ENGINEERING PROJECT APPRAISAL

ENGINEERING PROJECT APPRAISAL

The Evaluation of Alternative Development Schemes

Second Edition

Martin Rogers

Assistant Head of School
School of Civil and Building Services Engineering
Dublin Institute of Technology

Aidan Duffy

Lecturer
School of Civil and Building Services Engineering
and Dublin Energy Lab
Dublin Institute of Technology

WILEY-BLACKWELL

A John Wiley & Sons, Ltd., Publication

Registered Office
John Wiley & Sons, Ltd, The Atrium, Southern Gate, Chichester, West Sussex, PO19 8SQ, UK

Editorial Offices
9600 Garsington Road, Oxford, OX4 2DQ, UK
The Atrium, Southern Gate, Chichester, West Sussex, PO19 8SQ, UK
2121 State Avenue, Ames, Iowa 50014-8300, USA

For details of our global editorial offices, for customer services and for information about how to apply for permission to reuse the copyright material in this book please see our website at www.wiley.com/wiley-blackwell.

Library of Congress Cataloging-in-Publication Data

Rogers, Martin (Martin Gerard)
    Engineering project appraisal : the evaluation of alternative development schemes /
Martin Rogers, Aidan Duffy.
        p.   cm.
    Includes bibliographical references and index.
    ISBN 978-0-470-67299-0 (pbk. : alk. paper)
    1. Engineering–Costs.   2. Engineering–Management.   I. Duffy, Aidan.   II. Title.
    TA183.R55 2012
    658.4′04–dc23

                                                    2012003455

A catalogue record for this book is available from the British Library.

Wiley also publishes its books in a variety of electronic formats. Some content that appears in print may not be available in electronic books.

Cover design by Sandra Heath
Cover image courtesy of Shutterstock

Set in 10/13pt Times by SPi Publisher Services, Pondicherry, India

Printed in Malaysia by Ho Printing (M) Sdn Bhd

1   2012

Contents

Preface

Practising engineers nowadays require a broad range of skills. They must be aware of and understand the economic, environmental and social contexts within which a development project takes place, and be able to resolve problems that arise in these areas. Accredited professional engineering courses now require students to develop an awareness of the economic, financial, social and environmental factors of significance to development projects, along with an understanding of risk analysis and quality systems. To achieve this level of understanding, engineering project appraisal must form a core subject area within any course wishing to fulfil this educational objective. The advent of programmes such as the public–private partnership schemes requires professional engineers to be aware of a much broader range of issues related to the proposed scheme than merely the technical aspects of its design and construction. The overall implications associated with each project option must also be considered at the planning stage as part of the engineer's input to the project.

This textbook provides an introduction to the full breadth of evaluation techniques required for the assessment of competing engineering projects. The book is divided into two parts. An introduction to the topic of engineering project appraisal is given in Chapter 1. The remainder of Part I, spanning Chapters 2 to 10 of the book, initially covers the basic building blocks of economic appraisal, such as the time value of money, interest rates and time equivalence. It then proceeds to explain basic economic techniques, such as net present worth, internal rate of return and annual worth. The main application of these techniques to public project appraisal – Cost–Benefit Analysis (CBA) – is dealt with in detail, together with a number of related decision methods, such as Cost Effectiveness and Goal Achievement Matrix, all of which are derived from CBA but where the common aim is increased inclusiveness. Depreciation and taxation are also addressed. Value for money in construction projects and the economic analysis of renewable energy supply and energy efficient projects are also dealt with at the conclusion to the first part of the text.

The second part of the book, spanning Chapters 11 to 15, examines the appraisal techniques that are appropriate when factors other than purely economic ones require consideration. The text details three multicriteria models that are widely used in the planning and evaluation of engineering projects: the Simple Additive Weighting (SAW) Model, the Analytic Hierarchy Process (AHP) technique and Concordance Analysis. The procedures used by these models to deal with both risk and uncertainty are explained within the text. Previously, many textbooks in the area have made only brief reference to such models. In recent times, however, they have proved particularly useful in the evaluation of competing proposals in the transport, solid waste and water resources areas. The space given to them within this book reflects their growing importance as tools of engineering evaluation.

The economic and multicriteria methods should not be viewed as totally separate. Often, an initial economic evaluation undertaken for a set of competing project options can subsequently be assimilated into a wider evaluation where the economic scores constitute one criterion, viewed alongside other technical, environmental and social criteria within a multicriteria framework.

In an effort to make the book as useful as possible to both students and practising engineers, case studies and worked examples for the various economic and multicriteria techniques are given throughout the text. Within this second edition, additional worked examples are included within Chapters 2 and 3, with Chapter 7 containing two additional case studies, one from the water supply area and one from the sewer flooding alleviation area, to add to the existing case study from the highways area originally included within the first edition. Chapters 8 and 9, addressing value for money in the economic analysis of renewable energy supply, energy efficient projects and construction projects, respectively, are new chapters within the text, reflecting the growing importance of these topics within the planning, design and construction of engineering projects.

The book is seen as an essential text for both undergraduate and postgraduate students within professional engineering courses. It is also envisaged that students on planning and construction management courses will find the text useful.

Martin Rogers and Aidan Duffy
Dublin Institute of Technology

Introduction

Project appraisal

Project appraisal is a process of exploration, review and evaluation taken on by the decision maker as the alternative options for development are defined within the project planning process. It can also be expressed in terms of a number of mathematical techniques that simplify the comparison of project options on the basis of an agreed criterion or set of criteria. These techniques provide a rational and significant approach to evaluating diverse aspects of different alternatives and the ability of these alternatives to achieve a set objective. These aspects can be purely economic or can be more broadly set to encompass technical, environmental and social concerns as well. The primary objective is to aid in the process of making informed and rational choices regarding the most effective use of available scarce resources. In the context of the planning of engineering projects, it is concerned with establishing the priorities between competing project options by judging the real cost to society of resources. Its purpose is to judge the merits of each alternative based on a set of concerns that can be economic, technical, social or environmental (or any combination of these), depending on the nature of the evaluation.

Who are the decision makers? In the past, the decision whether to employ resources for one purpose rather than another lay with administrators, planners and financiers rather than engineers, who tended to concentrate their efforts on the design/construction aspects of the project in question. Nowadays, however, with engineers taking their place within project companies involved with the planning and financing of development projects, they are required to have a much broader range of skills. They are required both to be aware of and to understand the economic, environmental and social contexts within which a development project takes place, and to be able to resolve problems that arise in these areas.

Decisions within project appraisal have their basis in a number of fundamental concepts. They should be made among alternative courses of action, each of which is clearly and unambiguously defined. The decision itself should be based on the expected future outcomes arising from the various project options. It is desirable to have at least one if not several criteria of evaluation. These will allow judgements to be made between project options based on their relative intrinsic worth. Only criteria that demonstrate differences between the various options are of relevance to the decision maker. Any criterion where the options perform identically will not form the basis for making an informed choice.

Ultimately, it is the people involved who make the decisions. The techniques outlined within this text are only tools to assist in the moving forward from this process. The outputs that result from these techniques are valid only for the individual or group of individuals who chose the model in question for the particular purpose of interest to them. A different group may have selected a different type of model or may indeed take the same results and interpret them in a different manner. The final decision must only be arrived at after appropriate consultations have taken place between all actors involved in making the decision, with the output from the project appraisal technique helping to make sense of the information at their disposal.

Students of engineering must develop an awareness of the relevant economic, financial, social and political factors of significance to engineering development projects, along with an understanding of risk analysis and quality systems. This knowledge is a vital building block in an engineering student's education, given that the ability to analyse and solve engineering problems must include a capacity to make choices on the basis of environmental/commercial as well as engineering/technical constraints. The ultimate objective of project appraisal is to secure the greatest benefit from the available scarce resources.

Planning and decision making as primary functions of management

Management can be defined in terms of its four primary functions. It is the process of planning and decision making, organising, leading and controlling an organisation's human, financial, physical and information resources to achieve organisational goals in an efficient and effective manner. During the planning phase of a development project, its form and design are finalised. The subsequent construction/implementation phase requires the organisation of human and other resources required to complete it. Appropriate leadership ensures that available resources are used to their utmost potential in delivering the finished product in the most efficient and effective manner. Finally, control mechanisms must be put in place throughout all phases of the project's development to monitor actual progress against that which was originally planned and expected. This process highlights those areas where corrective action needs to be taken in order that the project can be completed in a form as close as possible to that envisaged in the original plan. It helps ensure the effectiveness and efficiency needed for the successful completion of the project in the form originally planned.

Planning is the first and most important function of management. All other functions flow directly on from it. In the context of the management of engineering projects, planning involves the determination of the type of scheme that will best meet the goals and objectives of the organisation in question. Decision making, as a core element of the planning process, involves selecting a course of action from a set of alternative schemes. It is thus the point within the engineering management process at which engineering project appraisal takes place. Decision making and planning are codependent – a plan cannot exist until a decision is made to commit resources to it.

The process of engineering management is action-orientated, with decision making at its centre. Use of project appraisal techniques will guide the manager in the making of these decisions. To set the context within which project appraisal takes place, the identity of the decision maker, the most appropriate type of decision making for the process in question and the environment of certainty/uncertainty/risk within which the decision is made, must all be determined. These topics are dealt with in detail in Chapter 1.

In reality, however, the behaviour of engineering managers is not adequately described by the four 'functions' of management referred to above. With respect to engineering decision making in particular, it is, in fact, a diverse and project-specific process. To be effective, it must take place within the context of almost continuous communication with relevant interested parties both within and outside the organisation. Engineers must, therefore, be able to communicate effectively, convincing their fellow workers that the selected course of action is the most appropriate one, resolving any conflicts that might arise and, if necessary, using their intuition.

A brief history of project appraisal

Engineering project appraisal has emerged from two completely separate streams of work. The economics-based methods addressed in Part 1 of this book are closely aligned with conventional microeconomics, where the economic behaviour of very small segments of the economy, such as individual firms or public/private organisations, are scrutinised. Engineering economics focuses on economic decision making within such individual organisational units. Interest in economics among engineers arose both from the obvious applicability of the laws of economics to the production and use of scarce resources and the desire on their part to make informed financial analyses of the effects of the implementation of projects they had developed and designed. *The Economic Theory of the Location of Railways* by Wellington (1887) was one of the earliest books on engineering economy. Written in the United States at a time when railway construction was of overriding importance to the economy, it was born out of the belief that engineers, when deciding on prospective locations for railway lines, paid scant regard to the costs and revenues the line would generate over its life-span. Wellington deduced that capitalised costs should be considered as a basis for selecting preferred lengths of rail lines or their curvature. By bringing this problem to light, Wellington captured the basic thrust of engineering economics. He believed that good engineering management required that those making strategic or tactical decisions should be aware of the economic consequences of their choices.

A second significant author within classical engineering economics was Eugene L. Grant, who, in his text *Principles of Engineering Economy* (Grant, 1930), discussed the importance of using compound interest calculations as a basis for comparing long-term investments in capital goods alongside the need for evaluating short-term investments. Riggs *et al.* (1996) emphasised the importance of engineering economics in the phrase 'those that manage money manage all'.

The second strand of thought from which engineering project appraisal has emerged, and one which is dealt with in Part 2 of the book, involves the examination of multicriteria-based methods of project analysis that go beyond the evaluation solely of the proposal's economic consequences. This class of decision methods was devised in order to allow the appraisal of projects in situations where other non-economic consequences needed to be introduced into the analysis. These have proved particularly appropriate in the civil engineering field, where complex development projects involving attributes that are diverse in nature and are often difficult to measure quantitatively let alone in monetary units are required to be evaluated. Work on these methods has proceeded on both sides of the Atlantic. In the United States, Keeney & Raiffa's *Decisions with Multiple Objectives* (Keeney & Raiffa, 1976) and Saaty's *The Analytic Hierarchy Process* (Saaty, 1980) introduced the theoretical basis for two multicriteria techniques that have been widely applied to engineering option choice problems. In Europe, Roy's ELECTRE Model (Roy, 1968) has been used over the past 30 years to solve decision problems in the transport, environmental and water engineering fields. In general, multicriteria decision methods offer a level of flexibility and inclusiveness that purely economics-based models tend to lack. On the downside, with some of the more complex multicriteria models, however, the numerical computation involved can be quite complex, unwieldy and inaccessible.

Summary

A practitioner within the field of engineering project appraisal will draw upon his or her combined knowledge of both engineering and decision modelling and will pick the appraisal tool, be it a purely economics-based or a multicriteria model, which he or she feels will be best suited to the problem under scrutiny and will most easily identify the correct course of action. There is still some debate among practitioners in the field regarding the theoretical basis for some of the methods referred to in this text. However, all the major evaluation methods outlined have shown themselves to be readily applicable to problems of option choice for engineering development projects. Such is the variety of methods open to the practitioner that the problem often lies in identifying from the wide variety of available methods that method which is most appropriate to the problem in hand. It is hoped that this text will go some way to guiding potential users of the models towards choosing the particular appraisal methodology which best suits their needs in terms of the quality and type of data available to be input into the model, the level of detail required in the final results output from it, and the time and resources at the decision maker's disposal for completing the decision process.

This book concerns itself with project appraisal in the broadest context. The assessments detailed here concentrate on the effect an engineering development has on society as a whole rather than on the project promoters themselves. Major engineering development projects, even if partially or wholly funded by private sector capital, must be assessed in terms of their effect on all those who come within its influence.

The aim of this book is to give civil engineers a basic technical knowledge of project appraisal, providing them with a platform which will allow them to participate as informed professionals within the planning process for any major infrastructure project. While the book concentrates on providing technical information on the appraisal techniques, it must be realised that the use of these in isolation will never achieve the results desired. All students of project appraisal must realise the importance of the political dimension inherent in such a selection process. Politics intrudes at every step in the decision process and at every level in the decision hierarchy. The politics of engineering project planning must be recognised and managed effectively. A more detailed discussion of the political decision-making process is given in Chapter 1.

This is not a comprehensive or advanced text on engineering project appraisal. The book cannot, through limitations on space, deal with all the complexities of the individual appraisal techniques detailed within the book. It is, nonetheless, hoped that it gives the reader a sufficiently broad knowledge of the range of assessment methods available to practitioners in the area, and will enable them to delve deeper if necessary into the technical complexities of any of the models outlined in the text and to participate fully, with professionals from other disciplines if necessary, in the planning and selection process for major infrastructure projects.

References

Grant, E.L. (1930) *Principles of Engineering Economy*. The Ronald Press, New York.

Keeney, R.L. & Raiffa, H. (1976) *Decisions with Multiple Objectives*. John Wiley & Sons, Inc., New York.

Riggs, J.L., Bedworth, D.D. & Randhawa, S.U. (1996) *Engineering Economics*. McGraw Hill, New York.

Roy, B. (1968) Classement et choix en présence de points de vue multiples (la méthode ELECTRE). *Revue Informatique et Recherche Operationnelle,* 2e Année, **8**, 57–75.

Saaty, T.L. (1980) *The Analytic Hierarchy Process*. McGraw-Hill, New York.

Wellington, A.M. (1887) *The Economic Theory of the Location of Railways*. John Wiley and Sons, Inc., New York.

PART 1

ECONOMICS-BASED PROJECT APPRAISAL TECHNIQUES

Chapter 1

Decision Making and Project Appraisal

1.1 Decision making context

Let us firstly discuss the identity of the decision maker. In answer to the question as to whether individuals or organisations make decisions, it is a widely held view that managerial decision making is essentially an individual process, but one which takes place within an organisational context. Therefore, while the decision maker is central to the process, any given decision made may influence other individuals and groups both within and outside the organisation, as well as having the potential to influence the surrounding economic, social and technical environment within which they all operate.

In the particular context of engineering project appraisal, complex decisions may need to be resolved involving not only the definition and evaluation of alternative actions, but also the resolution of how the chosen project should be physically undertaken. Such complex decisions, often involving the expenditure of vast amounts of money, are rarely taken by one single individual decision maker, such as a government minister, a technical expert or an administrator. Even if the final legal responsibility does lie with one specific individual, the decision will only be taken after consultation between this designated individual and other interested parties. For example, the final decision regarding whether a major highway project will proceed is the responsibility of the relevant government minister. However, his or her decision is made only after a consultation process with interested parties has been completed, usually by means of a formal public inquiry at which all affected parties are represented. Such a decision could in some cases be the ultimate responsibility of a collection of individual decision makers, such as a cabinet of government ministers or an elected or appointed body. Groups seeking to directly influence the decision maker, such as professional representative institutions or local community groups, could be directly affected by the decision. All these 'actors' are what Banville *et al.* (1993) call primary stakeholders in the decision process. They have a pre-eminent interest in the outcome of the process and will intercede to directly influence it. Also,

Engineering Project Appraisal: The Evaluation of Alternative Development Schemes, Second Edition.
Martin Rogers and Aidan Duffy.
© 2012 John Wiley & Sons, Ltd. Published 2012 by John Wiley & Sons, Ltd.

there are third parties to the decision, such as environmental and economic pressure groups that are affected only in general terms by the decision. Termed secondary stakeholders, they do not actively participate in making the decision. Their preferences, however, must be considered.

In such complex cases, it is usual for one of the primary stakeholders central to the decision process to be identified and designated as the decision maker. In the context of the appraisal, therefore, the decision is, in effect, reduced to an individual process. The diverse backgrounds and differing perspectives of the various stakeholders may mean that not all can benefit directly from the decision-making procedure. This chosen stakeholder, as the designated decision maker, then plays a critical part in the process. In some circumstances, however, he or she may only be a spokesperson for all the stakeholders, both primary and secondary. Whatever the relative influence of the various actors, the process requires that a decision maker be identified, even if the objectives specified by the chosen party are those commonly held or assumed to be commonly held by the entire group of stakeholders.

Although the actual process of decision making is generally carried out by the designated decision maker, in certain complex and/or problematic situations it is more usual for it to be undertaken by a separate party who is expert in the field of decision theory. This person, called the facilitator or the analyst, can work alone or as leader of a team. The function of the analyst is to explain the mechanics of the decision process to the decision maker, obtain all required input information and interpret the results, possibly with the use of decision models, in an easily understandable way.

For the purposes of this book, it will be assumed that the decision maker is an individual, responsible for each step in the decision process, with the ability to directly influence the decision-making procedure.

1.2   Techniques for decision making

A decision is only needed when there is a choice between different options. Such a choice can be made using either a non-analytic or an analytic technique. The first type is used for less important, relatively trivial decisions. The second type is required for more complex decisions involving the irreversible allocation of significant resources. These techniques justify greater input in terms of time and expense on the part of the decision maker.

1.2.1   Non-analytical decision making

Some decisions are made without conscious consideration, on the basis that they are perceived by the decision maker as being 'right'. These are intuitive in nature and reflect an ingrained belief held by the decision maker in relation to the situation under examination. There is, however, the danger that the decision environment may have changed and that new conditions could now prevail, resulting in the decision maker's intuition being misplaced and incorrect. For this reason, decisions based on intuition should only be used with extreme care, in matters where the outcome is of small consequence.

The other type of decision in this category – judgemental decisions – are more 'rational' or reasoned in their approach than the first type. They are appropriate only for those decisions that recur. The decision maker consciously reasons out the probable outcomes of the possible alternatives using his or her judgement, which has been developed from past experience and general knowledge. He or she selects the alternative that he or she believes will deliver the most desirable outcome. For a large organisation where the same types of decision tend to recur very frequently, these types of decision can be very useful. The similarity between these frequently occurring decision situations allows the effective use of 'programmed' decisions where, like a computer-based algorithm, the selection of options is highly structured and consists of an ordered sequence of clearly defined steps. An example of such a programmed decision is the use of a code of practice by a structural engineer to design a reinforced concrete building. Because the set of design decisions is standard for such a process, the code of practice provides a guide for the designer regarding the major decisions that should be made and the sequence in which they should be addressed. Professional judgement alone is inadequate for this decision process, as such a problem can be very complicated. Because the code of practice is used successfully by structural engineers on a daily basis to design reinforced concrete structures, they have the confidence that using this 'programme' as a framework for their design decision will result in a properly designed building. Such codes of practice are not static, unchanging documents, but are amended as technological advances dictate. In general terms, within this type of decision, the 'programme' must be altered to take account of situational changes, be they alterations in the economic, social or technological environment.

It is important, therefore, to distinguish between a programmed decision and a non-programmed decision. As previously defined, a programmed decision is applied to structured or routine problems, involving repetitive work and relying primarily on previously established criteria. Many of the problems at the lower levels of organisations are often routine and well defined, requiring less decision discretion and analysis. (For example, a relatively junior engineer in the organisation would be competent to carry out the structural design procedure referred to in the previous paragraph.) These are classified as 'non-analytical' decisions. Non-programmed decisions, on the other hand, are used for new, unstructured and ill-defined situations of a non-recurring nature, requiring substantial analysis on the part of the decision maker. Because of the unstructured nature of such decisions, managers, as they become more senior, are increasingly involved in these types of decisions (Figure 1.1).

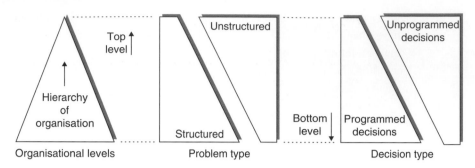

Figure 1.1 Types of problems and decisions at different levels of the organisation.

1.2.2   Analytical decision making

Non-programmed decisions are thus complicated in nature, involving a large number of factors where only correct actions will give rise to the desired results, and correct actions call for correct decisions carried out within an analytical framework. The probability of the correct choice being made in such situations is greatly increased by adopting a 'reasoned' or 'rational' approach that provides the appropriate analytical structure within which a coherent decision can be formulated.

1.2.3   Reasoned choice

The 'reasoned choice' model of individual or group decisions provides a technical foundation for non-programmed, non-recurring decisions (Zey, 1992). It comprises the following steps:

- *Recognising the problem.* The decision maker ascertains that a problem exists and that a decision must be reflected on.
- *Identifying goals.* The decision maker details the desired result or outcome of the process.
- *Generating and identifying options.* Different potential solutions are assembled prior to their evaluation.
- *Information search.* Characteristics of the alternative solutions are sought by the decision maker.
- *Assessing information on all options.* The information necessary for making a decision regarding the preferred option is gathered together and considered.
- *Selection of preferred option.* A preferred option is selected by the decision maker for implementation in the future.
- *Implementing the decision.* The chosen option is brought to completion.
- *Evaluation.* The decision is assessed after its implementation in order to evaluate it on the basis of its achieved results.

Clear rationality, where a judgement is arrived at following a sequence of deliberately followed logical steps, lies at the basis of this model for decision making.

1.2.4 Classical rational decision making

The principles of reasoned choice have been adapted into an analytic technique, called the rational approach, which has a specific application in the evaluation of project options at the planning stage of a proposed engineering scheme. The proper planning of a major engineering project requires a set of procedures to be devised which ensure that available resources are allocated as efficiently as possible in its subsequent design and construction. This involves deciding how the available resources, including manpower, physical materials and finance can best be used to achieve the desired objectives of the project developer. Systems analysis can provide such a framework of procedures in which the fundamental issues of design and management can be addressed (de Neufville & Stafford, 1974). Engineering systems analysis provides an orderly process in which all factors relevant to the design and construction of major engineering projects can be considered. Use of the process has the following direct impacts on the coherent and logical development of such a project:

- The process forces the developer/decision maker to make explicit the objectives of the proposed system, together with how these objectives can be measured. This has the effect of heightening the developer's awareness of his or her overall core objectives.
- It provides a framework in which alternative solutions will be readily generated as a means to selecting the most desired one.
- Appropriate methodologies for decision making will be proposed within the process for use in choosing between alternatives.
- It will predict the major demands which will be placed on the facility under examination through the interaction of the various technical, environmental and social criteria generated by the process. These demands are not always detected in advance.

The planning of major engineering projects is, therefore, a rational process. It involves a project's developer acting or deciding rationally in an attempt to reach some goal that cannot be attained without some action. He or she must have a clear awareness of alternative paths by which agreed goals can be achieved within the limitations of the existing environment, and must have both the information and the ability to analyse and evaluate options in light of the goals sought. Within the rational model, therefore, appropriate future action by the developer is determined by using the available scarce resources in such a way that his or her aims and objectives are maximised. It is a problem-solving process which involves closing the gap between the developer's objectives and the current situation by means of the developmental project in question, the 'objectives' being, for example, a more coherent transport infrastructure, a better quality rail service or a more efficient and cleaner water supply system.

The basic rational procedure can be represented by five fundamental steps. They constitute the foundation of a systematic analysis and are summarised in Table 1.1.

Table 1.1 Steps in the rational decision making process.

Step	Purpose
Definition of goals and objectives	To define and agree the overall purpose of proposed project
Formulation of criteria/measures of effectiveness	To establish standards of judging by which the options can be assessed in relative and absolute terms
Generation of alternatives	To generate as broad a range of feasible alternatives as possible
Evaluation of alternatives	To evaluate the relative merit of each option
Selection of preferred alternative/group of alternatives	To make a final decision on the adoption of the most favourable option as the chosen solution

Define goals and objectives

Goals can be seen as conceptual statements that set out in detail the intended long-term achievements of a proposed plan. They articulate the social values to be used within the planning process. Initially, they may only exist in outline form. Considerable data collection and evaluation may need to be undertaken and existing problems may need to be addressed before the goals can be precisely defined. Goals are, by their nature, abstract, and must therefore be translated into quantitatively based measurable objectives. These will form the basis for the criteria used within the process for evaluating alternative options. No appraisal process should proceed without an explicit statement of the objectives of the proposed undertaking. All analyses have a set of objectives as their basis. Much of the value of the planning process lies in the identification of a clear set of objectives.

The process will generate different classes of objectives that may be potentially conflicting. For example, within the planning of major transport infrastructure, the designer may have to reconcile the maximisation of economic and technical efficiency with the minimisation of social and environmental impact. These objectives will each have their own merits, and must be considered by their own individual set of criteria.

In an engineering context, the determination of broad objectives, such as the relief of traffic congestion in an urban area or changing the method by which domestic waste is disposed of, is seldom within the design engineer's sole remit. Their setting predominantly takes place at what is termed 'systems planning level' where input is mainly political in nature, with the help and advice of senior technical experts, some of whom will be professional engineers. The objectives serve to define the 'desired situation' that will transpire as a direct result of the construction of the proposed facilities.

Establish criteria

Defining the planning problem involves identifying the actual gap between the 'desired situation', as defined by the set of objectives derived, and the current situation, and assembling a range of measures designed to minimise or even close that gap. The ultimate aim of the process is thus to develop a grasp of the relative effectiveness with which these selected alternatives meet the derived set of objectives. Measures of performance, or criteria, must therefore be determined. They are used as 'standards of judging' in the case of the options being examined. Preferably, each criterion should be quantitatively assessed, but if, as with some social and environmental criteria, they cannot be assessed on any cardinal scale, it should nonetheless be possible to measure them qualitatively on some graded comparative scale.

The selection of criteria for the evaluation of alternatives is of crucial importance to the overall process because it can influence to a very great extent the final design. This selection process is also of value because it decides to a large degree the final option chosen. What may be seen as most desirable from the perspective of one set of criteria may be seen as much less so using another set of criteria. Thus the selection of the preferred design may hinge on the choice of the criteria for evaluation.

Identify alternative courses of action

Given that the ultimate end point of the process is to identify a preferred solution or group of solutions, it is logical that the decision maker should invest substantial effort in examining a broad range of feasible options. It would not be possible to subject all feasible options to a thorough analysis. Moreover, because resources for the analysis are never limitless, the decision maker must always be selective in the choice of options to be considered within the process. The decision maker must pay particular attention to identifying those alternatives that are shown to be most productive in achieving objectives, while ensuring that effort spent on the analysis of a given alternative does not exceed its anticipated benefits. This process should result in the drawing up of a set of alternative proposals, each of which would reasonably be expected to meet the objectives stated. There is seldom a plan for which reasonable alternatives do not exist.

Evaluate the alternatives

The relative merit of each option is determined on the basis of its performance against each of the chosen criteria. Each alternative is aligned with its effects, economic costs and benefits, environmental and social impacts, and functional effectiveness. This process is usually undertaken using some form of mathematical model. Selecting the appropriate model for the decision problem under consideration is a key step in the evaluation process. In the case of complex engineering projects where numerous alternatives exist and where so many variables and limitations need to be considered, it is at this point in the planning process that the application of decision-aid techniques becomes helpful. Ultimately, people make decisions. Computers, methodologies and other tools do not. But decision-aid techniques and models do assist engineers/planners in making sound and defendable choices.

Selection/recommendation

This is the point at which a single plan or shortlist of approved plans is adopted as most likely to bring about the objectives agreed at the start of the process. This is the real point of decision making, where a judgement is made on the basis of the results of the evaluation carried out in the previous step. As expressed above, because decisions are made by people, value judgements must be applied to the objectively derived results from the decision-aid model within the evaluation process. Political considerations may have to be allowed for, together with the distribution of the gains and losses for the preferred alternatives among a range of incident groups affected by the proposed facilities. The act of selection must, therefore, not be seen solely as a technical problem. This step within rational planning is the point in the process at which the final decision is actually taken.

1.2.5   *Behavioural decision making*

Although the reality of the decision situation may dictate otherwise, rationality assumes that, in order to arrive at the optimum solution for the planning problem under consideration, the decision maker must have:

- Complete information regarding the decision situation, that is why the decision is necessary, what stimulus initiated the process, and how it should be addressed.
- Complete information regarding all possible alternatives.
- A rational system for ordering alternatives in terms of their importance.
- A central goal that the final choice will be arrived at in such a way that maximises the economic benefit to the developer.

The basic, central assertion of this theory is that the decision maker, possessing complete knowledge of the problem, can, within the appraisal process, select the option which best meets the needs and objectives of the developer. This approach, termed optimisation, is strongly influenced by classical economics, and assumes that the decision maker is unerringly rational and devoid of personal preferences, motives and emotions. Since this is, in reality, unlikely to be the case, the behavioural model takes account of the imperfections likely to exist in the environment surrounding the planning process for engineering projects. Its originator, Herbert Simon (1976), recognised that full rationality did not accurately describe actual decision-making processes. In contrast to strict classical rationality, behavioural decision theory makes the following assumptions regarding the decision process:

- Decision makers have incomplete information in relation to the decision situation.
- Decision makers have incomplete information on all possible project options.
- Decision makers do not have the capacity or are not prepared to fully foresee the consequences of each option considered.

Simon notes that decision makers are, in reality, limited by their value systems, habits and skills as well as by less-than-perfect levels of knowledge and information.

He believes that, while decision makers seek to behave in a rational goal-oriented manner, their rationality has limits. They can be rational in striving to achieve a set of objectives only to the extent that:

- they have the ability to pursue a particular course of action;
- their concept of the end-point of the process is correct; and
- they are correctly informed regarding the conditions surrounding the choice.

Simon called this concept 'bounded rationality'. A decision maker is rational only within the boundaries laid down by the above limiting internal and external factors. Given that these limitations of information, time and certainty may, in practice, hinder a manager from being completely rational in his decision making, the manager may, as a result, decide to 'play it safe' rather than strive to arrive at the 'best' solution. Simon called this practice 'satisficing', where, rather than searching exhaustively for the best possible solution, a decision maker will search only until an option that meets some minimum standard of sufficiency is identified.

In the context of the planning of a major engineering project, decision makers may practise satisficing for a variety of reasons. A lack of willingness to ignore their own personal motives and objectives may lead to an inability on their part to continue the search after the first minimally acceptable option is identified. They may be unable to evaluate large numbers of options and/or criteria. Subjective considerations, such as the actual selection of criteria for evaluation, often intervene in decision situations. For all such reasons, the process of satisficing thus plays a major role in engineering decision making.

1.2.6   Irrational decision making

Both the classical and behavioural theories assume that the decision process involves at least some degree of rationality. Here, options are again generated and evaluated prior to the decision. In this instance, however, the decision maker is assumed to act in an irrational manner, with the final choice made prior to the initial generation of development options.

This model, termed the implicit favourite approach, was put forward by Soelberg (1967). It assumes that the decision maker does not search for the best option or even one that 'satisfices'. The process is only used as a vehicle for confirming that the initial favourite was the best option available, with spurious and sometimes irrelevant criteria of evaluation being invented to justify the final selection.

It is generally believed that unusual non-recurring decision problems will most often give rise to this type of solution in situations where the decision maker may not have ready-made rules and guidelines at his or her disposal for establishing and evaluating options. It has been found that the more political a decision, the more likely that the irrational model will be used. Political groupings may champion a particular option that they perceive as being to their own benefit. These groups will try to convince others of the chosen option's merits relative to the others under consideration. If the power position of the group pushing a particular option is strong enough, the opinions of others may not even be taken into consideration within the decision process.

1.2.7   Political involvement in the project planning process

In the context of a major engineering development project, the rational view of the planning process incorporates political involvement at two steps:

(1)   The determination of community goals is assumed to be the responsibility of political representatives.
(2)   The decision/selection process is usually viewed as primarily a political process, with elected representatives acting on the basis of information and advice from professional engineers and planners.

This perspective makes certain assumptions regarding the environment within which the decision is made:

* A set of community values and policies exist which is consistent with the goals and objectives of the proposed project.
* The project options are developed in response to rationally determined needs.
* The decision makers are primarily influenced by the rational evaluations of the various project options put forward by the technical experts.

Routine decisions, handed down on a day-to-day basis by those agencies responsible for the planning of engineering development projects, are generally resolved within a 'rational' framework. In many cases such decisions are taken by the professionals within the planning agency, with the political actors merely ratifying their actions. For extraordinary engineering planning decisions, Banks (1998) believes a form of irrational decision making, which he terms the 'political planning process', prevails in the case of 'one-off' extensive and complex engineering projects. The process is described as 'proposal oriented' rather than comprehensive, beginning with a specific development proposal rather than the definition of a broad set of goals and objectives that the chosen project must fulfil. Banks describes such a process as disjointed and confused, with different actors having different concerns and disagreement arising primarily out of people's lack of understanding of the decision problem. The process itself may be crisis oriented if the project being proposed is one of many such schemes within the political arena, in which case it will only be addressed if the problems which the proposal is intended to solve have reached crisis point.

Banks notes that the political planning process involves the following elements:

* A project proposal is made regarding a specific engineering development project. Specific projects such as the construction of a mass transit system for a given urban area or a toll-bridge connecting two major motorway networks could be proposed.

- The promoter of the proposal attempts to gather support for it through political influence, compromise or the manipulation of public opinion. Success in this regard could depend on the developer's ability both to gather political support from other parties in the planning process and to amend his proposal where necessary to gain additional support.
- A decisive action occurs, such as the decision of the planning authority or appeals board, to authorise a particular project. This may occur at either central or local government level.
- If the decision is favourable, the project is implemented. If it is unfavourable, the project is modified and then reintroduced at the first opportunity. Proposals of this type are rarely abandoned outright – it can fail many times, but it need only succeed once.

These steps are summarised in Table 1.2.

Banks' view is that the rational process can be incorporated into the political planning process as a means of persuasion. The professionals, such as planners and engineers, will tend to study the proposal within a structured rational framework, and the results of this work will be used within the overall political planning process to justify the project. The problem with this mixing of the two processes occurs where the two conflict with each other. For example, if a comprehensively rational decision process is followed by the professionals involved and results in a decision being reached that is incompatible with the more 'political' concerns of both local authority management and the members of the planning appeals board, it will lead to a divisive and unsatisfactory conclusion to the process.

Table 1.2 Steps in the political decision making process.

Step	Purpose
A proposal is made	To define the project as specific and non-choice based
The developer attempts to gather support for it	To generate political momentum in favour of the proposal
A decisive action occurs	To locate the point in the process at which approval/non-approval actually occurs
Resubmission if first submission rejected	If approval is not gained, the ability to continually amend and resubmit the proposal until consent is obtained.

1.3   Primacy of the rational model

The existence of 'non-rational' decision processes of the type outlined by Banks must be acknowledged. Such theories offer a useful insight into how, in a particular environment where political considerations tend to dominate, certain one-off, non-recurring project proposals gain approval via this process. However, for the purposes of this book, it will be assumed that the appraisal of engineering projects takes place within a format, overseen by planning specialists, where rational decision making, be it classical or behavioural, is the primary methodology at the basis of decision making. Within Banks' political planning process model, rational planning is seen as secondary and supplementary to the main process, used by the project promoters as a means of persuasion. Its use as a means of justifying proposals is seen primarily as a political asset rather than as a coherent logical tool for decision making. From the perspective of the professional engineer, it seems appropriate to assert the primacy of the rational model, on the basis of its logical foundation and its wide level of acceptance as an appropriate decision-making technique for use within this sphere of work. Within the rational model, the pivotal step is the evaluation or appraisal process where the relative merit of each proposal is determined. The main purpose of the succeeding chapters within this text is to explain the workings of various appraisal methodologies of direct use to planning engineers.

1.4   Decision-making conditions

While we can assume that decision making is, in effect, an individually-based process, and that it takes place, from the planning engineer's perspective, within a rational format, the environmental conditions surrounding it can vary markedly. Virtually all decisions are made under conditions of at least some uncertainty. The extent will range from relative certainty to great uncertainty. There may also be certain risks associated with making decisions. There are thus three categories of environmental conditions: certainty, risk and uncertainty.

1.4.1   Certainty

Very few decisions are made under conditions of certainty. This state, therefore, never truly exists. The complexity of an engineering project, together with the cyclical nature of the economic environment surrounding it, makes such a condition unattainable. It defines an idealised situation where all project alternatives and the conditions surrounding them are assumed to be known with complete certainty. Suppose an engineering contractor is awarded a project ahead of the other tendering companies on the basis of its bid. While this decision to award may appear to approach the condition of complete certainty, each of the contractors may have written non-identical cost increase clauses into their respective contracts so that the engineer making the decision to award may not be 100% certain of the relative conditions associated with each alternative bidder.

1.4.2   Risk

In a risk situation, the outcomes of the decision are random, with the probabilities of each outcome being known. Under these conditions, the availability of each project option and its potential pay-offs and costs are all associated with probability estimates. The probability in each case indicates the degree of likelihood of the outcome. The key element in decision making under a state of risk is the accurate determination of the probabilities associated with each project alternative. The probabilities can be determined objectively using either classical probability theory or statistical analysis, or subjectively using the experience and judgement of the decision maker. The values derived can then be used in a rationally based quantitative approach to decision making.

1.4.3   Uncertainty

In the context of an engineering project, the vast majority of decision making is carried out under conditions of uncertainty. Because of the complex and dynamic nature of the technology associated with present day engineering projects, the decision maker does not know what all the options are, the possible risks associated with each, or what the results or consequences of each will be. In such a situation, the decision maker has a limited database, is not certain that the data are completely reliable and is not sure whether the decision situation will change or not. Moreover, the interaction between the different variables may be extremely difficult, if not impossible, to evaluate.

Consider the environmental appraisal of an engineering development project. Because of the complexity of criteria relating to the estimation of noise and air pollution valuations, the database compiled by environmental specialists for each impact may be incomplete and there may be no guarantee that the values measured will not change with time. The accuracy and reliability of the data are therefore in question, and this uncertainty must be reflected in the decision process.

Under these conditions, if decision making is to be perceived as effective, the decision maker must seek to acquire as much relevant information as possible, approaching the situation from a logical and rational perspective. Explicit estimates of the levels of uncertainty associated with criterion estimates, together with judgement, intuition and professional experience will be of central importance to the decision making process.

Both the newness and the complexity of the rapidly changing technology associated with modern engineering development projects tend to induce uncertainty in their evaluation.

Many of the models outlined in this book as aids to the decision maker in the process of appraisal take explicit account of the levels of uncertainty associated with the relative evaluations of the competing proposals under consideration.

1.5   Project planning process

Accepting the importance of the rational model, certain steps within it are of particular importance in the context of examining an engineering development project. Assuming that such a proposed project will be planned in a logical manner, in an environment where some uncertainty/risk may exist, the three main steps in the process can be identified as:

- Identifying the project options.
- Identifying the criteria for evaluation.
- The appraisal process in which a preferred option is identified.

While the appraisal process may be the most important, the proper execution of the two preceding stages is of vital importance to the success of the overall process, as they provide an invaluable platform for effective appraisal. Let us look at each of these stages in some detail.

1.5.1   Identifying project options

A central objective of a given decision situation is the identification of feasible options. The term 'feasible' refers to any option that, upon preliminary evaluation, presents itself as a viable course of action, and one that can be brought to completion given the constraints imposed on the decision maker, such as lack of time, information and resources.

Finding sound feasible options is an important component of the decision process. The quality of the final outcome can never exceed that allowed by the best option examined. There are many procedures for both identifying and defining project options. These include:

- Drawing on the personal experience of the decision maker himself as well as other experts in the field.
- Making comparisons between the current decision problem and ones previously solved in a successful manner.
- Examining all relevant literature.

Some form of group brainstorming session can be quite effective in bringing viable options to light. Brainstorming consists of two main phases. Within the first, a group of people put forward, in a relaxed environment, as many ideas as possible relevant to the problem being considered. The main rule for this phase is that members of the group should avoid being critical of their own ideas or those of others, no matter how far-fetched. This non-critical phase is very difficult for engineers, given that they are trained to think analytically or in a judgemental mode (Martin, 1993). Success in this phase requires the engineer's judgemental mode to be 'shut down'. This phase, if properly done, will result in the emergence of a large number of widely differing options.

Proposed option vs. an accepted 'tried and tested' solution	Better	Worse
Construction cost	✓	
Maintenance Cost		✓
Visual appearance		✓
Technical innovation		✓

Table 1.3 Example of T-Chart.

The second phase requires the planning engineer to return to normal judgemental mode to select the best options from the total list, analysing each for technological and economic practicality. This is, in effect, a screening process that filters through the best options. One such method is to compare by means of a T-chart each new option with an existing, 'tried-and-tested' option that has frequently been used in previous similar projects (Riggs *et al.*, 1997). The chart contains a list of criteria which any acceptable option should satisfy. The option under examination is judged on the basis of whether it performs better or worse than the conventional option on each of the listed criteria. An example of a T-chart is given in Table 1.3.

In the example shown in Table 1.3, the proposed option would be rejected on the basis that, while it had a lower construction cost, its maintenance costs and visual appearance, together with its relatively limited degree of technical innovation, would eliminate it from further consideration.

The above example illustrates a very preliminary screening process. A more detailed, finer process would contain percentages rather than checkmarks. The level of filtering required would depend on the final number of project options the decision maker wishes to bring forward to the full evaluation stage.

1.5.2   Identifying attributes/criteria of evaluation

Attributes represent the characteristics associated with the essential features of a proposed development. Any given option being considered must perform positively with respect to these features if it is to have any hope of fulfilling the overall objectives for the project as laid down by the decision maker. Once these attributes can be measured or scaled in some way, they are termed criteria. Generating criteria will thus provide a means of evaluating the extent to which each option under consideration achieves the objectives set down. They provide a tool for the comparison of project options.

The characteristics of the decision making environment may vary substantially, and this may be reflected in the decision criteria used. Within a complex engineering development, criteria may vary from well defined quantitative attributes, such as economic and financial viability, to ones that are extremely difficult to define and quantify, such as morale and environmental welfare. Many of the more straightforward decision problems involve quantitative, monetary-based criteria that can be understood, defined and measured with relative ease. Many view criteria that can be expressed in monetary terms as the most important within a decision problem, given that selecting the most efficient option that will make the best use of limited resources is the primary concern of the decision maker. Grant *et al.* (1990) believe, therefore, that, where possible, all criteria should be expressed in monetary terms, and that the primary criterion should be monetary-based. Criteria that are not reducible to monetary figures are considered, but their role in the decision process is a secondary one.

This primary concern with monetary attributes or attributes that can be easily converted to monetary units lies at the basis of most texts on engineering economics. The ability to score the total performance of each project option on a single scale – usually a monetary value – has the great advantage that an 'optimum' solution can be found. The most often used method of project appraisal that uses the principle of optimisation is Cost–Benefit Analysis (CBA), which is one of a number of monetary-based methods described in detail later in the book. The decision process in such instances constitutes an economic evaluation. Non-economic-based consequences of the project are assumed to be of lesser importance to the decision maker.

However, more complex decision problems may involve attributes that prove difficult to define and measure. Examples of these are levels of passenger comfort and safety on a proposed transport system and the effects of a highway project on the cultural heritage of the area. These attributes are termed 'intangible' or 'qualitatively based'. Within an appraisal process where the measures of the relative effectiveness of the project options are not just economic but also possibly technical, social and environmental, a wide range of criteria, some monetary, some non-monetary but quantitative, and others purely qualitative must be taken into account. This can only be achieved within a multi-attribute or multicriteria decision-making format. Multicriteria decision models integrate quantitative and qualitative criteria to produce an aggregate performance measure using a 'compromise' technique which ranks or scores each option on the basis of a trade-off of its performances relative to the other options on each of the decision criteria.

Whether the criteria for assessment are purely economic or cover a wide range of objectives, the success of the evaluation process depends on being able to select and define criteria relevant to the appraisal process. Criteria can be developed by studying the relevant technical literature, by examining written details of similar decision situations or by asking the opinions of people with expertise in the relevant field. A number of formal methodologies for compiling a list of decision criteria exist.

Consumer surveys

Those who will be the eventual users of the development being proposed are one of the most logical sources of information regarding attributes to be used in an appraisal of competing options. For example, before deciding on a new mass transit system for a city, it is wise to ask the inhabitants what features of such a system they consider to be most important. However, certain care must be taken with the responses, as consumer tastes are open to manipulation and can change rapidly over time. Their stated behaviour towards the proposed development may vary greatly from how they actually react to the finished product.

Technical documents

Handbooks or government guidelines may exist in the relevant area on the attributes that must be taken into account when a development in a particular discipline of engineering is proposed. For example, in the water engineering area, the Battelle System (Dee, 1973) supplies a list of 78 economic, social and environmental criteria that must be considered when an option for a given water resources project is being assessed. Such information, used alongside the opinion of experts in the relevant field of engineering, can result in the compilation of a complete and exhaustive list of criteria relevant to the decision problem.

Delphi Method

The Delphi Method combines opinions into a reasoned and logical consensus. It requests and collates, in a systematic manner, opinions from experts regarding the correct decision criteria for appraising the proposed development. Initially, a precisely prepared questionnaire is given to a panel of experts from the professional specialties relevant to the problem. Replies to the questionnaires must contain answers with written supporting reasons for them. These reasons are summarised by a moderator/facilitator who gives them to the full panel for consideration. Full anonymity with regard to the source of each written response is maintained at all times. This iterative process is continued until the exchange of arguments and the transfer of knowledge results in a consensus being formed.

Overview

Whether the criteria are purely economic, or are more widely based, taking into account environmental and technical concerns, those finally selected for use in the appraisal process should be:

- measurable on some scale, be it quantitative or qualitative (measurability in monetary terms is only required for a purely economic appraisal);
- complete and exhaustive, covering all aspects of the decision problem;
- mutually exclusive, allowing the decision maker to view the criteria as separate entities, thereby avoiding 'double-counting';
- restricted to performance attributes of real importance to the decision problem.

1.5.3   Methods for engineering project appraisal

Assuming that the rational model forms a basis for this appraisal procedure, and that all options and decision criteria have been identified, the most important stage within this process is the actual common evaluation of the individual project options. For the engineer involved in assessing the relative merits of the individual proposals, a properly structured evaluation is central to the overall success of the appraisal. The remainder of this book concentrates on the models that might be used by the planning engineer to assess each of the options under consideration. These models fall into two categories as mentioned briefly above: the first based on optimisation, and the second based on compromise. They are distinguished by the set of rules they employ to make the decision.

A set of rules, which can also be called an evaluation method, is required to interpret the criterion valuations for each alternative considered. This set is a procedure that enables the pros and cons of alternative projects to be described in a logical framework so as to assess their various net benefits. They transform the facets of each proposal, as expressed within the agreed measure (or measures) of performance, into statements of its net social benefits.

The evaluation method must provide an insight into the formal relationships between the multiple aspects of alternatives as expressed in their performance on the decision criteria. The challenge is to develop an evaluation procedure appropriate for both the decision problem under consideration and the available information. It must be readily understandable to those involved in the decision process. The set of decision rules at the basis of the evaluation process is of vital importance.

If the conditions for the classical rational model are assumed to exist, then the chosen option emerging from this process can be designated the 'single best' of all the competing proposals. It is thus deemed the optimum choice. For the principle of optimisation to be at the basis of the decision taken, the decision maker must assume that the different objectives of the proposal, as stated through their relevant measures of performance, can be expressed in a common denominator or scale of measurement. This allows the loss in one objective to be directly evaluated against the gain in another. This idea of compensatory changes is central to many of the models used within engineering appraisal. The optimising principle is very elegant, providing a straightforward tool for the evaluation of alternative strategies on the basis of their economic benefit to society. In the case of CBA, the contribution of each alternative to the community is expressed in monetary terms.

Within the context of many engineering development projects, however, the decision maker may have limited knowledge regarding the decision situation, the available alternatives or the decision criteria to be used within the appraisal. In such instances, the optimising principle is rather limited, since the specification of a function expressing total benefit to society presumes the possession of complete information about all possible combinations of actions, about the relative trade-offs between actions, and about all constraints prevailing in the decision making process.

Given the somewhat limiting nature of these constraints on finding solutions to real-life and often complex engineering problems, certain circumstances exist where the so-called 'compromise' principle should be considered (Van Delft & Nijkamp, 1977). It stems from Simon's concept of 'bounded rationality' referred to earlier (Section 1.2.5), where the rational model must operate within the limitations imposed on the decision maker by lack of information in certain vital areas. It assumes the existence of a variety of decision criteria, not all measurable in a common denominator. The principle states that any viable solution has to reflect a compromise between various priorities, while the various discrepancies between actual outcomes and aspiration levels are traded off against each other by means of preference weights. The quality of each option can only be judged in relation to multiple priorities, so that a desired alternative is one that performs comparatively well according to these priorities. The compromise principle is particularly relevant for option evaluation/choice problems leading to multicriteria analyses. Given the potential complexity of the planning process for major engineering projects, such multicriteria methodologies can provide a useful resource for decision makers in the completion of their task.

Optimising methods

In situations where this analysis is predominantly an economic one, computations are performed on each of the alternatives in order to obtain one or more measures of worth for each. Engineering economics provides techniques that result in numerical values termed measures of economic worth. These, by definition, consider the time value of money, an important concept in engineering economics that estimates the change in worth of an amount of money over a given period of time. Some common measures of worth are:

- Net Present Value (NPV)
- Benefit/Cost Ratio (B/C)
- Internal Rate of Return (IRR)

In economic analysis, financial units (Euro/Pounds/Dollars) are used as the tangible basis of evaluation. With each of the above 'measure of worth' techniques, the fact that a quantity of money today is worth a different amount in the future is central to the evaluation.

Within the process of actual selection of the best option in economic terms, some criterion based on one of the above measures of worth is used to select the chosen proposal. When several ways exist to accomplish a given objective, the option with the lowest overall cost or highest overall net income is chosen. While intangible factors that cannot be expressed in monetary terms do play a part in an economic analysis, their role in the evaluation is, to a large extent, a secondary one. If, however, the options available have approximately the same equivalent cost/value, the non-economic and intangible factors may be used to select the best option.

Multicriteria methods

Within the context of the 'compromise' principle, multicriteria decision aid gives project planners some technical tools to enable them to solve a decision problem where several often conflicting and opposite points of view must be taken into account within the decision process (Rogers *et al.*, 1999). With such complex infra-structural planning problems, in many cases no single option exists which is the best in economic, technical and environmental terms. Furthermore, criteria from such diverse sources are seldom measurable in a common denominator. As a result, direct comparison of scores from different attributes becomes more complex. Hence the optimisation techniques available within operations research, referred to above, are not applicable to this problem type. The word 'optimisation' is inappropriate in the context of this type of decision problem. It may be virtually impossible to provide a truly scientific foundation for an optimal solution/decision. Multicriteria methods based on the compromise principle provide tools and procedures to help us attain the 'desired situation', as expressed in the set of objectives, in the presence of ambiguity and uncertainty. However refined our models may be, we must recognise that no amount of data will remove the fundamental uncertainties which surround any attempt to peer into the future. Multicriteria methods do not yield a single, 'objec-tively best' solution, but rather yield a kernel of preferred solutions or a general rank-ing of all options. They are the most readily applicable models to problems of option choice within civil engineering where it is virtually impossible to provide a scientific basis for an optimal solution. Solving such a multicriteria problem is, therefore, not searching for some kind of 'hidden truth', as Vincke (1992) put it, but rather helping the decision maker to master the complex data involved in a decision problem in such areas and advance towards a solution. This process involves compromise, and depends to a great extent both on the personality and experience of the decision maker and on the circumstances in which the decision-aiding process is taking place. However complete the information, the need for personal judgement and experience in the making of project planning decisions remains.

1.6   Example of a decision process

The project appraisal techniques described within this book are divided into two broad categories:

(1)   Purely monetary-based evaluations.
(2)   Multicriteria evaluations.

The methods in the first category require the monetary evaluation of all criteria relevant to the decision. The second category contains those methods that enable the evaluation of a potentially diverse range of attributes ranging from economic to social, environmental and technical criteria. Within this group, each criterion does not have to be measurable in monetary terms. Any scale that differentiates the perfor-mance of a number of options on the criterion in question, whether qualitative or quantitative, is permissible.

Before going into the details of these methods in the succeeding chapters, brief descriptions of typical decision problems solved using the two method types are given in the case studies in Sections 1.6.1 and 1.6.2. Both cases detail the data at the basis of the problem in question and outline the final decision taken. In neither case is the actual method for directly appraising the relative merit of the alternative proposals on the basis of the chosen decision criteria actually described. The methods of project appraisal detailed in the succeeding chapters of this text will attempt to provide the reader with the means of:

- selecting the appropriate appraisal methodology;
- collating all relevant information on the available options and their performance on each of the decision criteria chosen; and
- translating this information into a measure of the relative performance of the project options.

These three steps lie at the basis of project appraisal.

1.6.1   Case study 1: Economic analysis of alternative port access routes for a major city

A municipal authority wishes to evaluate the economic performance of a number of highway options for providing better access for heavy vehicles to the port area. The objective of the road is to reduce the negative effects on the city centre arising from the heavy vehicle traffic travelling to the port area from the outskirts of the city. Four options are assessed:

(1)   A 'do minimum' traffic management option involving the banning of trucks from some roads in the city centre (Option 1).
(2)   A new north–south tunnel connecting the existing orbital motorway system to the port area (Option 2).
(3)   A new east–west tunnel connecting the port to an existing dual carriageway (Option 3).
(4)   A new overground highway running east–west along an existing rail corridor adjacent to an existing canal (Option 4).

Each of the four options is assessed on the basis of the following nine economic consequences/criteria:

(1)   Car-user time savings
(2)   Heavy vehicle time savings
(3)   Public transport time savings
(4)   Car operating cost savings
(5)   Heavy vehicle operating cost savings
(6)   Accident cost savings
(7)   Capital costs
(8)   Maintenance costs
(9)   Operating costs.

All nine can be assessed in monetary terms. The first six are benefits and will thus have a positive monetary value, while the last three are costs and have a negative valuation.

For each option, its score on each of the economic criteria is discounted to a present worth. They are then added up to give an overall net present worth for the option in question. The following estimates of net worth for each of the options are obtained from the appraisal process:

- Option 1: –£40 million
- Option 2: +£100 million
- Option 3: +£50 million
- Option 4: +£66 million

On the basis of the economic evaluation, Option 2 is chosen as the best performing proposal.

1.6.2   Case study 2: Multicriteria analysis of alternative waste management strategies for a region

The regional government of a country in Western Europe wishes to devise a new waste management strategy, involving better use of existing incineration facilities and the possible construction of new ones. The objective of any new strategy is not only to put the process of waste management on a firmer economic footing, but also to have regard to the environmental and social effects of the strategy. The problem is thus a multicriteria one rather than purely economic. Five strategies are assessed:

(1)   Constructing a number of new waste incineration plants and importing waste from other countries to subsidise their construction (strategy 1).
(2)   Constructing a number of new waste incineration plants with no import of foreign waste (strategy 2).
(3)   Maximising the transportation of waste between existing facilities in the region (strategy 3).
(4)   Decentralising waste facilities to the outlying areas of the region (strategy 4).
(5)   Rationalising the number of existing waste management facilities (strategy 5).

Each of the five options is assessed on the basis of the following nine economic, technical, social/political and environmental factors:

(1)   Quantity of waste transported and distance travelled (environmental)
(2)   Energy use (environmental)
(3)   Impact of gas emissions (environmental)
(4)   Cost (economic)
(5)   Flexibility of strategy to possible increases in quantity of waste produced (technical)
(6)   Flexibility of strategy to possible decreases in quantity of waste produced (technical)

(7)   Level of overcapacity resulting from strategy (technical)
(8)   Level of local opposition to strategy (political)
(9)   Dependency of strategy on supply of imported waste (political)

Of the nine, one is assessed in monetary terms, six in quantitative, non-monetary units, and two on a qualitative scale. This diversity of assessment necessitated the use of a compromise-based rather than an optimising technique, with the overall performance of the strategy options based on their comparative performance on each of the decision criteria. An option's overall performance entailed trading-off its good performance on one criterion against its relatively weak performance on another. This analysis requires information on the relative importance to the decision makers of the nine criteria considered.

An analysis of the relative performance of the options on each of the decision criteria yields the following ranking:

- First: strategy 5
- Second: strategy 4
- Third: strategy 3
- Fourth: strategy 1
- Fifth: strategy 2

Strategy 5, involving the rationalising of existing incineration facilities, performs very well relative to the other options, with strategy 4 also scoring strongly. Both options are presented by the decision maker as viable solutions to the responsible government minister.

1.7   Summary

Within this first chapter we have defined project appraisal, identified the broadly rational planning framework within which it operates and outlined the potential level of certainty/uncertainty/risk associated with the environment within which it takes place. Two distinct types of appraisal systems are defined: one a purely economic analysis with its basis in classic rationality, and the other based in bounded rationality with the ability to encompass more broadly based environmental, technical and political concerns in addition to the basic economic factors. The first set of techniques — the optimising methods — makes assumptions regarding the level of completeness of information available and allows a precise measure of the relative performance of the different options on a common monetary scale. The second set of techniques – the compromise-based methods – makes far fewer demands on the quality of information available. However, this offers an evaluation procedure which, while being more inclusive than the first set, is more likely to supply a preferred option which performs 'well' relative to the others, rather than one which is identified as the optimal or 'single best' option.

1.8   Review of succeeding chapters

Part I of the book, comprising Chapters 1 to 10, details the basic tools and methodologies required by a decision maker to perform an 'optimising' economic appraisal. Chapter 2 specifies the basic tools required to carry out this process. The time value of money, interest rates and time equivalence are defined and explained in detail. Chapters 3 to 6 detail four methods for computing the economic worth of a stream of cash flows arising during the life of a project. Chapter 3 deals with the computation of present worth for such a stream. Life cycle cost analysis and payback period are also explained. Chapter 4 explains the computation of equivalent annual worth. The importance of the economic lives assigned to the options being compared within this computation is emphasised. Chapter 5 deals with rate of return computations both for a single project and when a number of competing options are being compared. In the case of competing projects, the way in which the method is employed will depend on whether the options are independent or mutually exclusive. Use of the benefit/cost ratio technique is also subject to these conditions. It is dealt with in Chapter 6 along with the topics of depreciation and taxation. Chapter 7 is of central importance within Part 1. It introduces Cost–Benefit Analysis (CBA), the main method of economic appraisal for public projects, and illustrates how the four basic methods for computing economic worth explained in the early chapters – present worth, rate of return, annual worth and benefit/cost ratio – can be used within this methodology. It outlines the process of identifying and valuing the costs and benefits on which the options are to be compared. The importance of a sensitivity analysis for ensuring the robustness of the final result is emphasised. This chapter also includes case studies from the areas of highway engineering, water supply and sewer flooding alleviation, together with an introduction to some of the techniques that can be used to assign a monetary valuation to non-economic criteria. It concludes with brief descriptions of the application of CBA to different areas of engineering. Chapter 8 details the economic analysis of renewable energy supply and energy efficient projects. Chapter 9 introduces the concept of Value for Money in construction projects. Chapter 10 outlines three further methods of economic evaluation: Cost Effectiveness Analysis, Planned Balance Sheet and Goal Achievement Matrix. While all are derivatives of the Cost–Benefit Methodology, they do allow the inclusion within their framework of non-monetary valuations. As a result, the optimising principle, which lies at the heart of CBA, is somewhat diminished within these techniques; linkages between them and the compromise-based methods dealt with in subsequent chapters are highlighted. These three methods thus form a bridge between the first and second sections of the book.

Part II of the book, comprising Chapters 11 to 15, puts forward a number of multicriteria models, all of which have their basis in the 'compromise principle'. Within Chapter 11, a number of simple multicriteria techniques such as the Dominance, Satisficing, Sequential Elimination and Attitude-Oriented Methods are explained in some detail. Chapter 12 is of central importance within Part II. It details the most widely used multicriteria method, the Simple Additive Weighting (SAW) Model.

Techniques allowing uncertainty to be incorporated into criterion scores within the SAW Model, together with the various systems for assigning importance weights to the criteria, are outlined. Environmental Checklists are important applications of the SAW Model, and three types are explained in the text. The chapter concludes with a case study, outlining the application of the model to choosing a transport strategy for a major urban centre. Chapter 13 explains the Analytic Hierarchy Process (AHP) Method, a widely used decision model in the United States. Its use of hierarchies together with a seven-point scale to calculate the priorities assigned to the different options under consideration is detailed. In conclusion, Chapter 14 deals with Concordance Techniques, a set of multicriteria models used extensively throughout Europe to resolve decision problems in areas including engineering development. Worked examples of the different types of decision methods are given throughout the text.

Figure 1.2 Bridge on Arklow bypass, Ireland (Source: Arup. Photographer: Studioworks).

References

Banks, J.L. (1998) *Introduction to Transportation Engineering*. McGraw-Hill International Editions, New York.

Banville, C., Landry, M., Martel, J.M. & Boulaire, C. (1993) A stakeholder's approach to MCDA. Working Paper 93–77, Centre de recherche sur l'aide à l'évaluation et à la décision dans les organisations (CRAEDO). University Laval, QC.

Dee, J. (1973) Environmental evaluation system for water resources planning. *Water Resources Research*, **9**, 523–535.

Grant, E.L., Ireson, W.G. & Leavenworth, R.S. (1990) *Principles of Engineering Economy*. John Wiley and Sons, Inc., New York.

Martin, J.C. (1993) The Successful Engineer: Personal and Professional Skills – A Sourcebook. McGraw Hill International Editions, New York.

de Neufville, R. & Stafford, J. (1974) *Systems Analysis for Engineers and Managers*. McGraw Hill, New York.

Riggs, J.L., Bedworth, D.D. & Randhawa, S.U. (1997) *Engineering Economics*. McGraw Hill International Editions, New York.

Rogers, M.G., Bruen, M.P. & Maystre, L.Y. (1999) *ELECTRE and Decision Support: Methods and Applications in Engineering and Infrastructure Investment*. Kluwer Academic Publishers, Boston, MA.

Simon, H.A. (1976) *Administrative Behaviour*. New York Free Press, New York.

Soelberg, P.O. (1967) Unprogrammed decision making. *Industrial Management Review*, **8**, 19–29.

Van Delft, A. & Nijkamp, P. (1977) *Multi-criteria Analysis and Regional Decision-Making*. Martinus Nijhoff, Leiden, The Netherlands.

Vincke, P. (1992) *Multi-criteria Decision Aid*. John Wiley & Sons Ltd, Chichester.

Zey, M. (1992) Criticisms of rational choice models. In: M. Zey (ed.) *Decision Making: Alternatives to Rational Choice Models*. Sage, Newbury Park, CA. pp. 9–31.

Chapter 2

Basic Tools for Economic Appraisal

2.1 Introduction

Before detailing the methodologies that allow a decision maker to perform an economic appraisal on a number of competing project proposals, it is necessary to understand a number of basic economic concepts:

- The Time Value of Money
- The Estimation of Interest
- Simple and Compound Interest
- Nominal and Effective Interest rates
- Continuous Compounding
- Time Equivalence
- Economic Computation

These can be seen as the essential tools required by the decision maker to undertake such an economic evaluation. Let us examine each in turn.

2.2 The time value of money

The economic evaluation of an engineering project involves the analysis of both costs and revenues, expressed in money terms, which occur at different times during the life of the project. In order to compare revenues received and costs paid out, some method must exist for estimating how the value of money changes with time. The concept that allows us to make such comparisons is the *interest rate*. It allows the cost to a person of having the use of a certain sum of money over a given period to be quantified. For example, a one-year project's cost is estimated at £1 million, incurred at the very start of the project by its developer. The developer will not have sufficient funds to meet this cost until 12 months after the starting date. At this stage, because of the time value of money, the amount to be paid by the developer will be greater than £1 million. This is achieved using the interest rate, and the value set for the rate will determine the final

Engineering Project Appraisal: The Evaluation of Alternative Development Schemes, Second Edition.
Martin Rogers and Aidan Duffy.
© 2012 John Wiley & Sons, Ltd. Published 2012 by John Wiley & Sons, Ltd.

amount that the developer will have to pay. If the interest rate is set at 10% per annum, the developer will have to pay £1.1 million in twelve months after the start of the project. If the rate is 20% per annum, the final total becomes £1.2 million.

Time value mechanics require the use of interest rates to translate monetary payments of differing amounts occurring at various times to a single equivalent monetary value. The setting of the interest rate is of great significance to this process. Future earned benefits from a project may, in absolute terms, appear to be much greater than the costs incurred which are incurred immediately, but, when reduced to the same time frame as the costs using the interest rate, the reality may be somewhat different. Within our economic environment, capital is the basic resource. It has the power to earn through its conversion into a physical development project providing goods and services to customers. For the person lending money to a project developer, interest is compensation, in the main for the loss of earnings that would have resulted from the lender using it for an alternative use. For the developer borrowing money, the loan represents an opportunity to undertake something immediately that would otherwise be delayed. From the perspective of the development of an engineering project, interest can be seen as the productive advantage gained from using a monetary resource efficiently. The interest rate is, in essence, a measure that gauges the productivity one should expect from the resource. The prevailing interest rate fixes the minimum level of productivity expected from the use of money over a given time frame, in other words, it sets the *time value of money*.

2.3   The estimation of interest

The quantitative measure of the time value of money is the *interest*, which allows the difference between the amount of money originally borrowed/invested and the final amount owed/accrued to be calculated. The original amount invested or loaned is termed the *principal*. If an amount of money is invested at time $t=t_0$, and this amount has accumulated with time, the interest at time $t=t_1$ would be:

$$\text{Interest} = (\text{total amount accrued})t_1 - (\text{original principal})t_0 \tag{2.1}$$

Conversely, if an amount of money is borrowed at time $t=t_0$, the interest would be:

$$\text{Interest} = (\text{total amount owed})t_1 - (\text{original principal})t_0 \tag{2.2}$$

In both situations, the amount of money that was borrowed or invested in the first instance has increased with time, and this increase over the original amount is denoted as the interest.

When this is expressed as a percentage of the original principal per unit time, the interest rate is obtained as follows:

$$\text{Interest rate}(\%) = \frac{\text{Interest per unit time}}{\text{original principal}} \times 100\% \tag{2.3}$$

The time unit employed within the formula must be defined. It is usually set at one year, but can also be set at as low as one month. This time frame is known as the *interest period*.

Example 2.1   Calculating interest

A construction company invests £10 000 upon the successful completion of a building project, and withdraws £10 650 exactly 12 months later.
   Estimate the following:

(i)   The interest accrued, and
(ii)   The percentage interest rate obtained on the original principal.

Solution

(i)   Using Equation 2.1
      Interest = £10650 − £10000 = £650
(ii)   Using Equation 2.3
      Interest rate = $(650 \div 10000) \times 100\% = 6.5\%$ per annum

---

2.4   Simple and compound interest

Simple and compound formulae are used to calculate sums of money accumulated over more than one interest period.

Where a simple interest rate is used, the interest earned is directly proportional to the principal involved, and the interest accrued over any preceding interest periods are ignored. Expressed as a formula, the total simple interest I earned through several time periods is estimated as follows:

$$I = P \times i \times n \tag{2.4}$$

where P is the principal, i is the interest rate and n is the number of time periods.

Since the principal is a fixed value, the amount of interest accrued is constant. Therefore, the total future amount of money is estimated as follows:

$$F = P + I$$
$$= P + P \times i \times n$$
$$= P \times (1 + (i \times n)) \tag{2.5}$$

where F is the future sum of money earned/to be paid.

Example 2.2   Simple interest computation

If a firm borrows £12 000 from a lending agency for four years at 5% per annum simple interest, how much money must they repay at the end of the term of the loan?

Contd

Example 2.2 Contd

Solution

Using Equation 2.5, where *P* is £12 000, *i* is 5% per annum and *n* is four years, the total future amount of money to be paid to the lending agency, F, is estimated as:

$$F = £12000 \times (1 + (0.05 \times 4))$$
$$= £14400$$

Details of this calculation are given in Table 2.1.

Table 2.1 Details of simple interest computation.

End of year	Amount borrowed (£)	Interest (£)	Amount due (£)	Amount paid (£)
0	12 000			
1	—	600	12 600	—
2	—	600	13 200	—
3	—	600	13 800	—
4	—	600	14 400	14 400

Where a compound interest rate is used, the interest accrued for each time period is estimated on the principal plus the total amount of interest accumulated in all pervious time periods. Compound interest takes into account the effect of the time value of money on the interest as well as the principal.

The interest for one time period is estimated as follows:

$$i = (\text{principal} + \text{all interest accrued within previous time intervals}) \times \text{interest rate}$$

The generalised compound interest formula for estimating the future sum of money earned/to be paid over *n* time periods is as follows:

$$F = P(1 + i)^n \tag{2.6}$$

Example 2.3   Compound interest computation

Taking the same interest rate problem as outlined in Example 2.2, but with a borrowing rate of 5% compound interest allowed for in this case, calculate the revised payment due at the end of the term.

Solution

Using Equation 2.6, where:

Contd

Example 2.3 Contd

$$F = £12\,000(1+0.05)^4$$
$$= £14\,586.08$$

Details of this calculation are given in Table 2.2.

Table 2.2 Details of compound interest computation.

End of year	Amount borrowed (£)	Interest (£)	Amount due (£)	Amount paid (£)
0	12 000			
1	—	600	12 600	—
2	—	630	13 230	—
3	—	661.50	13 891.50	—
4	—	694.58	14 586.08	14 586.08

2.5   Nominal and effective interest rates

2.5.1   *Nominal interest rates*

In most situations, the interest rate is given on an annual basis. However, cases may arise where it is required that the interest be compounded several times within one year. If one year is divided up into two halves, with an interest rate of 6% per half year, this interest rate is expressed as 12% compounded semi-annually. The 12% figure is called the nominal annual interest rate. The final value of £1000 at the end of one year, earning 12% compounded semi-annually is calculated as follows:

$$\text{Amount after 6 months } = P(1+i)$$
$$= £1000(1.06)$$
$$= £1060$$
$$\text{Amount after 12 months} = £1060(1.06)$$
$$= £1123.60$$

The effect of the nominal interest rate is to result in a higher sum than would be obtained were the 12% compounded annually (£1120 in this case). If this nominal interest rate of 12% were compounded monthly at 1% per month, using Equation 2.6, the final value of the £1000 would:

$$F_{compounded} = £1000(1+0.01)^{12}$$
$$= £1127$$

Therefore, for a given nominal interest rate, the more frequently compounded an initial sum of money is, the greater the final amount will be.

2.5.2   *Effective interest rates*

The confusion created by the use of nominal interest rates in the estimation of the final sum earned is eliminated by the introduction of *effective interest rates*. It is the interest charged over one year divided by the original amount borrowed/loaned. Taking the example immediately above, where a one-year loan is given at a nominal interest rate of 12% compounded monthly:

$$\text{Effective Interest Rate} = (£1127 - £1000) \div £1000$$
$$= 12.7\%$$

The generalised formula for obtaining the annual effective interest rate i_{ae}:

$$i_{ae} = \left(1 + \frac{r}{n}\right)^n - 1 \tag{2.7}$$

where r is the nominal annual interest rate and n is the number of times the interest rate is compounded within the year.

For the problem immediately above where r equals 12% and n equals 12:

$$i_{ae} = \left(1 + \frac{0.12}{12}\right)^{12} - 1$$
$$= (1 + 0.01)^{12} - 1 = 1.127 - 1 = 0.127 \text{ or } 12.7\%$$

2.6   Continuous compounding

As the compounding period becomes less and less, the number of compounding periods in one year increases. Where the interest rate is compounded continuously, the number of compounding periods, n, approaches infinity as the length of the periods become infinitesimally small.

As the ultimate limit for the number of periods for re-estimating interest within one year is reached, the effective interest rate formula in Equation 2.7 can be stated in a new form:

$$i_{inf} = \lim_{n \to \infty} \left(1 + \frac{r}{n}\right)^n - 1 \tag{2.8}$$

Given that the definition of the natural logarithm base is:

$$\lim_{m \to \infty} \left(1 + \frac{r}{m}\right)^m - 1 = e = 2.718$$

Using $r/n = 1/m$, which makes $n = mr$, the right hand side of Equation 2.8 can be rearranged as follows:

$$\lim_{n\to\infty}\left(1+\frac{r}{n}\right)^{n}-1=\lim_{m\to\infty}\left(1+\frac{1}{m}\right)^{mr}-1=\lim_{m\to\infty}\left[\left(1+\frac{1}{m}\right)^{m}\right]^{r}-1$$

$$i_{\infty}=e^{r}-1 \qquad\qquad\qquad\qquad (2.9)$$

Equation 2.9 is used to calculate the effective continuous interest rate. The periods for both i and r must be the same. (This was the case with Equation 2.7 where the period in question was one year.) For an annual nominal rate, r, of 12% per annum, the effective continuous rate per annum is:

$$i=e^{0.12}-1$$
$$=1.1275-1$$
$$=0.1275 \text{ or } 12.75\%$$

Example 2.4   Using the continuous compounding formula
If an organisation requires a return of 20% on its investment, calculate the minimum acceptable annual nominal rate if continuous compounding is assumed.

Solution
Using Equation 2.9, we know that the value of i per annum is 20%. Therefore:

$$0.2=e^{r}-1$$

Bringing r, the unknown value, to the right hand side of the equation:

$$e^{r} \quad =1.2$$
$$\ln(e^{r}) = \ln(1.2)$$
$$r \quad = 0.1823 \text{ or } 18.23\%$$

In certain situations, the justification for using continuous compounding within an economic analysis is that cash flows are modelled most accurately by assuming a continuous pattern of flows. In reality, however, most organisations use discrete compounding intervals in their studies for reasons of familiarity.

2.7   Time equivalence

The term equivalence is founded on two concepts already explained – the time value of money and interest rates. It implies that different amounts of money at different points in time are equal in terms of their economic value. For instance, if a simple rate of 10% per annum is assumed, £1000 today is *equivalent* to £1100 in a year's time. It means that £1000 today has exactly the same economic value as £1100 one year from today. (If the interest rate is changed, however, the equivalence in question no longer applies.) The earning power of money links time and earnings in order to determine time-equivalent money amounts. If a sum of £200 was lost today and discovered again five years later, its value would have remained the same but its earning power would have been foregone. On the other hand, £200 deposited at 10% compounded annually would have a value of £200$(1+0.1)^5$=£322.10 after five years. Therefore, £200 today is *equivalent* to £322.10 in five years' time, assuming an interest rate of 10% compounded annually, as an investor would find the £200 paid now and the promise of a payment of £322.10 in five years' time equally attractive.

If the money was not invested, £200 could be used to pay annual instalments of £100 over the next two years. Alternatively, if the £200 were deposited at 10%, £220 would be available at the end of the first twelve months. After paying the first £100, the remaining £120 would generate additional interest over the second twelve months of £12. Payment of the second £100 would leave £32.

The earning power of money is such that the £200 today is equivalent to £115.24 received at the end of each year for the next two years, that is:

$$200 = 115.24 / (1.1) + 115.24 / (1.1)^2$$

£200 today is also the equivalent of £52.76 received at the end of each year for the next five years, that is:

$$200 = 52.76 / (1.1) + 52.76 / (1.1)^2 + 52.76(1.1)^3 + 52.76(1.1)^4 + 52.76 / (1.1)^5$$

Equivalence is central to the economic analysis of engineering projects, as it permits cash flows, or different amounts of money accruing at different points in time, to be compared at one common discrete time location.

Amounts of money that differ in total magnitude but are made at different points in time may be equivalent to one another. To demonstrate this concept, let us assume that an organisation has borrowed £10 000 and wishes to have it totally repaid in 10 years. The interest rate is set at 10%. Table 2.3 illustrates three different pay plans, each equivalent to one another.

In each case the total amount of money paid back is different, totalling £10 000, £16 270, £20 000 and £25 940. The longer the period required for repayment, the greater the apparent difference would be. From the standpoint of the investor, however, the four plans are equivalent to each other, and represent a series of payments that will repay the present sum of £10 000 at a rate of 10% interest compounded annually.

Equivalence is of great importance to the economic appraisal of engineering projects. Such evaluations invariably entail the comparison of several project options proposed for meeting the stated planning objectives. If the four options examined required financial outlays as shown in Table 2.3, then all options would be equally desirable in economic terms assuming a prevailing interest rate of 10%. Their

Table 2.3 Equivalent sets of payments.

Year	Capital (£)	Plan A (£)	Plan B (£)	Plan C (£)
0	10 000	—	—	—
1	—	1 627	1 000	0
2	—	1 627	1 000	0
3	—	1 627	1 000	0
4	—	1 627	1 000	0
5	—	1 627	1 000	0
6	—	1 627	1 000	0
7	—	1 627	1 000	0
8	—	1 627	1 000	0
9	—	1 627	1 000	0
10	—	1 627	11 000	25 940
Total	10 000	16 270	20 000	25 940

equivalence would not be evident, however, by reference solely to the total outlays called for with each option. Their comparison only becomes clear when the differently phased payments are converted to an equivalent present worth.

Figure 2.1 illustrates the equivalence that exists between different amounts of money paid at different points in time. This graphical representation is also termed *cash flow diagramming*.

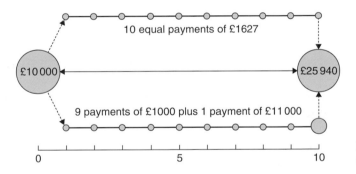

10 equal payments of £1627

£10 000

£25 940

9 payments of £1000 plus 1 payment of £11 000

0   5   10

Figure 2.1 Equivalent cash flows.

The appraisal of engineering projects invariably requires some conversion as a basis for sound judgement. Any comparison that omits the use of interest rates to convert different series of payments to an equivalent single one has the potential to be both vague and misleading.

2.8   Economic computation

2.8.1   *Symbols*

A number of engineering economy factors exists that are at the foundation of economic computation. These are concepts that lie at the heart of the economic analysis of any engineering project. In order to derive these factors, the following set of functional symbols must be identified:

$P =$ The present value of an amount of money, also called present worth. It is measured in monetary units (£).

$i =$ The interest rate per interest period (days, months or years).

$n =$ The number of interest periods.

$F =$ The future value of an amount of money, also called future worth. It is measured in monetary units (£).

$t =$ The time stated in periods (days, months or years).

$A =$ Each end-of-period payment in a uniform series progressing over the n periods in question. The entire set of consecutive payments taken together is equivalent to P at interest rate i, and is measured in monetary units per period. (£/year, £/month).

$G =$ An increase/decrease by a set amount each period of a series of end-of-period payments, representing a uniform arithmetic change in the magnitude of the amount from one time period to the next. The entire set of payments, taken together, again represent the equivalent of P at an interest rate i.

A one-year interest period is used in most situations. It is the standard for measurement, given that most organisations budget on a 12-month basis.

2.8.2   Formulae for single payments

> **Future worth factor:**
> To find: F
> Given: P
> Formula: $F = P(1 + i)^n$
> $\qquad\qquad = P \times \text{future worth factor}$

The future worth, F, of an amount of money that is accumulated after n periods from a single payment, P, when interest is compounded once per period is given in Equation 2.6 as follows:

$$F = P(1 + i)^n$$

The expression $(1 + i)^n$ is called the single-payment compound-amount factor, and represents the ratio of future to present worth. It is usually referred to as the *F/P factor*. It is represented by the symbol ($F/P,i\%,n$).

Example 2.5

If £5000 is invested today at 8% compounded annually, how much will it be worth in exactly six years.

Contd

Example 2.5 Contd

Solution

$i = 0.08, n = 6, P = 5000, F = ?$

$F = 5000(1.08)^6 = £7934.37$

> **Present worth factor:**
> To find: *P*
> Given: *F*
> *Formula: $P = F[1/(1+i)^n]$*
> $\qquad = F \times present\ worth\ factor$

Rearranging Equation 2.6 and solving *P* in terms of *F* gives:

$$P = F\left[\frac{1}{(1+i)^n}\right] \tag{2.10}$$

P, the present worth factor, determines the present worth of a given future amount *F* after *n* years at an interest rate *i*.

The term within brackets in Equation (2.10), which equals P/F, is called the present worth factor, and is simply the reciprocal of the compound-amount factor. It is the ratio of present worth to future value and is represented by the symbol (*P/F,i%,n*).

Example 2.6
How much must be invested today at 7% interest in order to accumulate £2800 in nine years?

Solution

$i = 0.07, n = 9, F = 2800, P = ?$

$P = 2800/(1.07)^9 = £1523.01$

2.8.3   *Uniform series formulae*

> **Sinking fund factor:**
> To find: *A*
> Given: *F*
> *Formula: $A = F\{i/[(1+i)^n - 1]\}$*
> $\qquad = F \times sinking\ fund\ factor$

A fund of money established in order to accumulate a desired future amount at the end of n periods of time, by means of a uniform series of payments made at the end of each period, is called a *sinking fund*. Each payment is a constant amount A. If the period is a year, A is termed an annuity.

If an amount A is invested at the end of each year for a total of n years, the final amount of money will be the sum of the individual compounded amounts. The sum of money invested at the end of the first year, through earning interest for the remaining $(n-1)$ years, will finally total $A(1+i)^{n-1}$. The payment at the end of the second year will total $A(1+i)^{n-2}$, with the third year's payment totalling $A(1+i)^{n-3}$. The last payment, made at the end of the nth year, earns no interest.

Each of the payments is collected to obtain the total future amount F as follows:

$$F = A(1+i)^{n-1} + A(1+i)^{n-2} + \ldots + A(1+i)^{n-(n-1)} + A(1+i)^{n-n}$$

$$F = A[(1+i)^{n-1} + (1+i)^{n-2} + \ldots + (1+i)^{n-(n-1)} + (1+i)^{n-n}] \tag{2.11}$$

Multiplying Equation 2.11 by $(1+i)$ gives the following expression:

$$F(1+i) = A[(1+i)^n + (1+i)^{n-1} + \ldots + (1+i)^{(n+1)-(n-1)} + (1+i)^{(n+1)-(n)}] \tag{2.12}$$

Subtracting Equation 2.11 from Equation 2.12:

$$F(1+i-1) = A[(1+i)^n - 1] \tag{2.13}$$

Bringing A over to the left-hand side of the equation:

$$A = F\left[\frac{i}{(1+i)^n - 1}\right] \tag{2.14}$$

The term $i/[(1+i)^n - 1]$ is called the *sinking fund factor* (A/F) and is represented by the symbol (A/F,i%,n).

Example 2.7

What amount of money must be deposited today at 5% per year for six years so that £1500 is accumulated on the date of the last deposit?

Solution

$i = 0.05, n = 6, F = 1500, A = ?$

$A = 1500\{0.05/[(1.05)^6 - 1]\} = £220.53$

Taking Equation 2.14, an expression for the future worth, F, can be formed as:

$$F = A \left[\frac{(1+i)^n - 1}{i} \right] \qquad (2.15)$$

> **Series compound-amount factor:**
> To find: F
> Given: A
> *Formula:* $F = A\{[(1+i)^n - 1]/i\}$
> $= A \times$ series compound amount factor

The term $[(1+i)^n - 1]/i$ is called the series *compound amount factor* (F/A) and is represented by the symbol ($F/A,i\%,n$).

Example 2.8
How much will accumulate in a fund at the end of 12 years if £110 is deposited at the end of each of these years? The interest rate is set at 8%.

Solution

$i = 0.08, n = 12, A = 110, F = ?$
$F = 110\{[(1.08)^{12} - 1]/0.08\} = £2.087.48$

> **Capital recovery factor:**
> To find: A
> Given: P
> Formula: $A = P\{[i(1+i)^n]/[(1+i)^n - 1]\}$
> $= P \times$ capital recovery factor

To find the uniform end-of-period payment A, which can be obtained for n years from a present investment P, substitute the value given for F in Equation 2.6 into Equation 2.14 as follows:

$$A = F \left[\frac{i}{(1+i)^n - 1} \right]$$

Since $F = P(1+i)^n$:

$$A = P(1+i)^n \times \left[\frac{i}{(1+i)^n - 1} \right]$$

$$A = P \times \left[\frac{i(1+i)^n}{(1+i)^n - 1} \right] \tag{2.16}$$

The expression $[i(1+i)^n]/[(1+i)^n - 1]$ is called the *capital recovery factor (A/P)* and is represented by the symbol $(A/P,i\%,n)$. It can be shown to be equal to the sinking fund factor plus the interest rate i.

Example 2.9

If £950 is invested today at 7%, what equal end-of-year withdrawals can be made for 15 years to completely empty the fund after the fifteenth and last withdrawal?

Solution

$i = 0.07, n = 15, P = 950, A = ?$

$A = 950(0.07(1.07)^{15})/((1.07)^{15} - 1) = £104.30$

> **Series present worth factor:**
> To find: P
> Given: A
> *Formula:* $P = A\{[(1+i)^n - 1]/[i(1+i)^n]\}$
> $\qquad\qquad = A \times series\ present\ worth\ factor$

Taking Equation 2.16, an expression for the present worth, P, can be formed as follows:

$$P = A \left[\frac{(1+i)^n - 1}{i(1+i)^n} \right] \tag{2.17}$$

The expression $[(1+i)^n - 1]/[i(1+i)^n]$ is called the *series present worth factor (P/A)* and is represented by the symbol $(P/A,i\%,n)$.

Example 2.10

What quantity of money would you need to deposit at 10% interest today in order to be able to withdraw £200 at the end of each year for the next four years, completely emptying the fund by the end after the fourth and last withdrawal?

Solution

$i = 0.10, n = 4, A = 200, P = ?$

$P = 200[(1.1)^4 - 1]/[(0.1(1.1)^4)] = £633.97$

2.8.4   Geometric series formulae

> **Arithmetic gradient conversion factor:**
> To find: A
> Given: G
> Formula: $A = G\{[1/i] - [n/((1+i)^n - 1)]\}$
> $\qquad\qquad = G \times$ arithmetic gradient conversion factor

Problems within engineering project appraisal frequently involve income or payments that increase or decrease each year by varying amounts. For instance, the running costs for a mass transit vehicle will tend to increase gradually for each successive year of its economic design life. Assuming that this increase/decrease is the same each year, the yearly increase/decrease is called the arithmetic gradient.

Given that the amount of money changes each year, the uniform series interest factors discussed above cannot be used, and new formulae must therefore be derived.

To achieve this, it is usual to assume that the cash flow occurring at the end of the first year is not part of the gradient series, but is treated separately as a *base amount*, denoted as A'. This is because the size of the base amount will, in all probability, be either larger or smaller than the gradient increase/decrease, G. The value of G may therefore be either positive or negative.

Assuming an increasing gradient, the pattern of the series is:

$$A', A'+G, A'+2G, \cdots, A'+(n-1)G$$

A decreasing arithmetic gradient yields the series:

$$A', A'-G, A'-2G, \cdots, A'-(n-1)G$$

If we ignore the base amount, the uniformly increasing gradient cash flow can be expressed for n years, as shown in Table 2.4. The amounts in the table can be seen as a set of payments that will sum to F after n years. This value can then be converted to a uniform series of payments by the product of F and (A/F) the sinking fund factor.

We assume an annual series of payments G commences at the second year,

Table 2.4 Example of uniform increasing gradient.

Year	1	2	3	4	......	$n-2$	$n-1$	n
Amount	0G	1G	2G	3G	....	—	$(n-2)G$	$(n-1)G$

another set of G commences at the end of the third year, going on until the nth year, when all terminate. By applying the series compound factor to each of these series of G every year, the compound amount at the end of the nth year can be estimated as follows:

$$F = G \times \left[\frac{(1+i)^{n-1} - 1}{i} + \frac{(1+i)^{n-2} - 1}{i} + \cdots + \frac{(1+i)^2 - 1}{i} + \frac{(1+i)^1}{i} \right]$$

$$= \frac{G}{i} \times \left[(1+i)^{n-1} + (1+i)^{n-2} + \cdots + (1+i)^2 + (1+i)^1 - (n-1) \right] \qquad (2.18)$$

$$= \frac{G}{i} \times \left[(1+i)^{n-1} + (1+i)^{n-2} + \cdots + (1+i)^2 + (1+i) + 1 \right] - \frac{nG}{i}$$

$$F(1+i) = \frac{G}{i} \times \left[(1+i)^n + (1+i)^{n-1} + \cdots + (1+i)^3 + (1+i)^2 + (1+i) \right]$$

$$- \frac{nG(1+i)}{i} \qquad (2.19)$$

Subtracting Equation 2.18 from Equation 2.19:

$$Fi = \frac{G}{i} \times [(1+i)^n - 1] - nG$$

$$Fi = G \times \left[\frac{(1+i)^n - 1}{i} \right] - nG \qquad (2.20)$$

$$F = \frac{G}{i} \times \left[\frac{(1+i)^n - 1}{i} \right] - \frac{nG}{i} \qquad (2.21)$$

To convert F to an annuity A, we must multiply F as expressed in Equation 2.21 by the sinking fund factor $i/[(1+i)^n - 1]$:

$$A = \frac{G}{i} - \frac{nG}{i} \times \left[\frac{i}{(1+i)^n - 1} \right]$$

$$A = G \times \left[\frac{1}{i} - \frac{n}{(1+i)^n - 1} \right] \qquad (2.22)$$

Example 2.11

A contractor, within a five-year contract, is awarded £2500 at the end of the first year, with the payments increasing geometrically by £500 over this initial amount for each of the next four years.

Contd

Example 2.11 Contd
  (i) Assuming an interest rate of 10%, what would the equivalent constant annual payment be?
  (ii) If the payments were geometrically decreasing by £500 from years 2 to 5, what would the equivalent constant payment be?

Solution (i)

$$i = 0.10, n = 5, A' = 2500, G = £500, A = ?$$

$$A = A' + G(A/G, I, n)$$

$$= £2500 + £500(1.81)$$

$$= £3405$$

Solution (ii)

$$A = £2500 - £500(1.81)$$

$$= £1595$$

Arithmetic gradient present worth factor:
To find: P
Given: G
Formula : $P = G \times 1/i \times \{[((1 + i)^n - 1)/(i(1+i)^n)] - [n/(1+i)^n]\}$
                 $= G \times$ arithmetic gradient present worth factor

---

The present worth of a uniform gradient G can be found by dividing Equation 2.22 by the capital recovery factor $(A/P, i\%, n)$ as expressed in Equation 2.16.

$$P = G \times \left[\frac{1}{i} - \frac{n}{(1+i)^n - 1} \right] \div G \times \left[\frac{i(1+i)^n}{(1+i)^n - 1} \right]$$

$$P = \frac{G}{i} \times \left[\frac{(1+i)^n - 1}{i(1+i)^n} - \frac{n}{(1+i)^n} \right] \tag{2.23}$$

Therefore, the annuity figure worked out using the equation $A' + A$, where A' is the base amount and A is estimated from Equation 2.22, is translated into a present worth by dividing it by the capital recovery factor

Example 2.12

Taking the previous example involving a base annuity of £2500 and a gradient of £500, the two cases, involving both increasing and decreasing gradients, can be converted to present worth values using the capital recovery factor (CRF):

$$\text{CRF} = [i(1+i)^n]/[(1+i)^n - 1] = [0.1(1.1)^5]/[(1.1)^5 - 1] = 0.2638$$
$$1/\text{CRF} = 3.791$$

Solution (i)

$$P = 3405 \times 3.791 = £12\,908.36$$

Solution (ii)

$$P = 1595 \times 3.791 = £6046.65$$

See Figure 2.2 for an illustration of problem (i)

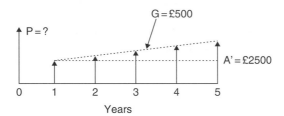

Figure 2.2 Illustration of Example 2.12.

2.8.5 Calculations involving unknown interest rates

In some situations, both present and future amounts of money are known, but the value of i which makes these two quantities equivalent needs to be estimated. Where a single payment is involved and the number of years, n, is known, Equation 2.10 can be re-arranged as follows to solve for i:

$$i = \left(\frac{F}{P}\right)^{1/n} - 1$$

Example 2.13

At what rate of interest will £2500 be worth £3506 in five years?

Solution

$$i = (3506/2500)^{1/5} = (1.4024)^{0.2} - 1 = 1.07 - 1$$
$$= 0.07 = 7\%$$

Where the payments are in the form of a uniform or geometric series, a trial-and-error method must be employed. For such problems, the use of interest tables commonly found in most engineering economics textbooks, and found at the back of this text, makes it relatively simple to solve the problem by interpolation.

Within interest tables, the values for the eight factors described in this chapter are estimated for interest rates that usually vary from 0.25% to 50% and for periods from one to sometimes up to 100 years. The values for each interest rate are usually detailed on a separate page. The tables are normally set up with the various factors ranged across the top of the page, with the number of periods down, n, down the left-hand column. The basic tables used normally refer to discrete compounding where the interest is compounded once within each interest period. For a given factor type, interest rate and value of n, the appropriate factor value can be located at the intersection of the row with the appropriate value of n and the column containing the values for the factor under examination.

Example 2.14
A firm has an investment policy that will yield £50 000 ten years from today. The company must pay £2000 per annum, the first payment falling due one year from now. What rate of return will the company have obtained on its investment?

Solution

$$F = A\{[(1+i)^n - 1]/i\}, \text{ therefore}$$
$$50\,000 = 2000((1+i)^n - 1)/i$$
$$[(1+i)^{10} - 1]/i = 25$$

Try 20%:

$$[(1.2)^{10} - 1]/0.2 = 25.95$$

Try 19%:

$$[(1.19)^{10} - 1]/0.19 = 24.7$$

The answer is approximately 19.24%.
(This can also be solved using interest tables. Going to the *F/A* factor, and going down the column until one finds the row where $n = 10$, the value of *F/A* for the interest rate on that page is located. The page with 20% compound interest factors gives a value of 25.95. The page with 18% gives a value of 23.52. Interpolating between these two values gives a value for i of 19.24%.)

2.8.6   Calculation of unknown years

There may be circumstances where both present and future amounts of money are known along with the value of i, but the number of years required for the present investment to accumulate to a future desired amount is unknown. Such problems are generally solved either by trial-and-error or by use of the interest tables.

> ### Example 2.15
> A developer receives £50 000 income from a toll road, now, £120 000 two years from now, and £140 000 four years from now. How many years from now will the total investment be worth £400 000, assuming a constant interest rate of 4%?
>
> *Solution*
>
> Assuming the investment will reach £400 000 in the nth year, the £50 000 (P_A) will have accumulated over n years, the £120 000 (P_B) over (n–2) years, and the £140 000 (P_C) over (n–4) years. Therefore:
>
> $$F = P_A(F/P,i,n) + P_B(F/P,i,n-2) + P_C(F/P,i,n-4)$$
> $$400\,000 = 50\,000(1.04)^n + 120\,000(1.04)^{n-2} + 140\,000(1.04)^{n-4}$$
>
> Try $n=10$ years:
>
> $$50\,000(1.48) + 120\,000(1.369) + 140\,000(1.265) = 415\,380$$
>
> Try $n=9$ years:
>
> $$50\,000(1.42) + 120\,000(1.32) + 140\,000(1.22) = 400\,200$$
>
> Therefore, answer $\cong 9$ years. (Interest tables can be used to help calculate the different values of F/P for the different values of n.)

2.8.7   Additional examples

Some economic appraisal problems can involve the use of more than one factor. Two such situations are encountered with cash flows involving a uniform or gradient series where payments/receipts commence beyond the end of the first period and where cash flows are a combination of a uniform series and randomly placed amounts. Bringing all amounts back to present values, and combining them either to give an overall present value or to express it as an equivalent series, can solve problems of this type.

Example 2.16
Determine the present value, P, and the equivalent uniform annual worth, A, over the next ten years, of five payments of £500 made from the end of year six through to the end of year ten, given an interest rate of 5%?

Solution

$$P_{t=0} = 500/(1.05)^{10} + 500/(1.05)^9 + 500/(1.05)^8 + 500/(1.05)^7 + 500/(1.05)^6$$
$$= 1696.13 \text{(present value of the five future payments at end of years}$$
$$ 6 \text{ to } 10)$$
$$A_{t=0-10} = 1696.13 \times [0.05(1.05)^{10}]/[(1.05)^{10} - 1]$$
$$= £219.66 \text{ (10 equal annual payments)}$$

Figure 2.3 illustrates the problem.

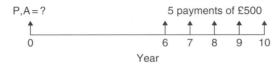

Figure 2.3 Illustration of Example 2.16.

Example 2.17
A developer will receive annual payments of £2500 at the end of each of the next eight years, together with additional lump sums of £8000 and £10 000 at the end of years 7 and 8 respectively. Assuming an interest rate of 8%, calculate the present worth of these payments?

Solution

$$P = 2500(P/A, 8\%, 8) + 8000(P/F, 8\%, 7) + 10\,000(P/F, 8\%, 8)$$
$$= 2500 \times [(1.08)^8 - 1/0.08(1.08)^8] + 8000/(1.08)^7 + 10\,000/(1.08)^8$$
$$= (2500 \times 5.7466) + (8000/1.714) + (10\,000/1.851)$$
$$= 14\,366.50 + 4667.44 + 5402.49$$
$$= £24\,436.43$$

Figure 2.4 illustrates the problem.

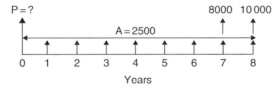

Figure 2.4 Illustration of Example 2.17.

Figure 2.5 M50 Interchange, Ireland (Source: Arup. Photographer: Peter Barrow).

2.9   Summary

The focal point of this chapter is the introduction of the concept of equivalence and interest rates, and their use within a set of engineering economics factors that allow calculations to be made regarding present, future, annual and gradient cash flows. The understanding and use of these formulae are critical to a decision maker wishing to undertake an economic appraisal of an engineering project and the ability to manipulate them is a powerful aid in the economic assessment process.

Chapter 3

Present Worth Evaluation

3.1 Introduction

The concept of present worth is a simple one. It denotes the value, in today's terms, of a given option. It is a form of analysis preferred by many engineering economists because it yields a single valuation that is less open to misinterpretation than other models. Within the context of project appraisal, once an understanding of the techniques for estimating the time value of money has been obtained, one is then in a position to evaluate an option or group of competing options on the basis of their economic performance. In addition to present worth, three other methods can be used for evaluating option performance:

(1) equivalent annual worth comparison,
(2) rate of return calculations, and
(3) benefit/cost ratio

All four require knowledge of certain assumptions that are inherent in the methods:

- All cash flows associated with each option must be known.
- All cash flows are in monetary units that are assumed to be of constant value.
- Interest rates are known. Any proposed project will be unattractive to an investor unless the investment can be recovered *with interest*. The rate of interest, i, should be the minimum rate of return that will be attractive to the investor in that situation (known as the *minimum attractive rate of return*).
- The comparison between project options is made on the basis of before-tax cash flows.
- All intangible/unquantifiable factors that cannot be valued in monetary terms are excluded from the evaluation.
- The availability of funds to finance the outlays associated with the option is not considered. It is assumed that funds will be available as and when required.

Each method will select the same most economically desirable option, if, in each case, the correct methodology is strictly adhered to.

Engineering Project Appraisal: The Evaluation of Alternative Development Schemes, Second Edition.
Martin Rogers and Aidan Duffy.
© 2012 John Wiley & Sons, Ltd. Published 2012 by John Wiley & Sons, Ltd.

The remainder of this chapter is devoted to the present worth method. The other three are dealt with in the succeeding three chapters.

3.2   Present worth – the comparison process

Present worth evaluation entails transforming all payments and receipts into a net equivalent monetary amount at time zero; that is, now. For the purposes of notation, receipts/revenues are treated as positive values on a given cash flow diagram, with payments/costs given negative values. There are generally three conditions under which the present worth evaluation method is used:

(1)   the options compared have equal lives,
(2)   the options under comparison have unequal lives, and
(3)   all options examined have unlimited lives

To execute a present worth analysis, certain elements must be known to the evaluator:

- Details of all cash flows, both in terms of the amounts of money involved and their timings. Knowledge of their timings requires an assessment spanning the life, n, of the proposal. In most situations, this refers to the economic life of the project, which is the period that maximises the proposal's net value, expressed as a present value, annual income or rate of return. At this point in time it is more profitable to replace it than continue using it. It thus represents the minimum-cost time of the proposal. It should not remain in service beyond this point.
- The interest rate at which the cash flows is to be discounted, with this value representing, at the very least, a stipulated minimum attractive rate of return for the investor.

3.2.1   Comparing options with equal lives

Estimates of present worth are made only between *coterminated* options. This ensures equivalent outcomes. Cotermination implies that the economic life of all options under consideration is the same. The lives of all assets will finish at the same point in time. If all project options have the same economic life, they are described as being of *equal service*.

If cotermination does not exist, the comparison is no longer being made on the basis of outcomes that are equivalent. For example, one piece of mechanical plant has a net cost in today's terms of £50 000 with an economic life of 5 years, while an alternative has a cost in present terms of £60 000 over a life of 8 years. In this situation a judgement based on the two present day valuations is not possible as, although the second costs £10 000 more, it is providing 3 extra years of serviceable life.

Once cotermination exists, the calculation is relatively straightforward, with the option delivering the highest net present worth deemed to be the most desirable.

Within the context of an appraisal, an engineering option such as a piece of mechanical plant or a fleet of buses or rail cars may still have an economic value which can be recouped at the end of their design life. This value is called the *residual value* or *salvage value*. This valuation is often factored into present worth estimates.

Example 3.1
Compare the present worth (PW) of the following two mechanical plant options, assuming an interest rate of 12% compounded annually. Initial purchase and annual maintenance costs are given, together with residual valuations. Details of the two options are given in Table 3.1.

Solution

$$PW_{Option A} = -50\,000 - 6000(P/A,12\%,15) + 8000(P/F,12\%,15)$$

$$= -50\,000 - 6000[(1.12)^{15} - 1]/[0.12(1.12)^{15}] + 8000[1/(1.12)^{15}]$$

$$= -50\,000 - 6000(6.81) + 8000(0.183)$$

$$= -89\,396 \,(\text{present value of net costs is } £89\,396)$$

$$PW_{Option B} = -70\,000 - 0(P/A,12\%,15) + 5000(P/F,12\%,15)$$

$$= -70\,000 + 5000(0.183)$$

$$= -69\,085 \,(\text{present value of net costs is } £69\,085)$$

Option B is selected because its present worth is less negative; that is, the present worth of its costs is less.

	Option A	Option B
Initial cost (£)	50 000	70 000
Annual maintenance cost (£)	6 000	0
Residual value (£)	8 000	5 000
Economic life (Years)	15	15

Table 3.1 Present worth comparison of options with equal lives.

3.2.2 Comparing options with unequal lives

The assumption of equal service or cotermination implies that the economic lives of all options considered within a present worth evaluation have a common endpoint. Situations will arise, however, where options with unequal economic lives must be compared. In such cases, the problem must be manipulated so that the options concerned can be compared over the same number of years. This process results in an even-handed comparison of all options, where the valuations of present worth estimated represent the payments and receipts associated with equal service. If the economic lives of the different options are not equalised, the resulting present worth evaluation will generally favour the option with the shorter life span, given that less costs will be incurred, even though it may not be the most economically desirable one.

The requirement for equal service can be met by comparing the options over a period equal to the *least common multiple* (*LCM*) for their economic lives. Within this approach, comparing options over the least common multiple value results in the extension of their cash flows to the same future point in time. Within this method, for a given option, the cash flow for one cycle is assumed to be repeated for the least common multiple number of years, thus allowing the comparison of all options over an equal time frame. For example, if two options are being compared, one with an economic life of 6 years, the other with one of 9 years. Both will be assumed to coterminate after 18 years, their least common multiple. At that stage, the first option will have gone through three complete cycles, with the second going through two.

Example 3.2

A developer has two design options for building a small storage warehouse. One involves the use of timber, the other uses structural steel. Financial data for the two plans are shown in Table 3.2. Compare the two options on the basis of their present worth, assuming an interest rate of 12%.

Solution

The LCM Method assumes that identical models having the same costs will replace the assets associated with a given option on a cyclical basis. Cotermination results from the comparison of costs over a period that can be divided evenly by the service lives of the options involved. For this example, the least common multiple is 30 years, with the timber option cycle being repeated three times (0–10, 10–20, 20–30) and the steel option cycle being repeated twice (0–15, 15–30).

$$PW_{timber} = 40\,000 + 6500(P/A,12,30) + (40\,000 - 5000)[(P/F,12,10)$$
$$+ (P/F,12,20)] - 5000(P/F,12,30)$$
$$= 40\,000 + 6500(8.055) + 35\,000(0.322 + 0.104) - 5000(0.033)$$
$$= 40\,000 + 52\,357.5 + 14\,910 - 165$$
$$= £107\,432.50\,(\text{negative worth})$$

$$PW_{steel} = 60\,000 + 3500(P/A,12,30) + (60\,000 - 0)[(P/F,12,15)]$$
$$= 60\,000 + 3500(8.055) + 60\,000(0.183)$$
$$= 60\,000 + 28\,192.50 + 10\,980$$
$$= £99\,172.50\,(\text{negative worth})$$

The steel structure is preferred because the present worth of 30 years of its service is less negative than that for the timber structure.

Figure 3.1 shows a cash flow diagram for options with different lives.

	Timber	Steel
Initial cost (£)	40 000	60 000
Annual Maintenance cost (£)	6500	3500
Residual value (£)	5000	0
Economic Life (Years)	10	15

Table 3.2 Present worth comparison of options with unequal lives.

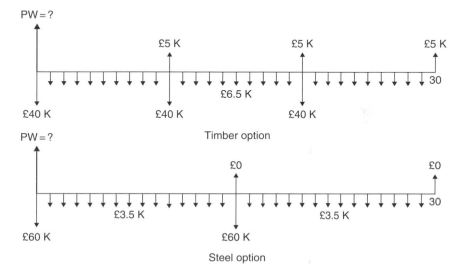

Figure 3.1 Cash flow diagram for options with different economic lives.

Another method for estimating the present worth of different-life options is the *planning horizon* approach. Within this, a time horizon is selected over which the analysis is to be carried out, and only those cash flows that occur within that chosen period are considered. Any other cash flows existing beyond this agreed point, whether payments or receipts, are ignored. The time horizon selected may be of relatively short duration. An estimate of the residual value at the end of the chosen period must be derived for all of the options studied. This methodology is used within replacement analysis, which is beyond the scope of this text.

While this method of present worth analysis for different-life options may be more straightforward and, indeed, more realistic than the LCW method, the latter technique is emphasised here because of its ability to give the reader a deeper understanding of the concept of equal service.

3.2.3 Comparing options with infinite lives

Capitalised cost (*CC*) is defined as the sum of the initial capital cost and the present worth of subsequent payments or receipts that are assumed to last forever. It is thus the present worth of a project option that is assumed to have an infinite sequence of cash flows, and thus provides extended service. It is used particularly to evaluate engineering projects, such as dams and tunnels, that provide extremely lengthy service.

The capitalised cost is estimated in the same manner as the present worth valuation, but with the number of years, n, set equal to infinity. As a result, the choice of interest rate determines, to a great extent, the final value obtained. From the previous chapter, the uniform series present worth factor ($P/A,i,n$) was defined by the formula:

$$P = A\left[\frac{(1+i)^n - 1}{i(1+i)^n}\right]$$

If above and below the line are divided by $(1+i)^n$, the equation becomes:

$$P = A\left[\frac{1 - (1 \div (1+i))^n}{i}\right]$$

As n approaches infinity, the term above the line becomes equal to one, yielding the following equation:

$$P = A/i \tag{3.1}$$

or:

$$(P/A,i,\infty) = 1/i$$

If P_0 represents the initial cost, the capitalised cost can be defined as:

$$\text{Capitalised cost} = P_0 + A/i \tag{3.2}$$

The comparison process of two or more options using capitalised costs involves:

- drawing up the cash flow diagram for two or more cycles of all recurring costs/ receipts, and indicating all once-off costs/receipts on it;
- calculating the present worth of all one-off costs/receipts;
- calculating the uniform annual worth (A) through one life cycle of all recurring costs/receipts;
- adding this valuation to all other uniform amounts occurring between the year 1 and infinity to give the total annual worth;
- dividing the value obtained by the interest rate, i, and summing this with the present value of all one-off costs/receipts to give the capitalised cost.

Example 3.3

The Dublin Port Authority intends to build a bridge spanning the River Liffey to the east of the city. There are two main options. The first is a high-level reinforced concrete bridge, spanning the river at its widest point, which has the advantage of connecting directly into the motorway system orbiting the city, but does not link in the local road network and is environmentally intrusive. The second option is a low-level steel bridge that crosses the river at a much narrower point but requires much greater additional road construction to connect into the motorway network. It links into the local network easily and is much less environmentally intrusive.

The high-level bridge has an initial construction cost of £40m, with maintenance costs estimated at £20 000 per annum. Its asphalt road surface would have to be re-laid every 9 years at a cost each time of £75 000. The low-level bridge and ancillary works are estimated at £35m, with maintenance costs estimated at £15 000 per annum. Its steel frame must be painted every 4 years at a cost of £220 000. Its road surface must be re-laid every 9 years at a cost of £45 000.

Compare the two options on the basis of their respective capitalised costs, assuming an interest rate of 5%.

Solution

(i) High-level bridge

Initial cost:

$$P_0 = -40\text{m}$$

Recurring maintenance cost (for total life):

$$A_1 = -20\,000$$

Recurring surfacing cost (every 9 years):

$$A_2 = -75\,000(\text{A/F}, 5\%, 9) = -75\,000(0.0907) = -6801$$

Total recurring costs:

$$A_T = A_1 + A_2 = -26\,801$$

Total capitalised costs:

$$\text{PW} = P_0 + A_T\,/i = -(40\,000\,000 + (26\,801.76\,/.05)) = -40\,536\,035$$

(ii) Low-level bridge

Initial cost:

$$P_0 = -35\text{m}$$

Contd

Example 3.3 Contd

Recurring maintenance cost (for total life):

$$A_1 = -15\,000$$

Recurring painting cost (every 4 years):

$$A_2 = -220\,000\ (A/F, 5\%, 4) = 220\,000\,(0.232) = 51\,040$$

Recurring resurfacing cost (every 9 years):

$$A_3 = -45\,000\,(A/F, 5\%, 9) = -45\,000\,(0.0907) = 4081$$

Total recurring costs:

$$A_T = A_1 + A_2 + A_3 = -(15\,000 + 51\,040 + 4081.5) = -70\,121$$

Total capitalised cost:

$$PW = P_0 + A_T/i = -\left(35\,000\,000 + 70\,121.50\,/\,0.05\right) = -36\,402\,430$$

The economic analysis favours construction of the low-level bridge as it has a lower capitalised cost.

3.2.4   Life-cycle cost analysis

One particularly well used application of the system of present worth computation is life-cycle costing. It has proved itself a readily usable methodology within engineering economics over the last half-century. It considers every cost associated with an engineering project over its entire economic life. Its purpose is to minimise the total cost incurred by the project developer by selecting the option not just on the basis of its initial cost but also on the basis of those additional expenses/revenues that occur subsequently over the life of the project. The costs incurred during the life of a project tend to follow a certain pattern. Initially, investigation and design costs are incurred, followed by acquisition and installation costs while the project is being put in place. During the course of its economic life, maintenance costs are incurred before possible disposal costs come into play at the end of the period. A negative cost, arising from the salvage value of the project at the termination point, may apply in some cases.

The factors *A/F* and *P/F* are applied within the life-cycle cost analysis, in the same way as they have been applied to the problems shown thus far within this chapter.

What sets this method apart, however, is the level of effort applied to including all types of cost likely to be incurred over the economic life of the project, which in the case of projects analysed is usually in the order of between 15 and 30 years. The option with the least life-cycle cost is identified as the best performing one, and is selected on that basis.

Example 3.4

The transport planner within a bus company has two proposed new routes for which there will be practically the same customer demand. The planner can only afford to proceed with one, and decides to make this choice on the basis of which one has the least life-cycle cost associated with it. An interest rate of 8% is assumed.

The following are the cost details for the two proposed routes:

(i) Route A

Five new medium-sized buses required. The buses will operate a total of 500 000 vehicle-kilometres per year, 25 000 vehicle-hours per year.

    Economic Life of each bus: 15 years
    Purchase costs: 5 buses at £220 000 per bus
    Operating costs (wages, benefits, etc.): £40 per vehicle-hour
    Maintenance costs: £0.15 per vehicle-kilometre
    Fuel costs: £0.45 per vehicle-kilometre
    Salvage value per bus after 15 years: £25 000

(ii) Route B

Nine new small-sized buses are required. The buses will operate a total of 450 000 vehicle-kilometres per year, 28 000 vehicle-hours per year.

    Economic Life of each bus: 15 years
    Purchase costs: 9 buses at £180 000 per bus
    Operating costs (wages, benefits, etc.): £35 per vehicle-hour
    Maintenance costs: £0.12 per vehicle-kilometre
    Fuel costs: £0.40 per vehicle-kilometre
    Salvage value per bus after 15 years: £15 000

Solution

(i) Route A

Purchase cost	= (£220 000/bus)(5 buses)	= £1 100 000
Operating cost	= (£40/veh-h)(25 000veh-h/yr)	= £1 000 000/yr
Maintenance cost	= (£0.15/veh-km)(500 000veh-km/yr)	= £75 000/yr
Fuel costs	= (£0.45/veh-km)(500 000veh-km/yr)	= £225 000/yr
Salvage value	= (£25 000/bus)(5 buses)	= £125 000

Contd

Examples 3.4 Contd
Present value of costs:

$$= 1100\,000 + (1000\,000 + 75\,000 + 225\,000)(P/A, 8\%, 15)$$
$$\quad - (125\,000)(P/F, 8\%, 15)$$
$$= 1100\,000 + (1\,300\,000)(8.559) - (125\,000)(0.315)$$
$$= £12\,187\,325$$

(ii) Route B

Purchase cost	= (£180 000/bus)(9 buses)	= £1 620 000
Operating cost	= (£35/veh-h)(28 000veh-h/yr)	= £980 000/yr
Maintenance cost	= (£0.12/veh-km)(450 000veh-km/yr)	= £54 000/yr
Fuel costs	= (£0.40/veh-km)(450 000veh-km/yr)	= £180 000/yr
Salvage value	= (£15 000/bus)(9 buses)	= £135 000

Present value of costs:

$$= 1\,620\,000 + (980\,000 + 54\,000 + 180\,000)(P/A, 8\%, 15)$$
$$\quad - (135\,000)(P/F, 8\%, 15)$$
$$= 1\,620\,000 + (1\,214\,000)(8.559) - (135\,000)(0.315)$$
$$= £11\,968\,101$$

The transport planner decides to opt for Route B as its life-cycle cost is lower.

3.2.5   Payback comparison method

This is an extremely simple procedure that can be used to make an estimate of the time it will take for the project to recoup its costs. It does not require information on an appropriate interest rate, but the lack of accuracy of the method requires that results from it should not be given the same weight as those from some of the other techniques outlined in this chapter. The method assumes that a given proposal will generate a stream of monies during its economic life and, at some point in time, the total value of this stream will exactly equal its initial cost. The time taken for this equalisation to occur is called the payback period. It is normally applied to small-scale projects where the timescale for equalisation is relatively short. The method itself does not concern itself with the performance of the project option after the payback period has occurred. Its analysis is thus not as complete as the others described earlier, and will, therefore, favour some projects more than others, often unfairly. It is thus best used as a back-up technique, used to supplement the information from one of the more comprehensive economic evaluation methods.

The payback period can be derived from the formula:

$$\text{Payback period } (n_p) = (C_0 \div \text{NAS}) \qquad (3.3)$$

where C_0 is the initial cost of the project and NAS is the net annual savings or net cash flow that will recover the investment after the payback period, n_p.

Equation 3.3 assumes that a zero discount value is being used. This is not always the case. If it is assumed that the net cash flows will be identical from year to year, and that these cash flows will be discounted to present values using a value $i \neq 0$, then the uniform series present worth factor (P/A) can be used within the equation:

$$0 = C_0 + \text{NAS}(P/A, i, n_p) \qquad (3.4)$$

Equation 3.4 is solved to obtain the correct value of n_p.

The method is, however, widely used in its simplified form, with the discount rate, i, set equal to zero, even though its final value may lead to incorrect judgements being made. If the discount rate, i, is set equal to zero in Equation 3.4. the following relationship is obtained:

$$0 = C_0 + \sum_{t=1}^{t=n_p} \text{NAS}_t \qquad (3.5)$$

Equation 3.5 reduces to $n_p = C_0/\text{NAS}$, exactly the same expression as given in Equation 3.3.

Given its shortcomings, why is it used by engineering economists so frequently? It is mainly because of its simplicity and the fact that it appears so logical. Also, it does address a question that it very important to a project's developer, as a relatively speedy payback will protect liquidity and release funds more quickly for investment in other ventures. This is particularly the case in times of recession when cash availability may be limited. Projects with a relatively short payback period can be attractive to a prospective developer. The short time frame tends is seen as lessening the risk associated with a venture.

Example 3.5

A contractor wishes to buy a new piece of site equipment and has a choice between two options, Machine A and Machine B. Machine B is more expensive than A, but is more elaborate and technologically advanced, and will have a longer serviceable life. Details of the costs and revenues associated with each option are given in Table 3.3.

Calculate the payback period and use the present worth method to justify the validity of your initial result.

Contd

Examples 3.5 Contd
Solution

(i) Machine A

Payback period $(n_p) = C_0/\text{NAS} = 24\,000/8750 = 2.74$

(ii) Machine B

Payback period $(n_p) = C_0/\text{NAS} = 52\,000/1400 = 3.71$

Therefore, on the basis of simple payback, Machine A is preferred as the initial outlay is recouped in almost one year less.

Let us now estimate the present worth of the two options. To do this more realistically, assume a discount rate of 7%. Given that Machine A has half the economic life of Machine B, two cycles of Machine A are assessed to bring the two economic life-spans to equality:

$$PW_A = -24\,000 - 24\,000\,(P/F,7,4) + 8750\,(P/A,7,8)$$
$$= -24\,000 - 24\,000\,(0.763) + 8750\,(5.97)$$
$$= £9925.50$$

$$PW_B = -52\,000 + 14\,000\,(P/A,7,8)$$
$$= -52\,000 + 14\,000\,(5.97)$$
$$= £31\,580$$

On the basis of its present worth evaluation, Machine B is the preferred option, as its net present worth is greater. Even with much smaller of larger discount rates, Machine B will continue to perform best using this method of analysis.

Note: This example illustrates why payback analysis should not be treated as a primary method of economic evaluation. In this case Option A, with the shorter economic life, repaid itself earlier but did not possess the later-in-life stream of benefits which accrued to Option B with its longer life span. This fact was emphasised by the more complete present worth comparison carried out on the two options.

	Machine A	Machine B
Initial cost (£)	24 000	52 000
Annual savings (£)	8750	14 000
Economic life (years)	5	10

Table 3.3 Comparison of two options using payback analysis.

3.2.6   Additional examples

Example 3.6
Compare the following two options on the basis of a present worth analysis. A discount rate of 10% applies. Details of the costs and revenues associated with both options are given in Table 3.4.

Solution

For this example, the least common multiple is 16 years, with Option B being repeated twice.

$$PW_{Option\ A} = -40\,000 - 7000\,(P/A,10,16) - 2500(P/F,10,5)$$
$$- 2500\,(P/F,10,10) - 2500\,(P/F,10,15) + 4000\,(P/F,10,16)$$
$$= -40\,000 - 7000\,(7.82) - 2500\,(0.621)$$
$$- 2500\,(0.386) - 2500\,(0.239) + 4000\,(0.218)$$
$$= -40\,000 - 54\,740 - 1553 - 965 - 598 + 872$$
$$= -£96\,984$$

$$PW_{Option\ B} = -20\,000 - 8000\,(P/A,10,6) - 20\,000\,(P/F,10,8) - 3000\,(P/F,10,4)$$
$$- 3000\,(P/F,10,12) + 2500\,(P/F,10,8) + 2500\,(P/F,10,16)$$
$$= -20\,000 - 8000\,(7.82) - 20\,000\,(0.467) - 3000\,(0.683)$$
$$- 3000\,(0.319) + 2500\,(0.467) + 2500\,(0.218)$$
$$= -20\,000 - 62\,560 - 9340 - 2049 - 957 + 1168 + 545$$
$$= -£93\,193$$

Therefore, Option B is preferred as it will cost less over the coterminated life of 16 years.

	Option A	Option B
Initial cost (£)	40 000	20 000
Annual operating cost (£)	7 000	8 000
Maintenance work every 4 years	–	3 000
Maintenance work every 5 years	2 500	–
Salvage value	4 000	2 500
Economic life (years)	16	8

Table 3.4
Comparison of two options using present worth analysis.

Example 3.7

Compare the following two options on the basis of a present worth analysis. A discount rate of 12% applies, with the index k going from 1 to 10. Details of the costs and revenues associated with each option are given in Table 3.5.

Solution

For this example, the least common multiple is 5 years, with Option A being repeated twice.

$$PW_{Option\,A} = -22\,000 - 2050\,(P/A, 12, 10) - 50\,(P/G, 12, 10) - 22\,000\,(P/F, 12, 5)$$

$$= -22\,000 - 2050\,(5.65) - 50\,(20.25) - 22\,000\,(0.567)$$

$$= -22\,000 - 11\,583 - 1013 - 12\,474$$

$$= -£47\,069$$

$$PW_{Option\,B} = -40\,000 - 1040\,(P/A, 12, 10) - 40\,(P/G, 12, 10)$$

$$= -40\,000 - 1040\,(5.65) - 40\,(20.25)$$

$$= -40\,000 - 5876 - 810$$

$$= -£46\,686$$

Therefore, Option B is preferred as it will cost less over the coterminated life of 10 years.

	Option A	Option B
Initial cost (£)	22 000	40 000
Annual operating cost (£)	2000 + 50k	1000 + 40k
Economic life (years)	5	10

Table 3.5 Comparison of two options using present worth analysis.

Example 3.8

Compare the following two options on the basis of their capitalised costs (CC). A discount rate of 10% applies. Details of the costs and revenues associated with each option are given in Table 3.6.

Solution

Using the A/P and A/F formulae detailed within Chapter 2:

$$AW_{Option\,A} = -10\,000 - 100\,000\,(A/P, 10, 5) + 5000\,(A/F, 10, 5)$$

$$= -10\,000 - 100\,000\,(0.264) + 5000\,(0.164)$$

$$= -10\,000 - 26\,380 + 819$$

$$= -£35\,561$$

Contd

Example 3.8 Contd

$$CC_{Option\,A} = -35\,561/0.1$$
$$= -£355\,611$$

$$CC_{Option\,B} = -280\,000 - 8000/0.1$$
$$= -£360\,000$$

Therefore, Option A is preferred to Option B on the basis that its capitalised cost is less.

	Option A	Option B
Initial cost (£)	100 000	280 000
Annual operating cost (£)	10 000	8 000
Salvage value	5 000	2 000
Economic life (years)	5	∞

Table 3.6 Comparison of two options using capitalised cost analysis.

Figure 3.2 British Airways cargo facility, UK (Source: W.S. Atkins).

3.3   Summary

All the main techniques within this chapter assess different project options through the conversion of cash flows into single-value present amounts. When the options being compared have different economic lives, manipulation of the figures allows a direct comparison to be made on the basis of equal service. Life-cycle cost analysis is seen as an important application of the present worth technique. Payback analysis is introduced to the reader as a method that can be used to supplement the main procedures outlined within the chapter.

Chapter 4

Equivalent Annual Worth Computations

4.1   Introduction

This method converts all payments and receipts arising from a project into an equivalent uniform annual amount of money. It produces results entirely compatible with those obtained from the present worth method outlined in Chapter 3. Its utility as a method stems from the tendency to view the performance of a project or an organisation over a twelve-month cycle, within which profits/losses are made and taxes are declared. The set of assumptions listed at the start of the previous chapter regarding the use of present worth apply equally to the use of this method.

4.2   The pattern of capital recovery

The capital recovery factor $(A/P, i, n)$ forms the basis for the computation of annual worth, as it converts single payments into equivalent annual amounts, be they costs measured as a negative cash flow or receipts which are assigned positive values. For a given initial investment, its value indicates the amount of money repaid or recovered together with the interest earned on that proportion of the initial amount still to be paid. Therefore, where a piece of machinery is being paid for over its economic life via a series of uniform payments, each payment will comprise less interest and more capital recovery as we progress towards the end of its lifespan.

For example, a contractor purchases an excavator for £50 000 that will be paid for over 5 years in equal annual payments at the end of each 12 months. Assuming a discount rate of 8%, the capital recovery factor (CRF) is:

$$CRF = i(1+i)^n / (1+i)^n - 1 = 0.08 (1.08)^5 / [(1.08)^5 - 1] = 0.2505$$

Engineering Project Appraisal: The Evaluation of Alternative Development Schemes, Second Edition.
Martin Rogers and Aidan Duffy.
© 2012 John Wiley & Sons, Ltd. Published 2012 by John Wiley & Sons, Ltd.

Therefore:

Annual repayment=50 000(0.2505)=£12 522.82

Given that interest is payable only on the outstanding amount, at the end of the first 12 months, when the first payment falls due, interest will be due on the entire £50 000; that is, £4000. Given that the first repayment is £12 552.82, all but £4000 of this will be available to pay off the total capital.

End of year 1
Total outstanding   = £50 000
Total payment   = £12 522.82
Interest due   = £4000 (8% of £50 000)
Capital recovered   = £8522.82
Capital unrecovered = £41 477.18

Thus, at the end of year 1, £8552.82 is recovered, leaving the unrecovered capital at £41 477.18. At the end of the second year, interest must be paid on this reduced amount. This interest will total £3318.17, leaving £9204.65 available from the payment to pay off the remaining capital.

End of year 2
Total outstanding   = £41 477.18
Total payment   = £12 522.82
Interest due   = £3318.17 (8% of £41 477.18)
Capital recovered   = £9204.65
Capital unrecovered = £32 272.53

By the time the last repayment is being made at the end of the fifth year, only £927.62 is paid as interest, with the rest of the £12 552,82 going towards paying off the total quantity of unrecovered capital. Thus, the relative proportions of the total payment going towards paying interest due falls from just over 30% at the end of the first year to just over 7% by the end of year 5.

Table 4.1 shows the variations in capital recovered and interest paid over the life of the project.

Table 4.1 Variations in capital recovered and interest paid over life of a project.

End of year	Capital outstanding (£)	Interest due (£)	Capital recovered (£)	Total annual payment (£)
0	50 000.00			
1	41 477.18	4 000.00	8 552.82	12 552.82
2	32 272.53	3 318.17	9 204.65	12 552.82
3	22 331.51	2 581.80	9 941.02	12 552.82
4	11 595.21	1 786.52	10 736.3	12 552.82
5	0.00	927.62	11 595.21	12 552.82

4.3   Modifying annual payments to include salvage value

If a salvage value is to be included in the equation for equivalent annual worth, the basic Equation 2.16 must be modified as follows:

Annual worth=the present cost (converted to a uniform annual payment using the capital recovery factor) + the salvage value (converted to a uniform annual amount using the sinking fund factor)

This equation can be written as:

$$AW = -P(A/P,i,n) + SV(A/F,i,n) \tag{4.1}$$

Let us look at the equations for the two factors contained in Equation 4.1 in some detail.

From Equation 2.14, the sinking fund factor can be expressed as:

$$(A/F,i,n) = \left[\frac{i}{(1+i)^n - 1}\right]$$

From Equation 2.16, the capital recovery factor can be expressed as:

$$(A/P,i,n) = \left[\frac{i(1+i)^n}{(1+i)^n - 1}\right]$$

Subtracting i from Equation 2.16 gives the following expression:

$$\frac{i(1+i)^n}{(1+i)^n - 1} - i = \frac{i(1+i)^n - i(1+i)^n + i}{(1+i)^n - 1} = \frac{i}{(1+i)^n - 1}$$

This expression, as illustrated by Equation 2.14, is the sinking fund factor $(A/F,i,n)$. Therefore, it can be deduced that:

$$\begin{aligned} AW &= -P(A/P,i,n) + SV[(A/P,i,n) - i] \\ &= -(P - SV)(A/P,i,n) - SV(i) \end{aligned} \tag{4.2}$$

This equation makes sense intuitively. Say, for example, a manufacturer buys a piece of equipment at a cost of £20 000, which, after its 5-year economic life, will have a savage value of £5000. In this case, the actual amount that must be recovered through annual payments is £15 000 (£20 000 – 5000), with the rest of the original cost being retrieved via the £5000 salvage value of the equipment. This amount is estimated within the $-(P–SV)(A/P,i,n)$ term in Equation 4.2. Since the buyer has no access to the £5000 until the end of the equipment's economic life, the interest forgone over this period represents unrecovered capital, and can thus be treated as a net cost to him. The $SV(i)$ term within Equation 4.2 represents the cost incurred by the buyer due to interest foregone over the machine's economic life.

Example 4.1

A contractor buys an excavator for £65 000. The operation and maintenance costs over its 8-year life span total £15 000 per year. The salvage value is put at £25 000. Assuming an interest rate of 9%, calculate the equivalent annual worth of the excavator.

Solution

$$AW = -(P - SV)(A/P, i, n) - SV(i) - annual\,maintenance\,cost$$
$$= -(65\,000 - 25\,000)(A/P, 9\%, 8) - 25\,000\,(0.09) - 15\,000$$
$$= -40\,000[0.09\,(1.09)^8 / (1.09)^8 - 1] - 2250 - 15\,000$$
$$= -40\,000(0.1807) - 2250 - 15\,000$$
$$= -£24\,478$$

Example 4.2

Find the annual worth of a piece of equipment which has a purchase cost of £12 000 and a salvage value at the end of its 5-year economic life of £2500. Its maintenance costs are £1000 in year 1, and rising by a further £250 every subsequent year. Assume an interest rate of 6%.

Solution

$$AW = -(12\,000 - 2500)(A/P, 6\%, 5) - 2500\,(0.06) - 1000$$
$$\quad - 250(A/G, 6\%, 5)$$
$$= -9500(0.2374) - 150 - 1000 - 250(1.8836)$$
$$= -£3876.20$$

This is the equivalent annual cost of the piece of equipment to the purchaser over the 5-year economic life. Figure 4.1 illustrates the problem.

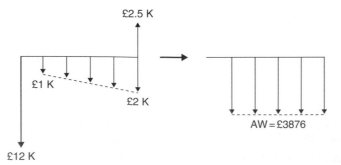

Figure 4.1 Conversion of purchase and maintenance costs plus salvage value into equivalent annual value.

4.4   Evaluating a single project

Thus far we have been calculating the annual worth of a particular purchase to the investor. The only revenue we have established is the salvage value of the purchase at the end of its economic life. As a result, therefore, the annual worth estimated has been negative, indicating a constant cost to the investor on an continuing annual basis. However, in reality, the purchase must reap some benefit to the investor that will recover the capital invested plus, possibly, some additional rate of return. When evaluating a single project, therefore, the only question to be answered is whether the purchase or project will generate a positive, neutral or negative cash flow. If the equivalent annual cash flow is positive, the project will yield the investor a net cash flow in addition to providing a rate of return, i, on the capital. As stated at the outset of Chapter 3, the value of i must be set at least equal to the minimum rate of return likely to be attractive to the investor. If the annual cash flow is zero, the project is one which will neither incur net costs nor generate net revenue, but is delivering exactly the rate of return, i, selected by the investor. If the annual cash flow is negative, no economic basis exists for the investor to proceed with the project, as annual net costs are being incurred at the minimum rate of return. The project is thus operating below the minimum attractive rate of return designated by the investor.

Example 4.3

A quarry owner buys five trucks at a cost of £30 000 per vehicle. He proposes to sell them in five years time at a salvage value of £8000. Total annual maintenance costs are estimated at £2000. Through their rental to building contractors, the quarry owner estimates that the revenue from the trucks will be £8500 per year in total. If the interest rate is set at 10% per annum, use AW to decide if the trucks should be bought.

Solution

$$AW = -(P - SV)(A/P, i, n) - SV(i) - annual\,maintenance\,cost$$
$$+ annual\,revenue$$
$$= -(30\,000 - 8000)(A/P, 10\%, 5) - 8000(0.10) - 2000 + 8500$$
$$= -22\,000(0.2638) - 800 - 2000 + 8500$$
$$= -£103.60$$

At the chosen discount rate, the proposal will generate a very small net negative cash flow. Given that the computed AW value is less than zero, the project should not proceed.

4.5   The comparison process

4.5.1   Introduction

Whereas the assessment process for a single project is based on whether $AW \geq 0$, selection of the best option from a number of possible projects is usually done on the basis of which one delivers the highest equivalent annual worth. Where a set of projects generates costs only, the one that generates the lowest equivalent annual cost is selected.

4.5.2   Equal life projects

Let us return to an option choice problem that was solved using the present value method in Example 3.1. This demonstrates that the same answer will be obtained regardless of which technique is used

> ### Example 4.4
> Assuming an interest rate of 12% compounded annually, compare the equivalent annual worth of two mechanical plant options. Option A has an initial cost of £50 000, an annual maintenance cost of £6000 and a residual value after 15 years of £8000. Option B has an initial cost of £70 000, no annual maintenance cost, and a residual value after 15 years of £5000.
>
> *Solution*
>
> $$AW_{Option\,A} = -(P - SV)(A/P, i, n) - SV(i) - annual\,maintenance\,cost$$
> $$= -(50\,000 - 8000)(A/P, 12\%, 15) - 8000(0.12) - 6000$$
> $$= -(42\,000)(0.1468) - 960 - 6000$$
> $$= -£13125.60$$
> $$AW_{Option\,B} = -(P - SV)(A/P, i, n) - SV(i) - annual\,maintenance\,cost$$
> $$= -(70\,000 - 5000)(A/P, 12\%, 15) - 5000(0.12) - 0$$
> $$= -(65\,000)(0.1468) - 600$$
> $$= -£10\,142$$
>
> Option B is selected because its annual worth is less negative; that is, the value of its equivalent annual costs is less than that for Option A.

4.5.3   Lease or buy?

This type of problem, where the choice in broad terms is between a large one-off payment and a number of smaller annual payments, is one that typically uses equivalent annual worth to find a solution.

Example 4.5

A factory owner wishes to decide whether to buy a photocopier for £8500 that can be sold after 3 years for £2000, or lease a comparable machine for £3750 per year. The organisation expects a return of 16% on investments. The maintenance costs for each option are assumed to be the same. What should be done?

Solution

$$AW_{Buy} = -(P - SV)(A/P, i, n) - SV(i)$$
$$= -(8500 - 2000)(A/P, 16\%, 3) - 2000(0.16)$$
$$= -(6500)(0.445) - 320$$
$$= -£3212.50$$
$$AW_{Lease} = -£3750$$

Since the equivalent annual cost for buying is less, the factory owner should proceed with this option.

4.5.4   Projects with different lives

Within an annual worth analysis of project options with different economic lives, only one life cycle for each option need be considered, since the annual worth will remain the same regardless of the number of life cycles analysed. Certain assumptions, however, which make the method consistent with the present worth methodology detailed in Chapter 3, must be observed:

- The options being compared will be required for the least common multiple of their economic lives.
- The cash flow in succeeding live cycles will be subject to the same discount rate as that assumed for the first.

Again, let us return to Example 3.2 involving the evaluation of options with different economic lives. Again the same answer is obtained as with the present worth method.

Example 4.6

A developer has two design options for building a small storage warehouse. One option involves the use of timber while the other uses structural steel. Option 1, the timber option, costs £40 000 initially, incurs an annual maintenance cost of £6500, and has a residual value after 10 years of £5000. Option 2, the steel option, involves an initial outlay of £60 000, costs £3500 in maintenance costs annually and at the end of its fifteen-year economic life has no salvage value.

Contd

Example 4.6 Contd

Compare the two options on the basis of their annual worth, assuming an interest rate of 12%.

Solution

$$AW_{Timber} = -(P - SV)(A/P, i, n) - SV(i) - annual\, maintenance\, cost$$
$$= -(40\,000 - 5000)(A/P, 12\%, 10) - 5000(0.12) - 6500$$
$$= -(35\,000)(0.177) - 600 - 6500$$
$$= -£13\,295$$
$$AW_{Steel} = -(P - SV)(A/P, i, n) - SV(i) - annual\, maintenance\, cost$$
$$= -(60\,000 - 0)(A/P, 12\%, 15) - 0(0.12) - 3500$$
$$= -(60\,000)(0.1468) - 3500$$
$$= -£12\,308$$

The steel option is proceeded with on the basis that it incurs a smaller equivalent annual cost.

Example 4.7

A contractor is purchasing steel-bending machinery. Two options are being considered. The first will cost £25 000, has a salvage value at the end of its 8-year life of £4500 and an annual maintenance and operating cost of £15 000. The second option has an initial cost of £75 000, a 12-year economic life, annual operating and maintenance costs of £9500, an upgrade cost after 5 years of £7000 and a salvage value of £30 000. If the required rate of return for the contractor is 20%, which machine should be chosen?

Solution

$$AW_{Option\,1} = -(P - SV)(A/P, i, n) - SV(i) - annual\, maintenance\, cost$$
$$= -(25\,000 - 4500)(0.2606) - 4500(0.2) - 15\,000$$
$$= -£21\,242.30$$
$$AW_{Option\,2} = -(P - SV)(A/P, i, n) - SV(i) - annual\, maintenance\, cost$$
$$= -[75\,000 + 7000\,(P/F, 20\%, 5) - 30\,000](A/P, 20\%, 12)$$
$$\quad - 30\,000(0.2) - 9500$$
$$= -[75\,000 + 7000\,(0.4019) - 30\,000](0.2253) - 6000 - 9500$$

Contd

Example 4.7 Contd

$$= -(75\,000 + 2813.30 - 30\,000)(0.2253) - 15\,500$$
$$= -(47\,813.30)(0.2253) - 15\,500$$
$$= -10\,770.66 - 15\,500$$
$$= -£26\,272.30$$

The first option should be selected on the basis that it incurs a smaller equivalent annual cost.

4.5.5   Projects with infinite lives

As with the present worth method, there are instances where an asset is treated as if it will last to infinity. In reality, this is never the case, as nothing man-made can last forever. Canals, dams, bridges and other large-scale projects can, in certain circumstances, come within this category because of their very long economic lives. Given that the difference between infinity and, say, 80 years is very small, we can assume such projects to have a perpetual life in economic terms.

$$(A/P, I, n) = \frac{i(1+i)^n}{(1+i)^n + 1}$$

As n approaches infinity, the term below the line approaches $(1+i)^n$. Therefore, the limit of the capital recovery factor as n goes to infinity is:

$$(A/P, I, n)_\infty = i \qquad\qquad (4.3)$$

Let us calculate the difference between capital recovery factor when one assumes an infinite life rather than a 80-year life, assuming a discount rate of 8%.

$$(A/P, I, n)_\infty = 0.08$$

$$(A/P, I, n)_{80} = \frac{0.08(1.08)^{80}}{(1.08)^{80} + 1} = 0.0802 \,(\text{a difference of } 0.25\%)$$

Therefore, for a project with, in effect, an infinitely long economic life, the capital recovery factor is replaced by the interest rate.

Again, the similarity between the assumption of infinite life for estimating annual worth and the capitalised cost method used with the present worth model is obvious. The infinite life project deemed best with the present worth method will again prevail using this technique. Let us take the bridge assessment example from Chapter 3 (Example 3.3).

Example 4.8

The Port Authority intends to build a bridge and has two main options, a high-level reinforced concrete bridge and a low-level steel bridge. The high-level bridge has an initial construction cost of £40m, with maintenance costs estimated at £20 000 per annum. Its asphalt road surface would have to be re-laid every 9 years at a cost each time of £75 000. The low-level bridge and ancillary works are estimated at £35m, with maintenance costs estimated at £15 000 per annum. Its steel frame must be painted every 4 years at a cost of £220 000. Its road surface must be re-laid every 9 years at a cost of £45 000.

Compare the two options on the basis of their respective equivalent annual worth, assuming an interest rate of 5%.

Solution

(i)   High-level bridge

$$AW_{Initial\ cost} = -40 \times 10^6 (0.05) = -2\,000\,000$$

Annual maintenance $= -20\,000$

$$AW_{Road\ surfacing\ every\ 9\ yr} = -75\,000\,(A/F, 5, 9)$$

$$= -75\,000\,(0.0907) = -6801.76$$

$$AW_{Total} = -2\,000\,000 - 20\,000 - 6801.76$$

$$= -2\,026\,801$$

(ii)   Low-level bridge

$$AW_{initial\ cost} = -35 \times 10^6 (0.05) = -1\,750\,000$$

Annual maintenance $= -15\,000$

$$AW_{Road\ surfacing\ every\ 9\ yr} = -45\,000\,(A/F, 5, 9)$$

$$= -45\,000\,(0.0907) = -4081.50$$

$$AW_{Painting\ every\ 4\ yr} = -220\,000\,(A/F, 5, 4)$$

$$= -220\,000\,(0.232) = -51\,040$$

$$AW_{Total} = -1\,750\,000 - 15\,000 - 4081.50 - 51\,040$$

$$= -1820\ 121.50$$

As with the capitalised cost evaluation in Chapter 3, the low-level option is selected.

Example 4.9

A developer is faced with two options for providing drainage for a site he/she wishes to develop. The first proposal involves the construction of a high quality reinforced concrete open culvert, which is assumed to have an infinite economic

Contd

Example 4.9 Contd

life. The second proposal suggests a steel culvert with an economic life of 10 years, which incurs an increasing maintenance cost over this period (the value x in the gradient detailed below ranges from 1 to 10). The detailed costings for both options are given in Table 4.2. A discount rate of 9% is assumed.

Solution

(i) Open culvert

$$AW_{construction\ cost} = -100\,000\,(0.09) = -9000$$
$$\text{Annual maintenance} = -1000$$
$$AW_{Total} = -£10\,000$$

(ii) Steel culvert

$$AW_{Construction\ cost} = -35\,000\,(A/P,9\%,10) = -35\,000\,(0.1558)$$
$$= -5453$$
$$\text{Annual maintenance} = -\left[3000 + 100\,(A/G,9\%,10)\right]$$
$$= -[3000 + 100\,(3.7978)]$$
$$= -3379.78$$
$$AW_{Salvage} = +9500\,(A/F,9\%,10) = -9500\,(0.0658)$$
$$= +625.29$$
$$AW_{Total} = -(5453 + 3379.78 - 625.29)$$
$$= -£8207.49$$

The steel culvert option is selected on the basis of its marginally smaller equivalent annual cost.

	Open culvert	Steel culvert
Construction cost (£)	100 000	35 000
Annual maintenance cost (£)	1000	3000 + £100(x−1)
Residual value (£)	0	9500
Economic life (years)	∞	10

Table 4.2 Comparison of open and steel culverts.

4.6   Summary

The equivalent annual worth (EAW) methodology outlined in this chapter is relatively straightforward to use, particularly when the competing options have different economic lives, because of the necessity to examine only one life cycle of the proposal. However, it must be emphasised that it will deliver an answer consistent

Figure 4.2 Burj al Arab Tower, UAE (Source: W.S. Atkins).

with the present worth method only on the basis that the underlying assumptions are being adhered to, namely that the options are being compared over the least common multiples of the estimated lives of the alternatives. For each option under consideration, the annual worth for one life cycle is assumed to be reproduced exactly for all succeeding cycles until the point where each option gives equal service.

Strengths of EAW
- EAW is particularly appropriate where options have different lives, as only one life cycle need be considered.

Limitations of EAW
- As with the NPV method, the final value computed is dependent on the discount rate chosen.
- The decision maker must assume that the EAW for each option is reproduced until the point of equal service for all is reached. The omission of this point from any answer given can leave it open to misinterpretation.

Chapter 5

Rate of Return Computation

5.1 Introduction

Within this chapter we calculate the rate of return for a proposal based on the equations we have used in the previous two chapters on present and annual worth. The methodology usually employed to achieve this is a trial-and-error one.

5.2 Minimum Acceptable Rate of Return (MARR)

In Chapters 3 and 4 we have assumed that the chosen interest rate is the minimum rate of return that will be attractive to the project promoter. This minimum acceptable rate of return is seen as the lowest level at which a project option under consideration remains economically desirable. It provides a basis for judging a proposed investment on the basis of whether it complies with this minimum standard of attractiveness. Different organisations will estimate the interest rate in different ways but it is generally accepted that it should be set no lower than the cost of capital. The extent to which it rises above this base level will depend on the economic policies of the organisation itself, along with its aspirations, objectives and general circumstances. If a developer is in a financially weak state, with a low credit rating, it may have to pay more for its money, so any investment must provide a very attractive return before it becomes feasible. For larger more established developers with major resources behind them, lower money costs allow less onerous returns to be required. In such circumstances the return from the proposal will be viewed relative to what would generally be expected from an investment of that magnitude. If the proposal generates a lower return than other similarly-scaled potential projects, investment in it represents an opportunity lost by the developer to use the invested capital to greater effect.

Engineering Project Appraisal: The Evaluation of Alternative Development Schemes, Second Edition.
Martin Rogers and Aidan Duffy.
© 2012 John Wiley & Sons, Ltd. Published 2012 by John Wiley & Sons, Ltd.

5.3   Internal Rate of Return (IRR)

The IRR is the most frequently used rate of return technique. It is computed by either of the methods used in the previous two chapters – present and annual worth. Within it, the present or annual worth is set equal to zero and the interest rate that allows this equality is calculated. For ease of calculation, the present worth valuation is used most frequently. In this case, IRR denotes the rate at which the computed present worth of all costs and the computed present worth of all receipts are equal. As in the previous two chapters, the effect of taxes on the discounted cash flows is not considered.

5.4   IRR for a single project

5.4.1   Calculating IRR for a single project using present worth

For a single stand-alone project, the rate of return is calculated by equating the present worth (PW) of costs and the present worth of receipts; that is:

$$PW_{costs} = PW_{receipts}$$

or

$$PW_{costs} - PW_{receipts} = 0$$

And the interest rate, $i*$, is derived which makes this equality possible.

Example 5.1

An office development is purchased by a developer for £700 000 and is expected to increase in value to £1.5 m in 6 years. During this period it can be rented to a tenant for £15 000 per year. What rate of return will the developer obtain on his investment within this time period?

Solution

$$PW_{costs} - PW_{receipts} = 0$$

which can be expressed as:

$$700\,000 - [15\,000(P/A, i*, 6) + 1\,500\,000(P/F, i*, 6)] = 0$$

As a preliminary check, it can be seen that the development has a positive rate of return because:

$$1\,500\,000 + 15\,000*6 = 1\,590\,000 \gg 700\,000\,(PW_{costs})$$

Even though this computation assumes the interest rate equals zero, given that the revenue far exceeds the costs, the project will produce a positive rate of return.

Contd

Example 5.1 Contd

Approximate value of i*

To gain an approximation of the value of i^*, the '72 rule' can be applied. This states that a sum doubles in value every $72/i$ years. Since the zero-interest estimate shows a more than doubling of capital over six years:

$$72/i = 6 \text{ or } i = 12$$

Trial-and-error method

Trying $i = 15\%$ as a first guess, we get the result:

$$\begin{aligned}
PW_{receipts} &= 15\,000(P/A,15,6) + 1\,500\,000(P/F,15,6) \\
&= 15\,000 \times [(1.15)^6 - 1]/[0.15(1.15)^6] + 1\,500\,000 \times 1/(1.15)^6 \\
&= 15\,000 \times 3.78 + 1\,500\,000 \times 0.432 \\
&= 704\,700 > 700\,000\,(+4700)
\end{aligned}$$

Trying $i = 16$:

$$\begin{aligned}
PW_{receipts} &= 15\,000 \times [(1.16)^6 - 1]/[0.16(1.16)^6] + 1\,500\,000 \times 1/(1.16)^6 \\
&= 15\,000 \times 3.684 + 1\,500\,000 \times 0.410 \\
&= 670\,260 < 700\,000\,(-29\,740)
\end{aligned}$$

The IRR is situated somewhere between 15 and 16%. By interpolation:

$$\begin{aligned}
i &= 15 + 4700/(4700 - (-29\,740)) \\
&= 15.14\% - \text{say}\,15.15\%
\end{aligned}$$

Interest tables can be used to minimise the error involved in the interpolation process. Figure 5.1 illustrates the cash flows involved in this problem.

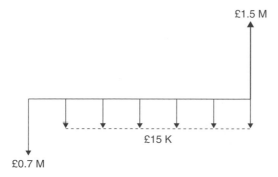

£1.5 M

£15 K

£0.7 M

Figure 5.1 Cash flows for office development.

There is no alternative to this trial-and-error procedure for manually determining the internal rate of return for all but the simplest cash flows. Spreadsheets greatly shorten this process, as they possess a 'rate of return' function that calculates automatically the correct value for a given set of costs and revenues.

5.4.2   Calculating IRR for a single project using annual worth

In this case the value of $i*$ at which the annual worth (AW) of the receipts is equal to the annual worth of the costs is determined; that is:

$$AW_{costs} - AW_{receipts} = 0$$

Example 5.2
Taking the figures from the previous example, use the annual worth method to compute the correct interest rate.

Solution
Using $i* = 15.15$:

$$
\begin{aligned}
AW_{costs} &= -700\,000(A/P,15.15,6) \\
&= -700\,000 \times [0.1515(1.1515)^6]/[(1.1515)^6 - 1] \\
&= -700\,000 \times [0.2653] \\
&= -185\,713 \\
AW_{receipts} &= 15\,000 + 1\,500\,000(A/F,15,6) \\
&= 15\,000 + 1\,500\,000[0.1515]/[(1.1515)^6 - 1] \\
&= 15\,000 + 1\,500\,000[0.11381] \\
&= 185\,715
\end{aligned}
$$

Again, since $AW_{costs} \cong AW_{receipts}$ for this value of $i*$, this value is designated the internal rate of return.

5.4.3   Single projects with more than one possible rate of return

Where the cash flows for a given project alternate from positive to negative, or vice versa, twice or more over its economic life, there will be more than one value of $i*$ for which the present or annual worth equation will equal zero. While this is a rare occurrence when evaluating a single project, a method must be found for dealing with it.

Example 5.3
The cash flows for a toll road project are indicated in Table 5.1 and Figure 5.2. For two years during the project the cloure of a major feeder road resulted in severe cash losses to the company. Because of these fluctuations, two values for rate of return exist, what are they?

Contd

Example 5.3 Contd

Solution

The nature of this cash flow would indicate that multiple roots exist for the solution of the present worth equation with the interest rate as the variable factor. To find the values of i, we must solve the following equation:

$$0 = 2000 + 750/(1+i)^1 - 4000/(1+i)^2 - 4000/(1+i)^3 + 1500(1+i)^4$$
$$+ 4000(1+i)^5$$

Figure 5.3 indicates that two values of i satisfy the above equation, 4.97 and 30.49%; that is:

$$0 = 2000 + 750/(1.0497)^1 - 4000/(1.0497)^2 - 4000/(1.0497)^3$$
$$+ 1500(1.0497)^4 + 4000(1.0497)^5$$

and

$$0 = 2000 + 750/(1.3049)^1 - 4000/(1.3049)^2 - 4000/(1.3049)^3$$
$$+ 1500(1.3049)^4 + 4000(1.3049)^5$$

Year	0	1	2	3	4	5
Cash flow (£)	2000	750	–4000	–4000	1500	4000

Table 5.1 Cash flows for toll road project.

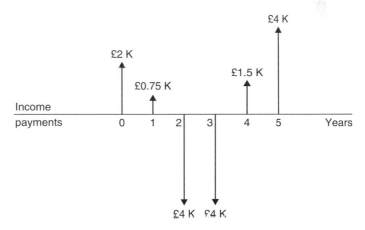

Figure 5.2 Diagrammatic representation of cash flows for toll road project.

5.4.4   *External Rate of Return (ERR)*

The external rate of return is a form of investment rate that is of particular use in overcoming the problem of a cash flow having multiple rates of return. The internal rate of return is the rate of return on the unrecovered sum of money where the cash

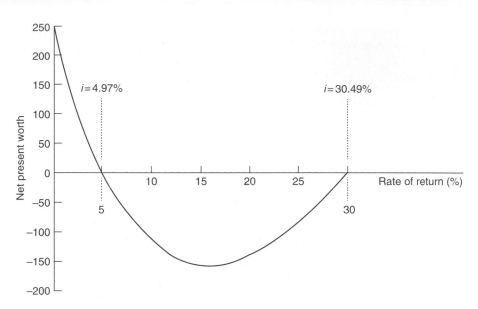

Figure 5.3 Illustration of multiple roots in cash flow problem.

flows are negative and the balance at the end of the cash flow period is zero. The external rate of return is used where cash flows are positive and is simply the return that it is possible to obtain from these available funds under existing market conditions. It therefore sets the rate one expects from the external market, and may constitute the minimum rate of return one would expect from any proposed engineering proposal. It may well be set at the minimum acceptable rate of return (MARR).

5.4.5   Historical External Rate of Return (HERR)

The historical external rate of return is one of the techniques that allow non-simple cash flow problems with multiple values of i, of the type outlined in Example 5.3, to be solved. It achieves this by assuming that any income/positive cash flows are invested at an explicit interest rate that constitutes the minimum attractive rate of return required by the investor.

The unknown rate of return (e') is obtained by discovering the point at which the future worth of all income, compounded at the minimum attractive rate of return becomes equal to the future worth of all outgoing payments/negative cash flows compounded at e'.

Example 5.4
Taking Example 5.3, but this time using the HERR method to estimate a single rate of return. A minimum acceptable rate of return/explicit interest rate of 7% is assumed.

Contd

Example 5.4 Contd

Solution

$$i = 7\%$$

$$\text{Future worth} = 2000(F/P,7\%,5) + 750(F/P,7\%,4) + 1500(F/P,7\%,1)$$
$$+ 4000 - 4000(F/P,e',3) + 4000(F/P,e',2)$$

$e' = 5\%$,	Future worth of cash flow = +£352.72
$e' = 6\%$,	Future worth of cash flow = +£134.75
$e' = 7\%$,	Future worth of cash flow = −£86.57
$e' = 6.611\%$,	Future worth of cash flow = −£0.08

(This estimation of the value of e' at which the future worth of incomes and payments become equal is shown in Table 5.2.)

This project should not be proceeded with, as the value of e' obtained under the HERR methodology (6.6%) is less than the minimum acceptable rate of return (7%).

Period (years)	Cash flow (£)	Future values (Period 5)		
0	2000	$2000(1.07)^5$	→	2805.10
1	750	$750(1.07)^4$	→	983.10
2	−4000	$-4000(1.066)^3$	→	−4846.92
3	−4000	$-4000(1.066)^2$	→	−4546.36
4	1500	$1500(1.07)^1$	→	1605.00
5	4000	$4000(1.1064)^0$	→	4000.00
		Total	→	−0.08

Table 5.2
HERR calculations.

5.4.6   *Project balance method*

The drawback of the historical external rate of return method is that the income from the re-investment of income payments is based on the absolute receipts rather than the cash flow balances. The project balance method is a more logical procedure, which applies either the internal rate of return (IRR) of the external rate of return (ERR) to the cumulative cash balance, depending on whether the balance itself is negative or positive.

Having ascertained that the cash flow is one which has multiple values of i^*, the current balance is estimated within each of the periods under consideration. If all balances are either zero or negative, it is assumed that all receipts will be used to pay off investment costs. Since they are being used internally, the IRR should be employed as the rate of return. If some balances are positive, however, some cash exists to earn income and an external rate of return, usually in the form of some minimum acceptable rate of return for investors, should be used. Assuming an external interest rate for the released funds arising from the positive net cash flows, a trial-and-error system is used to find the value of IRR at which the current balance at the end of the period for the project is zero.

Let us apply this process to the cash flows in Example 5.3, with the MARR again assumed to be at 7% and an initial arbitrary IRR of 10% (Table 5.3). Since the current balance is negative, the value of IRR is too high and must be reduced. By a trial-and-error based iterative approach we obtain a value of IRR = 6.144% at which the current balance is exactly zero:

$$IRR_{(guess)} = 10\% \quad \text{Current balance} = -£398.15$$

$$IRR_{(guess)} = 9\% \quad \text{Current balance} = -£292.90$$

$$IRR_{(guess)} = 8\% \quad \text{Current balance} = -£189.04$$

$$IRR_{(guess)} = 7\% \quad \text{Current balance} = -£86.57$$

$$IRR_{(guess)} = 6\% \quad \text{Current balance} = +£14.51$$

$$IRR_{(guess)} = 6.144\% \quad \text{Current balance} = +£0$$

Table 5.3 Project balance calculations.

End of period t (years)	Cash flow (£)	Current Balance (£)	Rate used in next period
0	2000	2000	MARR for t = 1
1	750	$2000(1.07)^1 + 750 = 2890$	MARR for t = 2
2	-4000	$2890(1.07)^1 - 4000 = -907.70$	IRR for t = 3
3	-4000	$-907.70(1.1)^1 - 4000 = -4998.47$	IRR for t = 4
4	1500	$-4998.47(1.1)^1 + 1500 = -3998.32$	IRR for t = 5
5	4000	$-3998.32(1.1)^1 + 4000 = -398.15$	

The HERR method resulted in an external return of 6.6%, but it must be remembered that the minimum attractive rate of return was applied to all positive cash flows, whereas in this method the MARR was applied in a more accurate way to positive cash balances only.

5.4.7 Unequal lives

The rate of return method of project appraisal can be used to assess projects with different economic lives. As with the previous methods, a least common multiple of the various lives can be determined, and the analysis carried out over this period.

Alternatively, the lives can be shortened to a common analysis timescale. If the least common multiple is employed, the calculations involved may become very cumbersome, and it may be advisable to employ one of the other methods of project evaluation (present worth or equivalent annual worth).

5.5   Incremental analysis

The above method of IRR estimation can be used to compare independent project alternatives, where approval of one option does not in any way affect the acceptance of another under consideration. Such a situation exists where there are no financial constraints and several or indeed all of the viable alternatives can be chosen.

However, a more usual situation arises where there is a limitation on capital and several competing options are competing for available finance. In such circumstances, proposed projects are deemed mutually exclusive. These, if ranked according to their IRR values, will not produce a result consistent with the rankings from a present worth or annual worth analysis. To achieve an answer in line with the present worth/ annual worth methods, mutually exclusive options must be analysed using *incremental IRR*. This methodology commences with the assumption that an acceptable low-investment option has been identified. Any higher investment option is then judged on the basis of the cash flow differences, termed incremental cash flows, between it and the acceptable low-investment one. The cash flow of the second option is thus equal to the cash flows of the first plus the incremental values. The level of return from the incremental cash flows are analysed relative to a minimum acceptable level (MARR); if it exceeds this threshold, the second option is deemed preferable to the first, otherwise the first remains the favoured selection. The process is continued until all options have been scrutinised using this technique and the best performing one identified.

Example 5.5 illustrates how the present worth method can give an answer at variance with a direct rather than incremental IRR comparison.

Example 5.5
The cash flows for two alternative urban renewal proposals are given in Table 5.4. Both have an initial cost of £125 000, but with Option A the revenues are relatively low initially and increase with time. With Option B, the reverse is the case, with high revenues initially but decreasing with time. They are mutually exclusive projects. Determine the present worth and IRR of both. Are both answers consistent? If not, how can consistency be maintained?

Solution
Present worth analysis
Assuming a minimum required rate of return of 11%, the present worth of the two options is:

$$PW_{Option\,A} = -125 + (12/(1.11)^1) + (45/(1.11)^2) + (75/(1.11)^3) + (110/(1.11)^4)$$
$$= -125 + 10.811 + 36.523 + 54.839 + 72.46$$
$$= 49.634 \times 1000 = £49\,630$$

Contd

Example 5.5 Contd

$$PW_{Option B} = -125 + (125/(1.11)^1) + (25/(1.11)^2) + (25/(1.11)^3) + (25/(1.11)^4)$$
$$= -125 + 112.61 + 20.29 + 18.28 + 16.468$$
$$= 42.649 \times 1000 = £42\,650$$

Thus, Option A ranks higher than Option B on the present worth analysis.
IRR Analysis (calculation of value of i at which PW = 0)
For Option A:

At $i = 23\%$, PW = £2860
At $i = 24\%$, PW = –£190
At $i = 23.935\%$, PW = £0

Therefore, by interpolation, $IRR_{Option A} = 23.94\%$.
For Option B:

At $i = 34\%$, PW = £350
At $i = 35\%$, PW = –£1000
At $i = 34.257\%$, PW = £0

Therefore, by interpolation, $IRR_{Option B} = 34.26\%$.
    Direct IRR comparison results in Option B being ranked ahead of Option A in this instance.
    To resolve this inconsistency, compare the two options using an incremental IRR analysis. Because both options involve the same initial outlay, and are thus equally low investment, either can be selected initially. In this instance we select Option B.
    The incremental cash flow for Option A relative to B is indicated in Table 5.5.

$$PW_{A-B} = -113 + (20/(1+i)^1) + (50/(1+i)^2) + (85/(1+i)^3)$$

At $i = 14.2\%$, PW = £0

Therefore, by interpolation, $IRR_{A-B} = 14.2\%$, which is termed the incremental rate of return. Since this rate of return is greater than the minimum acceptable value of 11%, Option A is deemed preferable to Option B, a result consistent with the present worth analysis.

Option	\multicolumn{5}{c}{Annual cash flow (£000)}
	0	1	2	3	4
A	−125	12	45	75	110
B	−125	125	25	25	25

Table 5.4 Cash flows for calculation of IRR and PW for two mutually exclusive urban renewal proposals.

Option difference	Annual cash flow (£000)					**Table 5.5** Incremental IRR (Option A relative to Option B).
	0	1	2	3	4	
A–B	0	–113	20	50	85	

5.5.1   *Using rate of return analysis for ranking multiple mutually exclusive options*

There are situations where a number of options are available and, while they are mutually exclusive, one wishes to rank them so that if the best performing one is for some reason no longer feasible, the next best one can be quickly identified and proceeded with and so on. In such circumstances, the present or annual worth computation readily yields the required information. However, many planners like the relative rates of return of the different options under investigation to be the deciding factor. This methodology is a little less straightforward than PW or AW, so care must be taken to ensure that it is applied correctly.

The first requirement when analysing a group of mutually exclusive options using rate of return is that at least one must result in a return equal to or greater than the minimum attractive rate (MARR). If none of the options equal the MARR, the 'do-nothing' option is selected. When, however, more than one of them attains this threshold, the one with the lowest investment cost is identified as the initially preferred option. It is compared with the qualifying option having the next lowest investment cost. If the incremental analysis shows that the return on the extra investment exceeds the minimum required rate of return (MARR), the higher cost option is assigned the first rank, otherwise the lowest cost option retains its first place. The prevailing option is then compared with the one with the next highest investment cost, and an incremental analysis again decides which of the two will win through. This analysis continues until all options have been subject to this comparison process and the prevailing one has been identified. This procedure is illustrated in Example 5.6.

> ### Example 5.6
> A developer is considering four different sites for an office development. The costs of construction vary mainly due to the cost of the site; the annual net cash flows differ due to differences in maintenance and upkeep costs. Details are given in Table 5.6. The minimum acceptable rate of return is 10%. Use the rate of return method to choose the optimum site in economic terms. The economic life of all options is 35 years.
>
> *Solution*
> Initially, the four sites are placed in order of increasing construction cost; that is:
>
>   Option 3 → Option 1 → Option 2 → Option 4
>
> *Contd*

Example 5.6 Contd

The overall rate of return for each option is then calculated. The results are also shown in Table 5.6. All the values are greater than the MARR (10%), thus all sites are viable options, and are preferable to the 'do-nothing' alternative.

Let us start with the lowest investment option (Site 3). Given that its ROR exceeds the MARR, it is deemed preferable to the 'do-nothing' option. It is therefore termed the 'defender' and it is then compared with the next lowest investment option (Site 1), termed the 'challenger'. The yearly incremental cost and cash flow of these two options is determined using the following equations:

Incremental cash flow = cash flow of 'challenger' – cash flow of 'defender

Incremental cost   = cost of 'challenger' – cost of 'defender'

For the comparison of Options 1 and 3, the following values were obtained:

$$\text{Incremental cash flow}_{(1-3)} = \text{cash flow (Site 1)} - \text{cash flow (Site 3)} = £1500$$
$$\text{Incremental cost}_{(1-3)} = \text{cost (Site 1)} - \text{cost (Site 3)} = -£5000$$
$$PW_{1-3} = -5 + (1.5 \times (P/A, i^*, 35))$$
$$= -5 + (1.5 \times ((1+i)^{35} - 1)/i(1+i)^{35}))$$
$$\text{At } i = 29\%, PW = £171.72$$
$$\text{At } i = 30\%, PW = -£0.51$$
$$\text{At } i = 29.99\%, PW = £0$$

Since, in this instance, $i^* = 29.99\%$, which is greater than the required MARR of 10%, Option 1 is deemed preferable to Option 3 as it is incrementally justified.

Option 1 now becomes the defender and is compared to Option 2.

$$\text{Incremental cash flow}_{(2-1)} = \text{cash flow (Site 2)} - \text{cash flow (Site 1)} = £7,000$$
$$\text{Incremental cost}_{(2-1)} = \text{cost (Site 2)} - \text{cost (Site 1)} = -£30,000$$
$$PW_{2-1} = -30 + (7 \times (P/A, i^*, 35))$$
$$PW_{2-1} = -30 + (7 \times ((1+i)^{35} - 1)/i(1+i)^{35}))$$
$$\text{At } i = 23\%, PW = £413.07$$
$$\text{At } i = 24\%, PW = -£849.01$$
$$\text{At } i = 23.32\%, PW = £0$$

Since, in this instance, $i^* = 23.32\%$, which is greater than the required MARR of 10%, Option 2 is deemed preferable to Option 1 as it is incrementally justified.

Contd

Example 5.6 Contd

Option 2 now becomes the defender and is compared to Option 4.

Incremental cash flow$_{(4-2)}$ = cash flow (Site 4) – cash flow (Site 2) = £6000

Incremental cost$_{(4-2)}$ = cost (Site 4) – cost (Site 2) = –£60 000

$$PW_{4-2} = -60 + (6 \times (P/A, i^*, 35))$$
$$= -60 + (6 \times ((1+i)^{35} - 1)/i(1+i)^{35}))$$

At $i = 9\%, PW = £3400.93$

At $i = 10\%, PW = -£2135$

At $i = 9.595\%, PW = £0$

Since, in this instance, $i^* = 9.595\%$, which is less than the required MARR of 10%, Option 2 is deemed preferable to Option 4 as it is incrementally justified.

To rank Option 4 relative to Options 1 and 3, let us examine it first relative to the one with the least investment cost (Option 3).

Incremental cash flow$_{(4-3)}$ = cash flow (Site 4) – cash flow (Site 3) = £14 500

Incremental cost$_{(4-3)}$ = cost (Site 4) – cost (Site 3) = –£95 000

$$PW_{4-3} = -95 + (14.5 \times (P/A, i^*, 35))$$
$$= -95 + (14.5 \times ((1+i)^{35} - 1)/i(1+i)^{35}))$$

At $i = 15\%, PW = £940.81$

At $i = 16\%, PW = -£4877.60$

At $i = 15.15\%, PW = £0$

Since, in this instance, $i^* = 15.15\%$, which is greater than the required MARR of 10%, Option 4 is deemed preferable to Option 3 as it is incrementally justified.

Let us now rank Option 4 relative to Options 1.

Incremental cash flow$_{(4-1)}$ = cash flow (Site 4) – cash flow (Site 1) = £13 000

Incremental cost$_{(4-1)}$ = cost (Site 4) – cost (Site 1) = –£90 000

$$PW_{4-1} = -90 + (13 \times (P/A, i^*, 35))$$
$$= -90 + (13 \times ((1+i)^{35} - 1)/i(1+i)^{35}))$$

At $i = 14\%, PW = £1910.59$

At $i = 15\%, PW = -£3984.10$

At $i = 14.31\%, PW = £0$

Since, in this instance, $i^* = 14.31\%$, which is greater than the required MARR of 10%, Option 4 is deemed preferable to Option 1.

The ranking of the four options plus 'do-minimum' is thus:

First: Option 2
Second: Option 4
Third: Option 1
Fourth: Option 3
Fifth: Do nothing

Table 5.6 Details of the four development projects (Example 5.6).

Site	1	2	3	4
Construction cost (£000)	−120	−150	−115	−210
Yearly cash flow (£000)	14	21	12.5	27
Overall rate of return (%)	11.44	13.85	10.5	12.66

To demonstrate the consistency between the results obtained by incremental IRR and by present worth, take Example 5.4 and use the present worth methodology to generate a ranking of all options.

Example 5.7

The present worth of the four options under consideration can be calculated as follows, assuming a discount rate of 10%:

Note: Calculation of uniform series present worth factor:

$$(P/A, 10\%, 35) = [(1+0.1)^{35} - 1] / [0.1(1+0.1)^{35}]$$
$$= 9.644$$

$$\begin{aligned}
\mathrm{PW}_{Option1} &= -120\,000 + 14\,000(P/A, 10\%, 35) \\
&= -120\,000 + 14\,000 \times 9.644 \\
&= £15\,018.23 \\
\mathrm{PW}_{Option2} &= -150\,000 + 21\,000(P/A, 10\%, 35) \\
&= -150\,000 + 21\,000 \times 9.644 \\
&= £52\,527.34 \\
\mathrm{PW}_{Option3} &= -115\,000 + 12\,500(P/A, 10\%, 35) \\
&= -115\,000 + 12\,500 \times 9.644 \\
&= £5551.99 \\
\mathrm{PW}_{Option4} &= -210\,000 + 27\,000(P/A, 10\%, 35) \\
&= -210\,000 + 27\,000 \times 9.644 \\
&= £50\,392,29
\end{aligned}$$

On the basis that the options are ranked in order of their increasing present worth, the same ranking as derived in the previous example using the incremental IRR method is obtained:

Option 2 → Option 4 → Option 1 → Option 3

Because all options have a positively valued net present worth, they are all preferable to the 'do-nothing' option that is therefore again ranked fifth and last.

5.5.2 Using IRR to analyse options with different lives

It is entirely possible to use the IRR methodology to analyse options with different life spans. It is tackled in the same way as with the present or annual worth techniques, where the least common multiple of the lives of the different options is obtained and this value is taken as the life value for all. Therefore, take the situation where three options are under consideration, with lives of 10, 15 and 20 years, respectively. The least common multiple of these three numbers is 30, so this value is taken as the study period, n, for all three.

Example 5.8

A developer is given three mutually exclusive options for a solid waste recycling mechanism. Each one varies in terms of initial investment cost, annual cash flow, and economic life. Details are given in Table 5.7. Assuming a MARR of 14%, select the best option using the rate of return technique.

Solution

Table 5.7 places the options in order of increasing investment cost. Initially, Option 1 is compared with the 'do-minimum' option. Since, from Table 5.7, the overall rate of return for Option 1 is less than the MARR, the 'do-nothing' is preferred over it and is designated the 'defender', with the next most expensive proposal, Option 2, assigned the 'challenger'. Since the overall rate of return for Option 2 is appreciably greater than the MARR, it is preferred. The final selection is thus between Option 2, the defender, and Option 3, the challenger.

Table 5.8 indicates the incremental cash flow analysis for these two options. Given that these two have different lives, the analysis is completed over the least common multiple of the two values, that is 12 years.

The incremental rate of return of (Option 3 – Option 2) is calculated as 19.85%, which is greater than the MARR of 14%. Therefore, Option 3 is chosen in preference to Option 2.

The results from the rate of return analysis can be confirmed by estimating the annual worth of each option by the technique indicated in Chapter 4.

$$AW_1 = -15\,000(A/P,14\%,3) + 6000 = -460.97$$
$$AW_2 = -17\,500(A/P,14\%,4) + 8000 = +1993.92$$
$$AW_3 = -22\,500(A/P,14\%,6) + 8000 = +2213.96$$

Thus Option 3 is confirmed as the best proposal.

Table 5.7 Cash flow details for three recycling options (Example 5.8).

Option	1	2	3
Construction cost (£000)	−15	−17.5	−22.5
Yearly cash flow (£000)	6	8	8
Economic life (years)	3	4	6
Overall rate of return (%)	9.7	29.4	27.1

Year	Cash flow (£000)			
	2	3	3–2 $(i^* = 0)$	3–2 $(i^* = 19.85)$
0	−17.5	−22.5	−5.0	−5.0
1	8.0	8.0	0	0
2	8.0	8.0	0	0
3	8.0	8.0	0	0
4	−9.5	8.0	+17.5	+8.48
5	8.0	8.0	0	0
6	8.0	−14.5	−22.5	−7.59
7	8.0	8.0	0	0
8	−9.5	8.0	+17.5	+4.11
9	8.0	8.0	0	0
10	8.0	8.0	0	0
11	8.0	8.0	0	0
12	8.0	8.0	0	0
Σ	+43.5	+51	+7.5	+0

Table 5.8 Incremental analysis of Option 3 minus Option 2 (Example 5.8).

Figure 5.4 Millenium Dome project, UK (Source: W.S. Atkins).

5.6   Summary

The rate of return technique enables a decision maker to identify the most economically desirable option in a manner totally consistent with the annual worth and present worth methods outlined in Chapters 3 and 4.

When more than one option can be chosen from those put forward for evaluation, the options concerned are deemed 'independent' and need only be compared, using their overall rate of return, with the 'do-nothing' option and its associated minimum acceptable rate of return. In this case, the options are not directly compared with each other.

However, the main point to be remembered is that when the options under consideration are mutually exclusive, that is proceeding with one eliminates the possibility of advancing with any of the others, these options must be compared two at a time, starting with the one having the lowest initial cost. The rate of return analysis of each pair is based on their incremental cash flow that identifies the dominant option within every pairwise comparison. Once every option, plus the alternative 'do-nothing' option, has been subject to an incremental analysis, the dominant option becomes apparent.

If incremental cash flows are not used, incorrect results, inconsistent with annual worth and present worth, will be obtained. It must be noted that this type of analysis is somewhat more difficult than the annual worth and present worth methods.

Strengths of IRR

- Unlike NPV or EAW, this method takes away the need for any discount rate to be assumed by the decision maker before starting the analysis.
- When used correctly, it is entirely consistent with the NPV and EAW methods.

Limitations of IRR

- If the project under assessment has large expenditures at the start and end of its economic life, multiple rates of return can be obtained.
- For the analysis of mutually exclusive proposals, the incremental technique can be unwieldy.

Chapter 6

Benefit/Cost Ratio, Depreciation and Taxation

6.1 Introduction

All three previous methods of evaluation dealt with in Chapters 3, 4 and 5 of this book – present worth, annual worth and rate of return – can be used to analyse the economic performance of a development project. The fourth and final method, benefit/cost ratio, if used properly, gives answers completely consistent with the previous three methods. However, it operates within the context where all cash flows must be interpreted as either costs or benefits, and the necessary calculations can only be done once these have been distinguished from each other. It is necessary, therefore, to define what these terms mean.

The chapter concludes with a discussion on the effects of depreciation and taxation on project assessments and comparisons.

6.2 Costs, benefits and disbenefits

The main difference between private as opposed to public projects is that the former examines the economic costs and benefits of the proposal from the narrow, profit-based point of view of the developer or, in the case of a publicly quoted company, its shareholders, while the latter must examine the effect of the project on society as a whole. For example, if a toll bridge is examined purely in terms of its performance as a private venture, its success is measured in terms of the extent to which the revenue from tolls outstrips the construction, maintenance and running costs of the facility. Given, however, that the bridge is part of a major national motorway network, the scheme should be more properly analysed as a public project, with the economic benefits accruing to the developers offset against the negative effects (or disbenefits) of the scheme, such as the impact on local property prices due to increased noise pollution and the disturbance to local land owners due to loss of land and enforced demolition of buildings. The project is thus seen as being of benefit to one segment of the community while being a disbenefit to another.

Engineering Project Appraisal: The Evaluation of Alternative Development Schemes, Second Edition.
Martin Rogers and Aidan Duffy.
© 2012 John Wiley & Sons, Ltd. Published 2012 by John Wiley & Sons, Ltd.

In the context of a public project, costs, benefits and disbenefits can be defined as follows:

(i)   *Costs*: The construction, operating and maintenance costs of the project to the community.
(ii) *Benefits*: The economic return or advantages accruing to members of the community arising from the project.
(iii) *Disbenefits*: The economic disadvantages incident on members of the community as a result of the project.

All are measured in monetary units.

6.3   Estimating the benefit/cost ratio for a single project

Before the benefit/cost ratio can be calculated, all values of benefits, disbenefits and costs must be converted to the same time equivalent. The unit can be the equivalent future, present or annual worth of each constituent. Having done this, the benefit/cost ratio is calculated using the equation:

$$\text{Benefit / cost ratio} = (\text{benefits} - \text{disbenefits}) \div \text{costs} \tag{6.1}$$

A benefit/cost ratio equal to or greater than one indicates an economically viable project. Disbenefits are always subtracted from benefits rather than added to costs. Unlike the economic techniques discussed in earlier chapters, costs are expressed as positive rather than negative values.

Example 6.1
Two alternative bridge proposals are under consideration by a local authority. Design X costs £10 m to build, has annual maintenance and operating costs of £300 000 and will result in an annual net benefit to motorists of £1.5 m. Design Y has construction costs of £14 m, has annual maintenance and operating costs of £300 000 and will result in annual benefits to motorists of £1.6 m. The economic life of the bridge project is set at 25 years, with the interest rate set at 9%.

Use the benefit/cost ratio to decide which option should be selected for construction.

Solution

(i) *Route X*

Annual worth of costs $= 10\,000\,000(A/P, 9\%, 25) + 300\,000$
                         $= (10\,000\,000 \times 0.1018) + 300\,000$
                         $= 1\,318\,062.51$
Annual worth of benefits $= 1\,500\,000$
Benefit / cost ratio $= 1.138$ (economically justified)

Contd

Example 6.1 Contd

(ii)   Route Y

Annual worth of costs $= 14\,000\,000(A/P, 9\%, 25) + 300\,000$
$= (14\,000\,000 \times 0.1018) + 300\,000$
$= 1\,725\,287.51$

Annual worth of benefits $= 1\,600\,000$

Benefit / cost ratio $= 0.93$ (not economically justified)

Route Y is not economically justified relative to the 'do-nothing' situation, as its ratio is less than one. Since the benefit/cost ratio for Route X exceeds one, it is the only economically justified course of action and should be proceeded with.

6.4   Comparing mutually exclusive options using incremental benefit/cost ratios

The above example treats the two options under consideration as independent. They are both compared relative to the 'do-nothing' situation, denoted by a benefit/cost ratio of unity. Where two or more options are mutually exclusive, an incremental analysis must be used in exactly the same way as outlined in Chapter 5.

Initially an overall benefit/cost ratio is estimated for each project option using Equation 6.1; those with values less than one are eliminated from further consideration. The option with a ratio greater than one and with the lowest capital cost is identified as the defender, and the eligible option with the next highest investment cost is set as the challenger. The one that cost more only becomes the preferred option if its incremental benefit/cost ratio exceeds unity; otherwise the lower cost proposal remains favoured. The procedure continues in exactly the same way as the rate of return technique until all options have been subject to a pairwise comparison, at which point a final selection is possible.

Example 6.2

Three alternative designs, A_1, A_2 and A_3, are under consideration for a major motorway interchange. For each option, the present value of the benefit of the interchange to the motorists (B), the construction cost (C), maintenance and operating cost (MO) and the salvage value (S) are given in Table 6.1. Given that the options are mutually exclusive, use the benefit/cost ratio to establish which option should be built, and confirm your answer by means of the NPV scores.

Solution

For each option, the overall benefit/cost ratio is calculated using the following formula:

Benefit/cost ratio $= B \div (C + MO - S)$

Contd

> ### *Example 6.2 Contd*
>
> The ratio values for all options are shown in Table 6.1. Given that all are greater than one, all are preferable to the 'do-nothing' situation and are viable options. To establish the preferred one an incremental analysis must be used. Start with the least cost option, A_1, which is designated the defender, with the next highest costing option, A_2, designated the challenger. Table 6.1 shows that the incremental B/C ratio of $(A_2 - A_1)$ is greater than one, therefore A_2 is preferred to A_1 and becomes the defender; A_3 then becomes the challenger. Table 6.1 indicates a B/C ratio less than unity for $(A_3 - A_2)$; therefore, A_2 is confirmed as the best option of the three.
>
> The Net Present Value of each option is estimated using the equation:
>
> $$NPV = B - (C + MO - S)$$
>
> The NPV scores for the three options confirm A_2, which has the highest NPV, as the option preferred over the other two.

Table 6.1 Costs, benefits, B/C ratios and NPVs for three motorway options.

	A_1	A_2	A_3	$(A_2 - A_1)$	$(A_3 - A_2)$
Construction cost (£m)	12	13	20	1	7
Maintenance/operating cost (£m)	7	20	17	13	−3
Salvage value (£m)	2	1	3	−1	2
Benefit (£m)	95	127	128	32	1
B/C Ratio	5.58	3.96	3.76	2.13	0.5
NPV	78	95	94	17	−1

6.5   Depreciation

Capital goods are items such as buildings and engineering equipment that are used within the development process. They have a finite life, called a 'service life', which is usually a large number of years. Their value can be eroded over a number of years by wear and tear or technical obsolescence. This devaluation is a cost to the particular development in question, reducing its earning potential but also reducing the taxation the developer is liable for. It can thus be an important factor in estimating a project's earning potential, as it is a tax-allowed deduction. Taxation is discussed in brief later in this chapter.

The two most straightforward methods for evaluating depreciation are the straight-line method and the declining balance method.

6.5.1   *Straight-line depreciation*

This is the most widely used method and the simplest to apply. The annual depreciation is a constant value. At the end of any given number of years after the introduction of an asset, its book value (BV) is the difference between its initial value and the product of the annual depreciation charge (DC) and its number of years in use. The depreciation charge and book value can be expressed in the forms:

$$DC_{(n)} = (P - S) \div N \tag{6.2}$$

$$BV_{(n)} = P - (n(P - S) \div N) \tag{6.3}$$

where P is the unadjusted monetary value of the asset, S is the salvage value of the asset at the end of its useful life, N is the useful life of the asset, n is the number of years the asset is in service, $DC_{(n)}$ is the annual depreciation charge in year n and $BV_{(n)}$ is the book value of asset at end of year n.

6.5.2   Declining balance depreciation

This is an accelerated write-off method. The annual depreciation charge is determined by multiplying the book value at the beginning of each year by a uniform percentage, termed R, expressed in decimal form. If R was set at 10%, the depreciation write off for any given annual period would be 10% of the book value at the beginning of that period. The depreciation charge is at its largest value in the first 12 months and decreases thereafter. The maximum rate of depreciation allowed is usually twice the straight-line rate. When this rate is used, the method is termed the *double-declining balance* (DDB). This can be expressed as:

$$R_{max} = 2/n \tag{6.4}$$

In other circumstances it may be limited to 1.5 times the base value, in which case R equals $1.5/n$.

The actual depreciation rate for each year, t, relative to the first cost, is:

$$R_t = R(1 - R)^{t-1} \tag{6.5}$$

The depreciation charge for year t, DC_t, is the uniform rate, R, times the book value at the end of the previous year:

$$DC_t = (R)BV_{t-1} \tag{6.6}$$

If the book value in year t is not known, the depreciation charge can be calculated as follows:

$$DC_t = (R)P(1 - R)^{t-1} \tag{6.7}$$

The book value in year t can be determined in two ways. Using P and R, the following equation is constructed:

$$BV_t = P(1 - R)^t \tag{6.8}$$

Also, BV_t can be determined by subtracting the current depreciation charge from the previous book value:

$$BV_t = BV_{t-1} - D_t \tag{6.9}$$

Within this technique, the book value approaches very close to, but never actually reaches, zero. There is an assumed salvage value, SV, after n years, equal to BV in year n:

$$\text{Assumed SV} = \text{BV}_n = P(1-R)^n \qquad (6.10)$$

Example 6.3

A contractor purchases excavation machinery at a cost of £7500. Its useful life is set at 5 years, after which it would be disposed of, with no estimated salvage value. Calculate the following values using the straight-line depreciation model:

  (i) The depreciation charge during year 1.
 (ii) The depreciation charge during year 2.
(iii) The book value of the machinery at the end of the third year.

Solution

For (i) and (ii)
The depreciation charges for the first and second year are, in this case, the same, and are calculated using Equation 6.2:

$$
\begin{aligned}
\text{DC}(1) = \text{DC}(2) &= (P-S) \div N \\
&= (7500-0) \div 5 \\
&= £1500
\end{aligned}
$$

For (iii)
The book value at the end of the third year is estimated using Equation 6.3:

$$
\begin{aligned}
\text{BV}_{(3)} &= 7500 - (3 \times (7500-0) \div 5) \\
&= £3000
\end{aligned}
$$

Example 6.4

Using the same data as contained in the previous example, use the double-declining balance model to calculate the three values required.

Solution

In this situation, the maximum depreciation rate is allowed for. Therefore:

$$\text{Depreciation rate}_{max} = R_{max} = 200\% \div R = 0.4$$

For (i)

$$
\begin{aligned}
\text{DC}(1) = \text{BV}(0) \times 0.4 &= P \times 0.4 \\
&= 7500 \times 0.4 \\
&= £3000
\end{aligned}
$$

Contd

Example 6.4 Contd
For (ii)

$$DC(2) = BV(1) \times 0.4$$
$$= (7500 - 3000) \times 0.4$$
$$= £1800$$

For (iii)

$$BV(3) = P(1 - R)^3$$
$$= 7500 \times (0.6)^3$$
$$= £1620$$

6.6   Taxation

All the economic evaluations in this and the previous three chapters assume before-tax cash flows. In most circumstances this assumption still results in adequate resolutions to the problems posed. When options being assessed in comparative terms are put forward to meet certain agreed objectives and are subject to identical tax regimes, the before-tax comparison yields the correct preference. Evaluations of public sector proposals rarely include the effects of taxation and are carried out using before-tax economic information.

There are certain circumstances where the effects of taxation can cause a change in the preference ranking of competing options. These include:

- deductions for interest payments, and
- differences in depreciation schedules.

In such situations, the main concern for the decision maker will be the net return from the proposal after taxes are accounted for. This constitutes the amount of capital that is actually available for use by the developer.

The following formula gives a method for adjusting the before-tax rate of return so that the after-tax rate can be derived:

$$IRR_{after-tax} = IRR_{before\ tax} (1 - effective\ rate\ of\ taxation) \qquad (6.11)$$

For example, a developer subject to a taxation rate of 25% estimates the before-tax rate of return to be 20%. An after-tax return of 15% can be estimated using Equation 6.2 as follows:

$$15 = 20 (1 - 0.25)$$

Use of Equation 6.11 assumes that assets in the evaluation are non-depreciable and no special tax provisions apply. Example 6.5 illustrates how the after-tax rate of return can be calculated with a depreciable asset.

Example 6.5

A contractor purchases a cube-crushing machine for £40 000. It is assumed to have a useful life of 5 years. Over this period it is estimated that it will save the contractor £20 000 per year, while its running costs will amount to £8000. It will be depreciated using the straight-line method and will be assumed to have no salvage value. The contractor is assumed to pay tax at 40%. The organisation wishes to ascertain whether this investment meets their required after-tax rate of return of 10%.

Calculate the following:

 (i) The before-tax internal rate of return for the machine.
 (ii) Its approximate after-tax rate of return assuming no depreciation.
(iii) Its after-tax rate of return assuming straight-line depreciation.

Solution

For (i)
Try $i = 15\%$ as a first guess:

$$PW_{costs} - PW_{receipts} = 0$$
$$40\,000 - (20\,000 - 8000)(P/A,15,5)$$
$$40\,000 - 12\,000 \times [(1.15)^5 - 1]/[0.15(1.15)^5]$$
$$40\,000 - 40\,225.86 = -£225.85$$

Try $i = 16\%$ as a second guess:

$$PW_{costs} - PW_{receipts} = 0$$
$$40\,000 - (20\,000 - 8000)(P/A,16,5)$$
$$40\,000 - 12\,000 \times [(1.16)^5 - 1]/[0.16(1.16)^5]$$
$$40\,000 - 39\,291.52 = +£708.48$$

By interpolation:

$$i = 15 + 225.85/(225.85 + 708.48)$$
$$= 15.24\%$$

Thus the before-tax rate of return is 15.24%.

For (ii)
Using Equation 6.11, an approximate value for the after-tax rate of return is obtained as follows:

$$IRR_{after\text{-}tax} = IRR_{before\,tax}(1 - 0.4)$$
$$= 15.24(0.6)$$
$$= 9.14\%, \text{which is below the company's threshold.}$$

Contd

Example 6.5 Contd

For (iii)

Computations on the after-tax cash flows that allow for straight-line depreciation are given in Table 6.2. Using the after-tax data in column 6 of Table 6.2, the present worth formula from which the rate of return can be estimated is:

$$PW = -40\,000 + 10\,400(P/A, i, 5) = 0$$

Try $i = 10\%$ as a first guess:

$$40\,000 - 10\,400 \times [(1.10)^5 - 1] / [0.10(1.10)^5]$$
$$40\,000 - 39\,424.18 = +£575.82$$

Try $i = 9\%$ as a second guess:

$$40\,000 - 10\,400 \times [(1.09)^5 - 1] / [0.09(1.09)^5]$$
$$40\,000 - 40\,452.37 = -£452.37$$

By interpolation:

$$i = 9 + 452.37 / (452.37 + 576.82)$$
$$= 9.44\%$$

Thus, the after-tax rate of return corrected for depreciation is, at 9.44%, just below the firm's minimum required rate.

Table 6.2 Calculation of after-tax cash flows.

Year	Before-tax cash flow (£)	Depreciation charge (£)	Taxable income (£) (1) − (3)	Taxes(£) 40% of (4)	After-tax cash flow (£) (2) − (5)
(1)	(2)	(3)	(4)	(5)	(6)
0	−40 000				−40 000
1	12 000	8000	4000	1600	10 400
2	12 000	8000	4000	1600	10 400
3	12 000	8000	4000	1600	10 400
4	12 000	8000	4000	1600	10 400
5	12 000	8000	4000	1600	10 400

6.7   Summary

6.7.1   *Benefit/cost ratio*

Again, as with the rate of return technique, the use of the incremental benefit/cost ratio depends on whether budgetary constraints exist; that is, whether the projects are independent or mutually exclusive. Where no constraints apply, the projects are independent, with the selection of one option not precluding the selection of others – all projects with a benefit/cost ratio greater than one are economically feasible. This was the case in Example 6.1, where only one of the options is feasible and is thus automatically selected.

Where the budget for project investment is constrained to the point where the available options are mutually exclusive, and where benefit/cost ratios are being used to select the preferred project, the selection must be made by increments, beginning with the lowest-costing option having a ratio greater than one. Each increment of additional project cost is only justified if the incremental benefit/cost ratio is greater than one.

Strengths of the benefit/cost ratio

- It is a very easily understood indicator of a project's overall desirability. If the value exceeds one, the proposal is economically worthwhile.

Limitations of the benefit/cost ratio

- As with the IRR method, where the proposals are mutually exclusive, a rather cumbersome incremental analysis will be required.
- It can give misleading results where the projects differ in size.
- The final value obtained is very sensitive to whether an attribute is defined as a cost or a benefit.

6.7.2 Depreciation

Two simple methods for explicitly taking account of the depreciating value of an asset within an economic analysis are detailed. It is shown how allowing for

Figure 6.1 Sellafield development, UK (Source: W.S. Atkins).

depreciation can affect the after-tax economic performance of a given project, as making an allowance for depreciation will cause a reduction in tax liability, thus boosting the after-tax economic return of the proposal.

The more complex MACRS (Modified Accelerated Cost Recovery System) Model is not covered within this text. It is the approved asset-depreciation system within the United States. The US government provides depreciation rates in tabular form and updates them regularly. The recovery period is specified according to the class of the property in question (manufacturing office, residential etc.). The system always depreciates to zero. MACRS depreciation rates are presented for one year longer than the stated recovery period, n. This is due to the built-in half-year convention imposed by the system where all property is placed in service at the mid-point of the tax year of installation.

Chapter 7

Cost–Benefit Analysis of Public Projects

7.1 Introduction

Cost–benefit analysis (CBA) seeks to measure the net social benefit of a development project. It thus attempts to select those projects for which there is a high surplus of social benefit over cost. It forms a framework within which one can assess a course of action based on the costs it will incur and the benefits it will bring, balancing one against the other. Primarily, it is a 'tool box' to guide decisions on development proposals. It is a highly organised and quantitative process. To facilitate answering the question 'is the benefit worth the cost?' both must be measurable in some common unit of measurement. Given that costs have tended to be expressed in monetary values, this method of measurement has been adopted as the unit of assessment for cost–benefit analysis. Much of the work of cost–benefit analysis lies in attempting to express all relevant costs and benefits in monetary terms.

The two main types of decision that cost–benefit analysis guides are, firstly, 'yes/no' decisions, where it is determined whether a development proposal should be undertaken or not, and, secondly, 'either/or' decisions, where one of several proposed options is selected for implementation or where there is a choice between two or more alternative paths to achieving some predetermined goal.

Cost–benefit analysis is seldom used in isolation as a decision making tool. Because it is basically economic in nature, other techniques such as Environmental Impact Assessment or Risk Assessment are often used alongside cost–benefit analysis. However, as will be seen later in this chapter in the case of some of the environmental and social effects of relevance to infrastructure development projects, criteria that do not seen to be economic in nature, and whose value may not appear to be readily expressible in monetary terms, can be brought into the cost–benefit analysis framework using special techniques such as contingency valuation and hedonic pricing. The decision maker may come to the conclusion, however, that while including such criteria within the evaluation may be technically possible, such a course of action may not necessarily be desirable. The criteria may sit better being

Engineering Project Appraisal: The Evaluation of Alternative Development Schemes, Second Edition.
Martin Rogers and Aidan Duffy.
© 2012 John Wiley & Sons, Ltd. Published 2012 by John Wiley & Sons, Ltd.

considered explicitly in another framework alongside cost–benefit analysis rather than being subsumed within it. The decision maker may see such a course of action as being more transparent.

When one states that cost–benefit analysis estimates the net social benefit of a proposal, the word 'social' is used to concentrate on the effects, both good and bad, that a proposed project will have on society as a whole. It is important to distinguish it from a financial analysis, where the emphasis is on estimating the revenues derived and the costs incurred by the person or persons sponsoring the enterprise, usually termed the developer. Taking the construction of a toll bridge as an example, the financial analysis would concern itself only with the balance sheet details of the promoter – how much the project is likely to cost in terms of initial construction and subsequent maintenance over its economic life set against the likely toll receipts over the equivalent period. A cost–benefit analysis of the project would have a far wider remit, estimating benefits to the bridge users in terms of time and fuel savings together with the reduced likelihood of occurrence of accidents, and setting these against the loss of benefit to local landowners resulting from compulsory land acquisition in addition to the developer's construction and maintenance costs.

Cost–benefit analysis is the main method of economic appraisal for a major engineering project. It enables us to decide whether a certain course of action is worthwhile from the perspective of the community as a whole. Within it we actually assess the costs incurred by the undertaking of the project and comparing them with the benefits that will accrue from it. The chapters in the book thus far have given us the tools to examine all the costs and benefits on a time equivalent basis, and have introduced a number of techniques that can be used as indicators of the net benefit of the project, resulting in the final answer being delivered as a net present value, an equivalent annual worth, an internal rate of return or a benefit/cost ratio. What this chapter concentrates on is the process of computing the individual relevant costs and benefits of the engineering development project in question. These values constitute the raw material to which the various tools and indicators are applied to arrive at a final answer.

7.2   Historical background to cost–benefit analysis

Cost–benefit analysis was developed as a methodology for evaluating proposals within a single sector. Its origins can be traced back to a classic paper on the utility of public works by Dupuit (1844), written originally in the French language. The technique was first institutionalised in the United States of America in the first half of the twentieth century with the advent of the Rivers and Harbours Act (1902), which required that any evaluation of a given development option must take explicit account of navigation benefits arising from the proposal and that these should be set against project costs (Prest and Turvey, 1966). Following on from this, the US Flood Control Act (1936) deemed that a flood control project could only receive financial support from the United States Federal Government if benefits were shown to exceed costs. The US National Resources Planning Board initiated work aimed at guaranteeing the full implementation of this legislation. This effort resulted in the preparation of a general primer, known as the 'Green Book', by a federal interagency committee composed of the major water resource agencies (US Federal Interagency River Basin Committee, 1950). This report detailed the general principles of economic analysis as they were to be applied to the formulation and evaluation of federally funded water resource projects. It thus formed the basis for the application of cost–benefit analysis to water resource proposals, where options were assessed on the basis of one criterion – their economic efficiency. This single criterion was termed the maximisation of the net project contribution to the United States national income (Little, 1957).

The cost–benefit analysis model derived its conceptual basis from the theory of the firm, and the basic endeavour of the firm within the private sector to maximise its profits. Given the need to choose a course of action from among a number of viable options, the profit maximising developer is assumed to compare the profitability of different project options by establishing the profit levels of each over a set period, estimated on the basis of the monetary revenues and costs forecast to accrue to the developer, and linking these to the capital invested in the proposal. The developer then chooses the most profitable option. Basic economic theory dictates that the maximisation of private sector profit in an economy allowing perfect competition leads to 'optimal community welfare', where people in society make the best possible use of their resources. In a similar way, cost–benefit analysis can be viewed as a process within which a public utility, acting as the developer of the project, allocates its resources so that the most economically desirable or 'profitable' project option is selected in order to maximise economic efficiency. Profit in this instance is defined as the difference between total benefits and total costs. However, care must be taken in extending the private allocation model to the public sector. The two will only equate if certain conditions prevail. These conditions include the ability to determine costs and benefits in a competitive market and the absence of any external effects that cannot be accounted for within the process of evaluation.

Within the United States, the application of the principles of cost–benefit analysis to the assessment of public sector projects raised some controversial issues both with the Government executive and the legislature during the 1950s, with a number of economists developing an academic interest in these matters. In 1958, three influential texts on cost–benefit analysis by Eckstein, Krutilla and Eckstein and McKean were released. In 1965, Dorfman released an extensive report applying cost–benefit analysis to developments outside the water resources sector. Since the 1960s use of the technique spread beyond the United States and was used extensively to aid option choice in the transport, urban planning and environmental management sectors. Over the past forty years the method has been developed and extended in order that its approach to the valuation of costs and benefits might recognise as effectively as possible that the vast majority of public policy decisions have multiple objectives reflecting social goals that are much broader than the mere maximisation of economic efficiency.

7.3   Theoretical basis for cost–benefit analysis

Cost–benefit analysis is a methodology for assessing the economic desirability of competing engineering project options, estimating the extent to which each contributes towards the improvement of economic welfare of the entire community in question. But how is economic welfare measured in order that an improvement in it can be identified? The answer lies in the Pareto Optimum, a concept put forward by Vilfredo Pareto for measuring changes in social welfare. To explain it, define two distinct states, State A and State B, with the former consisting of the community in question **without** the engineering project under examination in existence, and the latter state describing the community **with** the project up and running. Pareto stated that State B was economically preferable to State A if the project in question resulted in at least one person being made better off and no one being made worse off. In this situation the project is deemed a Pareto improvement.

Thus, an engineering project will be deemed a welfare improvement on the 'do-nothing' situation if it improves one individual's economic well-being and results in a dis-improvement for no one. This rule, while being powerful in theoretical terms, is somewhat limited in its application, as all engineering projects are bound, by their very nature, to create 'losers' as well as 'winners' in economic terms, making at least one person worse off as a result of their development. This limitation is surmounted by use of a modification to the Pareto rule, called the Kaldor–Hicks criterion, which introduced the concept of compensation. It assumes that a project creates winners and losers, but states that its development can still constitute a welfare improvement if those that gain from the project are in a position to compensate those that lose out and still remain better off as a result of it. The actual process of compensation need not physically take place. It is the possibility of compensation occurring that is important, and the fact that the net 'overall' welfare improvement would still exist after it has taken place.

Cost–benefit analysis takes the rather abstract notion of improvements in economic welfare resulting from competing engineering projects, as expressed by Kaldor–Hicks, and provides the decision maker with a set of guidelines that will enable these projects to be appraised so that it can be decided which option is more desirable in economic terms. It does so, in most circumstances, using the yardstick of money. This stems from the fact that, in order to pose the question 'is the benefit from the proposed project worth the cost' in a quantitative form, one must have the cost and benefit expressed in a common unit of measurement, and money is the most convenient unit for use within this context. The benefits accruing to those that gain and the costs incurred by those that lose are thus all assessed in monetary terms. For a major public project, those that gain will be the users of the development in question, while the main loser is the organisation/public body financing the venture. If the monetary valuation of its economic benefits outweighs the full cost of the development, the project is deemed to be economically viable. Cost–benefit analysis is the vehicle within which this calculation is made. If a number of options are being compared, then the one resulting in the largest net positive welfare in monetary terms will be selected.

Much of the work of carrying out a cost–benefit analysis involves identifying the various costs and benefits relevant to a particular project and then putting them into the common unit of money. When they occur at different times, the various methods described in previous chapters for translating them into time equivalent values can be used. When the various costs and benefits are different in nature, various techniques are required to value them. These are described in the following sections. Before we examine this, however, we must first identify the relevant costs and benefits.

7.4   The procedure of cost–benefit analysis

There is a sequence that is followed when undertaking a cost–benefit analysis. While it is not a rigid or immovable set of steps, it forms a representation of how the method should be properly undertaken. The relative importance of the various steps within the process will vary depending on the type and scale of project being evaluated. The steps can be listed as:

- Identifying the main project options that the decision makers wish to be assessed.
- Identifying all costs/benefits/disbenefits relevant to the economic analysis.
- Placing valuations on all the costs/benefits/disbenefits identified.
- Assessing and comparing the cost–benefit performance of the options.
- Carrying out a sensitivity analysis.
- Making a final decision.

These stages can now be examined individually.

7.5   Identifying the main project options

This is a fundamental step in the cost–benefit analysis process where all the decision makers compile a list of all relevant feasible options that they wish to be assessed. It is usual to include a 'do-nothing' option within the analysis in order to gauge those evaluated against the baseline scenario of doing nothing. This ensures that, in addition to the various options being compared in relative terms, these are also seen to be economically justified in overall terms, in other words their benefits exceed their costs.

7.6   Identifying costs and benefits

This is a vital step in the cost–benefit analysis process, as costs and benefits must be identified as part of the process of valuing them. A proportion of them is directly related to the basic purpose of the proposed development, and is thus relatively obvious and straightforward, but others relating to unavoidable side effects are less obvious. The way to achieve proper and comprehensive identification of both costs and benefits has its foundation in the clear definition of each course of action. This is usually attained by defining the *with project* scenario and the *without project* scenario. Both are projections into the future and thus have levels of uncertainty associated with them. The latter scenario may not necessarily be equivalent to the 'do-nothing' option and may not equate to a continuation of the present situation, although it will probably be quite close to it. The definition and description of the without project scenario should be such that it constitutes an entirely feasible and credible course of action, one that might in reality materialise if the proposal does not obtain approval. For example, if a water treatment plant is not approved, an alternative course of action is a 'do-minimum' refurbishment of existing facilities. Once the two scenarios are described in detail, the full lists of costs and benefits associated with both must be drawn up. On the assumption that some initial estimations of costs and benefits for both situations are possible, any impact whose valuation is identical for both scenarios can be discarded, as no change in either cost or benefit will result from that impact. It will not form a basis for telling the two apart. Those costs and benefits that vary across the two scenarios form the basis for the evaluation.

For an engineering development project, the costs are easily identified. They constitute the initial construction cost plus any maintenance and operating costs that occur during its economic life. Identifying the benefits is less straightforward. This is done on the basis of the net output that becomes available to the economy of the community as a result of the project. For example, when assessing the benefits to a community from the construction of a light rail scheme, these are seen to accrue to the actual users of the rail system, mainly through the savings in time to commuters that result from the improvement in transport facilities. These savings, which can be given a monetary value, did not exist prior to the project's completion. An irrigation project might result in a lowering in crop production costs to local farmers and a consequent increase in farm revenues. This extra income would not have materialised without the project being built. These are thus the sources of the extra benefits accruing to sections of the community as a direct result of the construction of such projects, and are estimated against the benchmark of the 'without project' situation.

In practice, decision makers often examine previous cost–benefit analyses for projects similar in type and scale to the one under examination in order to build up a list of benefits likely to be directly relevant to the construction of the proposed project. Care must be taken, however, to ensure that the historical cases being used for this comparison are, upon close examination, truly similar both in structure and in nature to the proposal for which the costs and benefits are being identified.

7.7   Placing valuations on all costs and benefits/disbenefits

Having identified the costs and benefits that should be included within a cost–benefit analysis, one is then faced with the tasks of assigning monetary valuations to each of these so that the net benefit can be estimated using any one of a number of indicators.

Given that we are measuring either the costs to society of producing the inputs required to bring the project to fruition or the economic benefits to sections of society as a result of the project's outputs, the values placed on them must reflect as closely as possible the *real* or *actual* costs incurred or benefits derived. Such valuations are relatively straightforward where market prices exist for these inputs/outputs, as this is seen as the most accurate reflection of its real worth to society. For example, if the construction of a highway involves the demolition of a number of dwellings, the cost incurred by society, that is the money due to the owners for the loss of their property, is equal to the market value of these properties.

However, many inputs/outputs that we might wish to value within a cost–benefit analysis often have no direct market value, or market imperfections may make an accurate prediction of its direct value extremely difficult. In such situations, techniques such as shadow pricing and the hedonic price approach can be used to indirectly estimate the market value of an input or output.

7.7.1   Shadow pricing

The principle at the basis of the need for shadow pricing is opportunity cost. When looking for the monetary valuation of an item to be included in a cost–benefit analysis, it is its 'opportunity cost' that we seek to obtain. The opportunity cost of an input/output is defined as that which one must give up or sacrifice in order to obtain it. The opportunity cost for a Government of spending £100m on building an underground rail transport system for a city is that other public projects, valued together at this same amount, will now not be built and people will have to forego the benefits that would have been available from them. In other words, the benefits derived from spending £100m on public sector housing or improvements in the telecommunications network will now not be realised because of the construction of the underground rail system. In strict economic terms, the opportunity cost of using resources is the benefit foregone in the *best* alternative use rather than just any alternative use. In many situations, the market price will closely reflect its opportunity cost to society. However, in some circumstances, in the context of a major engineering project of significance to a wide community of people, it may be the case that market prices do not provide an appropriate measure of the opportunity costs associated with inputs and/or outputs relevant to the project. In such circumstances it may be judicious to substitute shadow prices for market prices.

Take, for example, the estimation of the benefits of a newly constructed highway. The main beneficiaries of such a scheme are the road users who gain economically by means of reduced costs. Yet there is no direct market that allows this benefit to be accurately estimated. The users may not be charged directly for road travel and, even if they are, the amount levied may not reflect the true market value of the benefit accruing to the motorists. To estimate as accurately as possible the true economic value of the benefits to the road users, and thus an amount that each individual driver would be willing to pay if required by the road authority, three shadow benefits arising from the construction of the new road are estimated: time savings for motorists and the reduction both in vehicle operating costs and in accident costs. Accurate market-based valuations of these three proxy benefits can be achieved. These three added together give an accurate economic valuation of the overall benefit to road users. The case study shown in Section 7.11 illustrates how these three shadow benefits are estimated.

On the cost side, the financial cost of construction may not equate with its true economic cost, and to get from one to the other a factor, known as the shadow price factor (SPF), is used. Say a land reclamation project is being built in a rural area using unskilled farm labour whose work might be seasonal in nature. Therefore, the opportunity cost to the developer of using this unskilled labour is the value of farm work foregone by the labourer. This assumes that the land reclamation project took place during the working season for the labourer. If it did not, then the opportunity cost of the worker would be low. This is overcome by taking the estimate of the earning power of the labourer during high season and multiplying it by an shadow price factor, say 0.5, to get an accurate economic estimate of the cost to the community of the labourer in question.

7.7.2   Non-market valuation of costs/benefits

In the context of an engineering development project, there are numerous costs/benefits/disbenefits, particularly in the environmental and social areas, for which no market exists within which it can be valued. Normal methods for assigning monetary values to such inputs/outputs cannot therefore be used. In these circumstances, two options are open to the decision maker. The first consists of using a 'revisionist' cost–benefit analysis technique or a multicriteria model as a separate decision framework, into which the particular effect can be readily introduced and assessed. It can be used, and its results examined, in parallel with those of the more economics-based cost–benefit analysis. These alternative decision frameworks are outlined in the following chapters An alternative approach lies in using one of a number of techniques, each at different stages of development, to place economic values on impacts where this previously was not possible. Once the monetary value for the impact has been established, it can then be included fully within the cost–benefit analysis. Some of these valuation techniques are:

- Techniques using surrogate market prices
  - Property values (the hedonic price approach)
  - Travel cost
- Contingent valuation methods
  - Bidding games
  - Take-it-or-leave-it experiments
  - Delphi technique

These methods are explained further in Section 7.15.

7.8   Assessing and comparing the cost–benefit performance of options

An engineering project is often complex and long term, with the costs and benefits associated with it occurring over a long time, termed the *life* of the project – a parameter we have dealt with in the earlier chapters of the book. It sets a limit on the period over which the costs and benefits are estimated, as all must occur within this time slot, be it 25, 35 or even 50 years or more. It is related, in principle, to the expected lifetime of the project under analysis. Given that many engineering developments (dams, bridges, tunnels, power stations) have the potential to be in service for a very long period, it may seem impossible to set a limit on the life of the project with any degree of certainty. In practice, however, this may not give rise to serious problems in the evaluation, as the loss of accuracy that results from limiting the life of a project to 35–40 years instead of continuing the computation far beyond this point in time is marginal, as shown by some of the examples in earlier chapters on projects with assumed infinite lives. The shortened analysis can be justified on the basis that, in time equivalent terms, substantial costs and/or benefits are unlikely to arise in the latter years of the project. If they are predicted, the life may well have to be extended. Truncating the analysis can also be justified on the basis of the uncertainty with which costs and benefits that occur beyond a certain time horizon can be predicted. Where this technique is applied after a relatively small number of years, the project may well have to be assigned a substantial residual or salvage value, reflecting the significant benefits still to be accrued from the project or, conversely, costs still liable to be incurred by it (a residual value can be negative, as say for a nuclear power station yet to be de-commissioned). The difficulty in assigning a meaningful residual value to a project after so few years in commission results in this solution being rather unsatisfactory. It is far more advisable to extend the evaluation to a future point in time where the residual value is extremely small relative to its initial value.

In addition to this, the costs and benefits occur at different times over this time horizon. Because of this, they cannot be directly combined until they are reduced to a common time frame. This is achieved using another parameter introduced earlier, the *discount rate*, which translates all costs and benefits to time equivalent values. The actual value used is the social discount rate, given that the decision maker is interested in the benefits and costs to society as a whole rather than to any individual or group of individuals. The setting of this rate is quite a complex process and is somewhat beyond the scope of this text. It is important to point out, however, that it is not the same as the market interest rate available to all private borrowers. It is a collective discount rate reflecting a project of benefit to a large number of people and spanning a time frame greater than one full generation. The long-term collectively-based rate should be relatively low, almost certainly in single figures and probably below 7%. The prevailing market interest rate for individual borrowers may well be substantially greater than this figure.

Having set the two parameters of project life and discount rate, the time frame for the comparison has been set together with the discount rate that allows all costs and benefits to be directly compared at the same point in time. Next, the decision maker must choose the actual mechanism for comparing and analysing the costs and benefits in order to arrive at a final answer for the net benefit of each of the project options under consideration. Any of the four techniques described earlier in the text can be used for this purpose:

(1)   Net Present Value (NPV)
(2)   Equivalent Annual Worth (EAW)
(3)   Internal Rate of Return (IRR)
(4)   Benefit/Cost Ratio (B/C)

The NPV will estimate the economic worth of the project in terms of the present worth of the total net benefits, with the EAW giving the equivalent annual worth of the total net benefits, all calculated over the entire designated project life and discounted at the agreed rate. The IRR will yield, for each option under consideration, the rate at which the net present value for it equals zero, with the B/C ratio based on the ratio of the present value of the benefits to the present value of the costs. For the last two methods, if the options under consideration are mutually exclusive, an incremental analysis must be carried out to establish the best performing one in economic terms.

All methods depend on discounting to arrive at a final answer. All, if used correctly, should give answers entirely consistent with each other, but the specific technique to be used varies with the circumstances. Thus, while the chosen technique is, to a certain extent, down to the preference of the decision maker, it is, nonetheless, dependent on the type of the decision to be taken within the analysis. If the decision is whether or not to proceed with a given project, the result from the chosen technique is compared with some predetermined threshold value in order to decide whether the project is economically justified. Once a discount rate/minimal acceptable rate of return is set, any of the above methods will give the same result. Assuming a discount rate of 9%, the project will be economically acceptable if the NPV or EAW of the net benefits at 9% exceed zero, if the IRR is above 9% or if the B/C ratio at 9% exceeds unity.

In the case of an independent project, all techniques yield the same result, the critical question being the choice of discount rate. In choosing between mutually exclusive projects, the most straightforward method involves choosing the option with the maximum NPV/EAW of net benefits.

There may, however, be situations where it is required to rank in order a number of projects. If, for example, there is a set quantity of resources available for developing a certain category of project, the decision maker wishes to have a sequence in which these projects should be approved and constructed until the allotted resources are exhausted. In these cases, ranking based on NPV is of no assistance, since high-cost projects deemed acceptable based on borderline NPV scores would be given priority over low-cost projects yielding far better benefits. A correct course of action would be to rank the different project options based on their benefit/cost ratio, with the one with the highest B/C score given the rank 1, the second highest score given the rank 2, and so on.

Selecting a criterion for deciding between project options can be contentious. Some decision makers are used to incorporating certain techniques in their analyses and are loath to change. IRR is rarely mentioned in the preceding paragraph, yet a number of National Governments have a preference for it. This inclination towards it by some decision makers is, to some extent, based on the fact that many have a background in banking and thus have an innate familiarity with this criterion, together with the perception that its use does not require a discount rate to be assumed or agreed. The latter statement is, in fact, incorrect, as, particularly when evaluating a single project, IRR must be compared with some agreed discount rate.

Other supplementary methods of analysis discussed in earlier chapters such as life-cycle cost analysis and payback period could also be used to analyse project options. Least-cost analysis is of use with mutually exclusive projects where benefits to all are known to be broadly the same and do not form part of the evaluation. Evaluating the costs alone, therefore, the option with the lowest discounted cost valuation, which can also be termed the least negative NPV, is selected for development.

7.9   Sensitivity analysis

Any of the analytical techniques referred to immediately above assume that the costs and benefits used within it have been estimated to the highest possible level of certainty. However, it is inevitable with projects whose evaluation extends over a large number of years that some uncertainty is associated with these estimates. Furthermore, because no perfect way exists for estimating any of the specific parameters in the study, some error must exist in the way the figures have been derived. For some parameters, such as construction and operating cost estimates, where it may be possible to increase the accuracy in their computation, the expense and time required may make the exercise prohibitive. Furthermore, for certain parameters, such as the discount rate, no one true 'correct' value exists.

In an effort to gauge the extent to which any variation in the values of certain individual parameters will affect the final answer, the decision maker will carry out a sensitivity analysis in which the effect of varying these values on the baseline cost–benefit result is determined. This process is called sensitivity testing. It is a transparent and flexible procedure in which the parameters are varied in an arbitrary manner, one-by-one, in an effort to ascertain the extent to which the economic indicator is altered as a result. For any given parameter, a number of incremental changes are made and the final indicator value is computed each time, noting the degree of change from its baseline value.

For a typical cost–benefit analysis case, for a given indicator, a value for it is calculated using the baseline benefit and cost inputs. Then, one of the parameters, be it a component of a cost or benefit, is chosen and gradually varied, enabling the effect of these variations on the overall cost–benefit analysis indicator score to be assessed. Similar computations can be undertaken for all relevant parameters. Such tests are straightforward and easily performed. Their use lies in giving the decision maker a feel for the relative importance of the various parameters to the final cost–benefit analysis score as shown by the indicator in question (NPV, EAW, IRR or B/C). An important aspect of this type of test is finding, in the case of a cost input, the level of increase above the baseline estimate at which the indicator hits a critical/threshold value, such as an NPV of zero or a benefit/cost ratio of unity. Conversely, individual benefit components can be lowered to establish the point at which this causes the onset of a critical indicator score. In cases where the decision maker is aware that two components of a cost or a benefit are correlated, it may be preferable to vary them together rather than individually. Such a process does, however, make the final analysis less transparent and straightforward.

Another useful test involves varying the point in time at which the benefits of a project come on stream. If the baseline figures assume that benefits commence at the end of year 2 – the point at which construction has finished and the project goes into commission – the sensitivity test might involve delaying this start-up point to the end of year 3 and seeing what effect this delay has on the overall cost–benefit analysis score. This can be an important indicator, as delays in construction by a year or more may have a drastic effect on the economic viability of the proposed project.

Figure 7.1 takes one of the benefits of a rail scheme – time savings – and gives a graphical representation of how increases and decreases in the value of this benefit affect the overall benefit/cost ratio. For the hypothetical case shown for illustrative purposes, the ratio is seen to decrease from its initial baseline value of 1.55 to unity when the time benefits are reduced to 54% of their initial valuation. Figure 7.2 gives a graphical representation of the effect a delay in the starting point of such benefits,

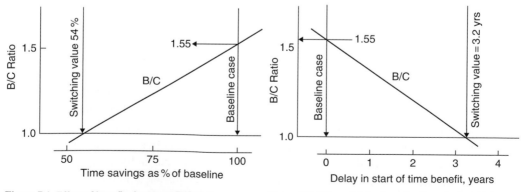

Figure 7.1 Effect of benefit change on B/C Ratio.   **Figure 7.2** Effect of delay in start of benefits on B/C Ratio.

due possibly to construction problems, has on the B/C ratio, indicating that a delay of 3.2 years in the start of the receipt of time saving benefits within the life of the project also reduces the ratio to unity.

The case study detailed later in this chapter (Section 7.11) includes a sensitivity analysis of the initial baseline result.

7.10   Final decision

The decision maker may not rely on the results of the cost–benefit analysis alone when making a final recommendation regarding which project option if any to proceed with. There must be an awareness of the criticisms that exist regarding the rather narrow focus of such a purely economic analysis and the omission of parameters that cannot be assessed in monetary terms – criticisms which have led to the introduction of some of the non-economic methods of project evaluation explained in subsequent chapters. It is wise not to use the results of cost–benefit analysis in isolation to justify a proposed course of action, as opponents will concentrate on criticism of the basic assumptions at the heart of the economic analysis, some of which may, to some extent, be justified on the grounds of a lack of equity inherent in a market-based analysis.

Professionals such as engineers, planners and architects may find the rational and logical basis of cost–benefit analysis appealing to the point where undue emphasis may be placed on its findings. This must, however, be avoided and its results seen in conjunction with an evaluation that takes into account the social and environmental consequences of the various options open to the decision maker. It should not be used as a decision making tool in isolation. If it is, its drawbacks outlined may result in it being ignored by political forces in the decision making process.

7.11   Case study: the cost–benefit analysis of a highway improvement project

7.11.1   Background to cost–benefit analysis for transport projects

The case study detailed is taken from the transport field. Before the details of this case study are given, some background information on the application of cost–benefit analysis to the assessment of transport infrastructure is provided.

The application of cost–benefit for project assessment in the transport area is made more complicated by the wide array of benefits associated with a given transport initiative, some easier to translate into monetary values than others. Many of the benefits of improvements to transport projects equate to decreases in cost. The primary grouping that contains this type of economic gain is termed user benefits. Benefits of this type accrue to those who will actively use the proposed installation. This grouping includes:

- reductions in vehicle operating costs;
- savings in time;
- reduction in the frequency of accidents.

This is the only grouping examined in the case study below. Other more comprehensive studies assess a secondary grouping of benefits, those accruing to 'non-users' of the proposed facility. These include:

- Positive or negative changes in the environment felt by those people situated either near the new facility or near the existing facility from which the new one will divert traffic. These can be measured in terms of the changes in impacts such as air pollution, noise or visual intrusion/obstruction.
- Savings in time by travellers not using the new installation. Where a new urban transit system is being assessed, the savings in time experienced by car users due to the reduced congestion on the road, resulting from some drivers switching their mode of travel, is assessed as a benefit to 'non-users' of the transit system.
- The loss or improvement of recreational facilities used by local inhabitants, or the improvement or deterioration in access to these facilities.

The costs associated with a proposed transport installation can fall into similar categories. However, in most evaluations, construction costs incurred during the initial building phase, followed by maintenance costs incurred on a continuing basis throughout the life of the project, are sufficient to consider.

Within the case study examined, the three primary user benefits listed above are estimated relative to the *without project* or 'do-nothing' situation. Let us examine each of these benefits in some detail.

Reductions in vehicle operating costs

This constitutes the most direct benefit derived from a new or upgraded transport installation. It is often the most important one and the one easiest to measure in money terms. While the users are the initial beneficiaries of these reductions, circumstances dictated by government policies, competition or the drive to maximise profits might lead to other groups within the broader community having a share in the ultimate benefit.

The levels of reduction in transport operating costs vary widely. Road and rail improvements generally lead to lower operating costs. Shipping port improvements may lead to savings in the cost of cargo handling, and overall shipping costs may be reduced as a result of the ability of the upgraded port to cater for larger more economical ships. In the case of a highway, the new upgraded project leads to lower levels of congestion and higher speeds than on the existing roadway, resulting in lower fuel consumption and lower maintenance costs due to the reduced wear and tear on the vehicles.

In the following highway case study, a formula is used which directly relates vehicle operating costs to speed. Costs included are both fuel and non-fuel based. The higher speeds possible on the new road relative to the existing road lead to monetary savings for each road user.

Savings in time

The upgrading of a transport installation will invariably reduce travel time as well as improve the reliability of transport services. For transport users, time has some connection with money. The degree of correlation between the two depends primarily on the manner in which the opportunities made possible by the increased availability of time are used.

In general, analyses of the value of time savings within the cost–benefit analysis framework focus on distinguishing between travel for work and travel for non-work purposes. Non-work time includes leisure travel and travel to and from work. Within developed economies, the value of working time is related to the average industrial wage plus added fringe benefits, on the assumption that time saved will be diverted to other productive uses. There is no broad agreement among economic evaluation experts regarding the valuation of non-work time. Because there is no direct market available that might provide the appropriate value, values must be deduced from the choices members of the public make that involve differences in time. Studies carried out in industrialised countries have indicated that travellers' value non-working time at between 20 and 35% the value attributed to working time (Adler, 1987). Less developed countries may, however, set the valuation at a lower percentage. In the case study presented, an average value for time savings is used which supplies a single value covering both 'workers' and non-workers' using the highway.

Reduction in the frequency of accidents

Assessing the economic benefit of accident reduction entails two steps. In the case of a highway, this requires comparison of the accident rate on the existing unimproved highway with that of other highways elsewhere in the country (or abroad) constructed according to the higher standard of the proposed new road. The higher the standard of construction of a highway, the lower its accident rate will be.

The second step involves the monetary valuation of the accident reduction. Three types of damage should be considered:

(i)   property damage;
(ii) personal injuries arising from serious accidents;
(iii) fatal accidents.

Property damage to vehicles involved in accidents is the most easily measured in monetary terms. Reduced breakage of cargo can also be a significant benefit in proposed rail-based and seaport installations. Valuations can be obtained directly from the extent of claims on insurance policies.

The cost of serious but non-fatal accidents is much more difficult to assess. Medical costs, the cost of lost output and personal pain and suffering constitute a large proportion of the total valuation.

There is major disagreement on which method is most appropriate for estimating the economic cost of a fatal accident to society. In recent times, stated preference survey techniques have been employed to estimate this valuation.

In the case study detailed, an average cost per accident, covering fatal and non-fatal, is employed, with damage costs also accounted for within the final estimated value.

7.11.2   *Introduction to the case study*

It is proposed to upgrade an existing single carriageway to a dual carriageway, with safety improvements being made at the major intersections. The time scale for the construction of the new roadway is 2 years, with traffic benefiting from the improvements from the start of the third year.

It is assumed that the main benefits accruing from the project are:

● time savings to motorists using the new road;
● accident savings arising from the higher design standard of the proposed roadway;
● vehicle operating cost savings due to higher travelling speeds attainable on the new route.

The initial construction phase is set at 2 years, with all construction costs incurred within this time frame. The subsequent operational life is set at 12 years, with annual maintenance costs being incurred over this period.

It is required to estimate the economic return from the proposed scheme using the following indicators:

(i)   Net Present Value (NPV)
(ii) Return on Investment
(iii) Benefit/Cost Ratio

7.11.3   *Data*

The data required to carry out the economic analysis for the above highway project are given in Table 7.1. The predicted flows on the route over the 12-year life of the project, together with the yearly construction and maintenance costs, are given in Table 7.2. It can be seen that the actual costs incurred consist of the construction costs that are paid out during the initial two-year building phase, with maintenance costs incurred during each of the subsequent 12 years during which the road is assumed to be in service.

Table 7.1 Data required for economic analysis of highway project.

Description	Data
Accident Rates	0.75 per 10^6 vehicle-kilometres (existing road)
	0.25 per 10^6 vehicle-kilometres (proposed road)
Average cost of an accident	£7500
Average value of vehicle time savings	£1.75 per hour
Average speed on existing road	45 km/hour
Average speed on proposed road	85 km/hour
Discount rate	8%
Average vehicle operating cost	$((2+25/v+0.00001v^2) \div 100)$ £ per km

Year	Predicted flow (10^6 veh-km/yr)	Construction cost (£)	Operating cost (£)
1	—	10 000 000	—
2	—	5 000 000	—
3	200		250 000
4	210		250 000
5	220		250 000
6	230		250 000
7	240		250 000
8	250		250 000
9	260		250 000
10	270		250 000
11	280		250 000
12	290		250 000
13	300		250 000
14	310		250 000

Table 7.2 Costs and flows for each year of operation of the highway.

7.11.4   Solution

Accident savings

The reduction in accident costs per year due to the improved road can be calculated as follows:

$$= (\text{accident rate on existing road} - \text{accident rate on proposed road})$$
$$\times \text{ accident cost} \times \text{flow}$$
$$= (0.75 - 0.25) \times 7500 \times \text{yearly flow}$$

For example, in year 3, the predicted flow is 200×10^6 vehicle-kilometres. Therefore, the accident savings for year 3 is calculated as follows:

$$\text{Accident savings}_{(\text{Year 3})} = (0.75 - 0.25) \times 7500 \times 200$$
$$= £750\,000$$

The accident savings for each of the subsequent 11 years are calculated using the relevant flows in Table 7.1. The resulting figures are shown in Table 7.3.

Operating cost savings

The difference in vehicle operating costs as a result of the new road can be calculated per vehicle as follows:

$$\text{Vehicle operating cost}_{(\text{Existing road})} = (2 + 25/45 + (0.00001 \times 45^2)) \div 100$$
$$= \pounds 0.025758 \text{ per km per vehicle}$$
$$= \pounds 0.25758 \text{ per vehicle-km}$$
$$\text{Vehicle operating cost}_{(\text{Proposed road})} = (2 + 25/85 + (0.00001 \times 85^2)) \div 100$$
$$= \pounds 0.0236637 \text{ per km per vehicle}$$

For example, in year 3, the predicted flow is 200 vehicle-kilometres. Therefore the savings in vehicle operating costs can be estimated as follows:

$$\text{Vehicle operation savings}_{(\text{Year 3})} = (0.025758 - 0.0236637) \times 200 \times 10^6$$
$$= \pounds 418876$$

The vehicle operating cost savings for each of the subsequent 11 years are calculated, again using the relevant flows in Table 7.2. The resulting figures are shown in Table 7.3.

Journey time savings

This saving is based on the shorter time taken by each vehicle to travel along the proposed new road as opposed to travelling along the existing road. This time saving multiplied by the value of time for the occupants of the vehicle allows a monetary value of the time savings to be computed.

For example, again in year 3, the value of time savings can be computed as follows:

Time taken for vehicle to travel existing road/km	$= 1/45$
	$= 0.022222 \text{ h/km}$
Time taken for vehicle to travel proposed road/km	$= 1/85 \text{ km/h}$
	$= 0.011765 \text{ h/km}$
Time saved per vehicle travelling new road	$= 0.010458 \text{ h/veh-km}$
Value of time saved per vehicle kilometre	$= 1.75 \times 0.010458$
	$= \pounds 0.0183/\text{veh-km}$

Table 7.3 Estimation of total discounted benefits.

Year	Accident cost savings (£)	Operating cost savings (£)	Travel time savings (£)	Total benefits (£)	Discounted benefits (£)
1	—	—	—	—	—
2	—	—	—	—	—
3	750 000	418 876	3 660 131	4 829 007	3 833 421
4	787 500	439 820	3 843 187	5 070 457	3 726 937
5	825 000	460 763	4 026 144	5 311 907	3 615 195
6	862 500	481 707	4 209 151	5 553 358	3 499 557
7	900 000	502 651	4 392 157	5 794 808	3 381 215
8	937 500	523 595	4 575 163	6 036 258	3 261 202
9	975 000	544 538	4 758 170	6 277 708	3 140 417
10	1 012 500	565 483	4 941 176	6 519 159	3 019 632
11	1 050 000	586 426	5 124 183	6 760 609	2 899 509
12	1 087 500	607 369	5 307 190	7 002 059	2 780 614
13	1 125 000	628 314	5 490 196	7 243 510	2 663 424
14	1 162 500	649 257	5 673 203	7 484 960	2 548 337
					$\Sigma = 38\ 369\ 460$

Again taking year 3 where the predicted flow is 200×10^6 vehicle-kilometres, the monetary value of the time savings for this period is calculated as follows:

$$\text{Time savings}_{(\text{Year 3})} = 0.0183 \times 200 \times 10^6$$
$$= £3\,660\,000$$

Discounting of costs and benefits

It is assumed that all construction costs are incurred in two payments at the end of each of the first two years, with all benefits accruing and all operating/maintenance costs being incurred at the end of the subsequent 12 years. To compare these monetary amounts, they must be converted to present value amounts; that is, their value at time zero. Take for example the construction cost payments in years 1 and 2 and the total benefits in year 3.

In Year 1, the total construction cost, incurred at the end of the period, is £1 m. The general formula for estimating the discounted cost–benefit analysis to present value is:

Net benefit / cost discounted to year n = net benefit / cost $\div (1 + r)^n$

where r is the discount rate, set at 0.08 (8%). The present value of the £1 m is therefore calculated as follows:

Discounted cost$_{(Year 1)}$ = $10\,000\,000 \div (1.08)^1$ = £9 259 259

In Year 2, the total construction cost, incurred at the end of the period, is £0.5 m. The present value of this sum is estimated as:

Discounted cost$_{(Year 2)}$ = $500\,000 \div (1.08)^2$ = £4 286 694

In Year 3, the total benefit accruing to motorists was estimated as £4 828 991, again incurred at the end of the period. The present value of this sum was estimated as:

Discounted benefit$_{(Year 3)}$ = $4\,829\,007 \div (1.08)^3$ = £3 833 421

These discount calculations are undertaken for all periods in Tables 7.3 and 7.4.

Net Present Value (NPV)

This measure of cost–benefit is obtained by subtracting the sum of the discounted costs from the sum of the discounted benefits:

NPV = £38 369 460 –£15 161 196

    = £23 208 264

To be economically viable, the net present value must exceed zero, demonstrating that the project results in a net monetary benefit to society discounted to time zero. In this case the net present value, at over £23 million, ranks high in terms of economic attractiveness.

Benefit/Cost Ratio (B/C)

This measure of cost–benefit is obtained by dividing the sum of the discounted costs into the sum of the discounted benefits:

Year	Construction and/or maintenance costs (£)	Discounted costs (£)
1	10 000 000	9 259 259
2	5 000 000	4 286 694
3	250 000	198 458
4	250 000	183 758
5	250 000	170 146
6	250 000	157 543
7	250 000	145 873
8	250 000	135 067
9	250 000	125 062
10	250 000	115 798
11	250 000	107 221
12	250 000	99 278
13	250 000	91 924
14	250 000	85 115
		Σ = 15 161 196

Table 7.4 Estimation of discounted costs.

$$B/C = 38\,369\,460 \div 15\,161\,196$$
$$= 2.53$$

To be economically viable, the benefit/cost ratio must be greater than one, demonstrating that the sum of the project's discounted benefits exceeds its discounted costs. The project examined has a ratio of between 2 and 3, emphasising its high economic rating.

Internal Rate of Return (IRR)

This measure of cost–benefit is obtained estimating the interest rate at which the sum of the discounted benefits becomes equal to the sum of the discounted costs. In this instance, this value was found to be 28.1%; that is:

$$IRR = 28.1\%$$

For economic viability, the value obtained must exceed some minimum acceptable rate of return set by the relevant government department. It would be most likely set at approximately 10%. It can be seen that, in the case of this highway project, the internal rate of return far exceeds 10%, thus reinforcing the economic attractiveness of this project.

Sensitivity analysis

The sensitivity testing took two forms. Firstly, the effect of reductions in the time savings on the economic performance of the proposal was gauged. These results are given in Table 7.5. These show that time savings have to be reduced to 20% of their baseline valuations before the benefit/cost ratio reduces to unity. The second test examined the effect of a delay in the commencement of all benefits beyond the point at which they are assumed to 'kick in' within the base study, at the start of year 3. This test shows that, even if the benefits only commence from the beginning of year 7, the benefit/cost ratio is still 1.86 (Table 7.6). These results reinforce the robustness of the economic performance of the proposal.

Table 7.5 Effect of reduction in time savings on Benefit/Cost Ratio.

Time savings as % of baseline	Benefit/ Cost Ratio
75.0	2.05
50.0	1.57
25.0	1.09
20.2	1.00

Table 7.6 Effect of delay in start of benefits on Benefit/Cost Ratio.

Delay in start of all benefits (years)	Benefit/Cost Ratio
1	2.34
2	2.16
3	2.01
4	1.86

Overall conclusions

The three indicators used within the analysis all point to the strong performance of this highway project. In absolute terms they give basic economic justification to its construction and provide enough information for it to be compared with other similar infrastructure projects that may come under consideration and which have been appraised using the same set of basic economic indicators.

Within the United Kingdom, the economic analysis of road schemes has been standardised within the *COBA* computer program and its user manual (COBA, 1996). Both have been updated on a regular basis since the early 1970s. *COBA* incorporates standard unit costs for travel time, vehicle operating costs and accident costs affecting users of the proposed highway. The method also provides for construction costs, land and property costs and future costs, including maintenance work and the cost of traffic delays resulting from construction. It contains inbuilt default values that decision makers can replace with case-specific values if available.

7.12   Case study: water supply scheme in a developing country

7.12.1   *Background to cost/benefit analysis of water projects*

Effective water resource planning is essential to the functioning of all sectors of an economy. Agriculture, industry and households all rely on the availability of a secure supply of clean water and the removal and treatment of wastewater, as well as the control of flooding events. Water is a limited resource and its abstraction, treatment, distribution and disposal typically represent relatively large capital investments at a national level. The effective economic appraisal of water resource projects is, therefore, important when deciding how to allocate these scarce resources.

The economic assessment of water resource projects, although similar to that of other large infrastructural projects, merits special consideration for a number of reasons, namely: they involve long investment horizons; are often subject to significant uncertainty; and may unfairly confer advantages on particular groups in society (This is the issue of 'equity'). For example, water infrastructure such as dams, reservoirs and water supply networks can represent very long-term investments, ranging from several decades to centuries and, therefore, the treatment of distant benefits and cash flows much be carefully considered. Some water resource projects must pay particular attention to uncertainty. For example, a project involving river water abstraction for domestic water supply must consider stochastic factors such as demographic trends and catchment development scenarios as well as inter-seasonal and long-term variations in rainfall patterns. Whereas project appraisal is concerned with the efficient allocation of resources, the distributional effects on society are often ignored. For example, should a water company favour a flood relief scheme in an affluent area where the benefits are greater rather than those in a poorer area? In general, however, the economic assessment of water projects employs the same methods used in other engineering projects.

The cost–benefit analysis of water infrastructure projects involves five main steps (UKWIR, 2007):

(1)   defining objectives;
(2)   establishing a baseline position against which alternatives can be measured;
(3)   identifying and valuing costs and benefits;
(4)   undertaking the costs–benefit analysis;
(5)   considering distributional impacts.

The objectives of water resource project cost–benefit analyses are dependent on the perspective of the body undertaking the study. Assessments may be undertaken by:

● governments, in order to provide evidence for international, national or regional water policies;
● water companies or local authorities to prioritise water investment programmes; or
● water companies or local authorities to identify the best projects among a number of alternatives.

The net costs or benefits of a water resources project must be judged against a credible baseline position. This is normally taken as the 'business as usual' (BAU) position, which considers future costs and benefits where no changes are made to existing practices and policies.

The values of costs and benefits can be direct (i.e. directly used by an individual, such as the avoided cost of buying bottled drinking water) or indirect (i.e. secondary effects not directly related to the decision to invest, such as increased tourism in a seaside area with improved water quality).

Private costs are direct monetary costs which are incurred by the project and include capital costs, operation and maintenance O&M costs and day-to-day costs associated with the project. These are typically easy to value through the application of either top-down or bottom-up techniques. The former comprise empirical cost models based on historic tender and operational data and the latter on budgets prepared by quantity surveyors and price books. Non-private or external costs include those costs imposed on society, rather than those imposed on the project investor. External social costs might include traffic disruption, loss of amenity or negative health effects. Hidden implementation costs may also arise and might include the cost of developing new regulations or training costs needed to comply with new standards. The social cost of energy-related carbon emissions is an increasingly important topic in water resources planning; this is considered in more detail in Chapter 8.

The economic value of a project benefit can be viewed as the extent to which society will sacrifice other goods or services to obtain or keep it. Social benefits include improved health (lower morbidity and mortality rates) and improved amenity as well as environmental protection or enhancement. However, because there are often no markets for benefits accruing from water projects, market-based valuation approaches are often not useful. Alternative approaches include opportunity costing, hedonic pricing (particularly useful for flood risk to properties), travel cost methods and contingent valuation (stated preference) methods using Willingness to Pay (WTP) and Willingness to Accept (WTA) techniques. An important approach is the Benefit Transfer Method where ex-post economic values from comparable projects elsewhere are used. An alternative is to adopt direct-use values. For example, where an environmental resource is lost as part of a water project (e.g. drainage of wetlands or damming a valley), its value can be inferred by summing its agricultural, fishing and tourism values. In addition, indirect-use values such as biodiversity provision and run-off control can also be included. Other benefits to consider include the 'option values' of a resource which is maintained for future use and the 'non-use value' of an environmental resource to society due to its existence (e.g. public desire to maintain wilderness even if it is never visited).

When costs and benefits have been identified and quantified, analysis involves choosing an appropriate discounting approach and considering uncertainty. The choice of discount rate depends on the investor perspective: society and government rates (for example, 4% is recommended by the Irish Government) will differ from those of private sector organisations (typically in the region of 5–8%). Declining discount rates are recommended for very long-term projects in order to give greater importance to the welfare of future generations (Groom *et al.*, 2005); the UK Treasury recommends the declining discount rates shown in Table 7.7.

Table 7.7 Declining discount rates recommended by the UK Treasury (HM Treasury, 2003).

Period, years	0–30	37–75	76–125	126–200	201–300	300+
Discount rate, %	3.5	3.0	2.5	2.0	1.5	1.0

The issue of uncertainty can be addressed using a variety of approaches: qualitative approaches, probability-weighting scenarios, sensitivity analysis and Monte Carlo analysis. These are not considered in this text but are dealt with in detail elsewhere (UKWIR, 2007). A significant constraint in uncertainty analysis is the absence of data on the probability of an event occurring (particularly extreme events, such as dam failure).

Once the cost–benefit analysis has been completed, and where a decision has been taken to proceed with a particular project or policy, the final step (where appropriate) is to assess the equity of project. Equity can be considered to be a measure of how fairly the costs and benefits of the decision are shared across different groups in society. Therefore, it is important to identify who benefits and who loses and, depending on the ethical stance adopted, favour projects which promote equity or consider only efficiency. This may involve qualitative analysis where different groupings (for example, under the headings of age, income, education or location) are designated as winners and losers. Alternative, more complex quantitative analysis can be undertaken on the distributional impacts of the project on the appropriate groups.

7.12.2   Introduction to the case study

A new water supply system is being considered for a town in a developing country which has a current population (2010) of approximately 50 000 people and is expected to grow by 3% per annum. Currently residents obtain clean water in two ways: (i) by taking it from wells and carrying it back to their homes; and (ii) by purchasing it from water sellers. A survey of households indicated that, on average, 80% of water is sourced in the former manner and the remainder is purchased at a price equivalent to $1.98/m³. It takes approximately 68 minutes to obtain each cubic metre of water and, when combined with the opportunity cost of labour (taken as an unskilled wage) of $0.30/h, the cost of fetching well water is $0.34/m³ (68 min/60 min × $0.30). The cost to the local government of providing water at the well (operation and maintenance) is estimated to be $0.02/m³, giving a total cost of well water of $0.36/m³. Approximately 75% of this is the opportunity cost of labour. Assuming real national income growth of 2.5% per annum, this cost is inflated by 1.9% per annum (75% × 2.5%).

The survey also established that the average consumption of water is 74.8 litres per capita per day (lcd). However, similar previous projects have found that switching from well to mains water supply is likely to result in an increase in water consumption of 23% to 92 lcd. Furthermore, per capita water demand is projected to rise by 0.5% per annum without the project.

The water supply scheme will involve the abstraction of raw water from a nearby watercourse, water treatment, storage, pumping and distribution to households. The programme for the implementation of the scheme is ten years, by which time it is projected to meet the needs of 80% of the population (an additional 8% added per annum) and produce a total of 1.9 million cubic metres per year (Mm³/yr). The projected lifespan of the project is 50 years and all capital will be funded at a rate of 7%. Revenues will comprise connection charges of $50/connection and a water tariff starting at $0.39/m³, increasing at a real rate of 2% per annum for 10 years and rising in line with inflation thereafter (i.e. no real price increases). The price charged by the local government for raw water abstraction is $0.02/m³.

7.12.3   Solution

Business as usual

In the business as usual case, the population grows at 3% per annum from a base of 50 000 in year 1. The existing cost of water is the volume-weighted average of the costs of water obtained from the well and from vendors; this is $0.68 (Table 7.8).

Water source	Proportion (%)	Cost ($/m³)	Weighted cost ($/m³)
Well	80	0.36	0.288
Purchased	20	1.98	0.396
Weighted average			*0.684*

Table 7.8 Existing weighted cost of water.

Future demand is the product of the future population, which is expected to grow at 3% annually, and per capita water consumption, which will grow from a base of 74.8 lcd at 0.5% per annum; water costs rise at 1.9% per annum. Based on these figures, the demographic and water demand forecast as well as unit water costs are shown in Table 7.9.

Demand with the project

It is assumed that population growth is the same with or without the project. Per capita water demand with the project is expected to rise by 0.7% per annum up to a maximum of 100 lcd, resulting in an annual demand of 1.87 Mm³ by 2019, at which stage it meets the water needs of 80% of the town's population for that year. Peak output of 1.90 Mm³/a is achieved in 2022. Water demand with the water supply scheme is shown in Table 7.10.

Table 7.9 Base case projections for population, water demands and unit water costs for selected years.

		Year					
		2010	2019	2029	2039	2049	2059
Population	('000s)	50.0	65.2	87.7	117.8	158.4	212.8
Water demand per capita	(l/d)	74.8	78.2	82.2	86.4	90.9	95.5
Total water demand	(Mm³/annum)	1.37	1.86	2.63	3.72	5.25	7.42
Water cost	($/m³)	0.68	0.81	0.98	1.18	1.43	1.72

Table 7.10 Projections for population and water demands with the water supply project.

		Year					
		2010	2019	2029	2039	2049	2059
Town population	('000s)	50.0	65.2	87.7	117.8	158.4	212.8
Population served by scheme	('000s)	4.0	52.2	52.2	52.2	52.2	52.2
Proportion of population served by scheme	(%)	8	80	60	44	33	25
Water demand per capita by scheme customers	(l/d)	92.0	98.0	100.0	100.0	100.0	100.0
Total demand	(Mm³/annum)	0.13	1.87	1.90	1.90	1.90	1.90

Project NPV for the investor

Project revenues include water sales and connection fees, the latter equally spaced over the 10-year implementation period. Water sales are the product of the $0.39/m³ tariff (increasing at 2% per annum for 10 years) and the total volume of water consumed.

Project costs include capital costs, which are phased over the construction period and are shown in Table 7.11; 20% of these costs are unskilled labour. Operation and maintenance (O&M) costs include raw water costs, chemicals, power, insurances and

Table 7.11 Capital cost phasing for the water supply project ($ '000s).

Item	| 2010	2011	2012	2013	2014	2015	2016	2017	2018	2019	Task Totals
						Year					
Source development	0.72			0.72				0.36			1.80
Treatment plant	0.18			0.06				0.06			0.30
Storage facilities	0.42			0.42				1.26			2.10
Pump station	0.48			0.16				0.16			0.80
Pipework	0.32	0.32	0.12	0.12	0.12	0.12	0.12	0.12	0.12	0.12	1.60
Design and procurement	0.18	0.01	0.01	0.01	0.01	0.01	0.01	0.01	0.01	0.01	0.30
Other	0.05	0.05	0.05	0.05	0.05	0.05	0.05	0.05	0.05	0.05	0.50
Annual Totals	*2.35*	*0.38*	*0.18*	*1.54*	*0.18*	*0.18*	*0.18*	*2.02*	*0.18*	*0.18*	*7.40*

All costs in $m.

Table 7.12 Project costs, revenues, cash flows and discounted cash flows ($ '000s) for selected years.

Year	2010	2019	2029	2039	2049	2059
Costs						
Capital	2.35	0.18	0.00	0.00	0.00	0.00
O&M	0.15	0.20	0.20	0.20	0.20	0.20
Raw water	0.00	0.05	0.05	0.05	0.05	0.05
Revenues						
Connections	0.20	0.33	0.00	0.00	0.00	0.00
Water sales	0.05	0.74	0.74	0.74	0.74	0.74
Net cash flow	−2.25	0.64	0.49	0.49	0.49	0.49
Discounted net cash flow	−2.25	0.35	0.14	0.07	0.04	0.02
NPV	*0.43*					

All figures in $m.

labour. Raw water costs are $0.02/m^3$ and assume that one third of the water sold is lost during treatment and distribution. The remaining O&M costs are estimated to be $150 000 per annum initially, rising to $200 000 per annum after four years, 30% of which are unskilled labour costs.

Table 7.12 gives costs, revenues, cash flows and discounted cash flows for selected project years. The project yields a positive NPV of $430 000 over its lifespan, indicating that, under the assumptions used, the project is worthwhile for the investors. However, the small positive NPV and significant uncertainties regarding key parameters warrant careful consideration of this figure, possibly using sensitivity analysis.

Cost–benefit analysis

The cost of well water has been calculated above as $0.68/m^3$ rising 1.9% per annum. The benefit to society of the scheme is, therefore, the product of the avoided cost of well water and the project water consumed. In addition, the salaries paid to workers on the project can be viewed as a benefit to society where the worker was previously unemployed. Here it is assumed that 50% of labour is drawn from unproductive sources (a shadow price factor of 0.5 is used) and that 20% of the project and 30% of O&M costs are unskilled labour costs. Project capital costs are, therefore, factored by 0.9 (i.e. $1 - 0.5 \times 20\%$) and O&M costs by 0.85 ($1 - 0.5 \times 30\%$).

The costs include the capital and operational costs of the project as well as connection costs and mains water costs for those connected to the scheme. An external cost arises because the price charged for raw water does not cover the real opportunity cost of using the water for alternative irrigated agricultural purposes, valued at $130 000 per annum. This cost is phased equally over the 10-year project implementation period to reflect the phased uptake of raw water.

The cost–benefit analysis was undertaken using a fixed discount rate of 5%; the results are shown in Table 7.13. The project gives a positive NPV to society of $15.7 m and benefit/cost ratio of 1.72, greater than the threshold of one, signalling that the benefits outweigh the costs.

Although the equity effects of the project are not considered as part of this case study, some observations regarding the distributional impact of the project include:

- consumers connected to the water distribution project will gain through lower water costs, although the benefits to more affluent households which purchase water from vendors will be greater than those from poorer households where water is taken directly from wells;
- unskilled workers will benefit since there will be more employment and income for them;
- farmers will lose due to the reduction in the availability of raw water for irrigation and farming.

Table 7.13 Cost–benefit analysis of the water project-costs and benefits for selected years.

Year		2010	2019	2029	2039	2049	2059
Benefits							
Project water supplied	(Mm^3/annum)	0.13	2.49	2.54	2.54	2.54	2.54
Unit value of project water	($/m^3)	0.68	0.81	0.98	1.18	1.43	1.72
Total value of project water	(m$/annum)	0.09	2.02	2.48	3.00	3.62	4.37
Costs							
Capital costs (factored by 0.9)	(m$/annum)	2.12	0.17	0.00	0.00	0.00	0.00
O&M costs (factored by 0.85)	(m$/annum)	0.13	0.17	0.17	0.17	0.17	0.17
Cost of water consumed	(m$/annum)	0.05	0.74	0.74	0.74	0.74	0.74
Foregone benefit of water	(m$/annum)	0.01	0.13	0.13	0.13	0.13	0.13
Total costs	(m$/annum)	2.31	1.21	1.04	1.04	1.04	1.04
Net benefits	(m$/annum)	−2.22	0.81	1.44	1.96	2.58	3.33
Discounted benefits	(m$/annum)	−2.22	0.52	0.57	0.48	0.38	0.30
NPV	(m$)	15.70					
C/BR		1.72					

In conclusion, given that it is attractive both to an investor (subject to further sensitivity analysis) and society, the project should proceed. However, policies should be identified which redistribute the gains experienced by both workers and consumers to farmers.

7.13   Case study: cost–benefit analysis of sewer flooding alleviation

7.13.1   Introduction to the case study

A housing estate comprising 124 properties has been subject to flooding during heavy rainfall events. A survey of the estate established that the flooding was due to the inability of the sewage system to carry the required hydraulic load during severe rainfall events; a total of 25 low-lying properties have experienced both internal and external flooding as a result. Field analysis and computer-based hydraulic modelling indicate that 14 houses will flood for an event with an exceedence probability of 0.1 (the probability of one flood of at least this magnitude occurring in any one year, or a 1 in 10 year storm), and a further 11 houses during a 0.2 probability event (1 in 5 year storm). These risks are deemed unacceptably high.

The objective of the study is to undertake a cost–benefit analysis of a scheme which would reduce the probability of sewerage flooding to an acceptable level, deemed to be a 0.02 probability event (1 in 50 year storm) based on computer simulation.

7.13.2   Solution

The solution involves both increasing the capacity of the local sewer network by upsizing pipework and installing attenuation tanks. The cost of the project is established using a bottom-up approach and total costs are estimated to be £1.58 m (Table 7.14). Social costs are regarded as being minimal: no major traffic disruption is anticipated because the project will take place in a quiet area and no businesses will be affected by the work.

The benefits of the scheme include the property damage (the dwelling and contents) and personal losses (the loss of items of emotional value, for example) that are avoided due to the flooding event. The former are quantifiable but the latter are not and are therefore ignored for the purposes of the study. The quantifiable benefits can be valued as the product of their monetary value and the change in the probability of a flooding event as a result of the upgrade works. The UK's The Water Services Regulation Authority (OFWAT) has published a series of look-up tables that estimate the present values of the amounts which are worthwhile spending on sewer flooding alleviation projects (Green and Wilson, 2004). Here, a bottom-up approach (using surveyors) was used to estimate flood damage costs to dwellings and a market-based approach was used to identify content costs. Net present values are based on a declining

Table 7.14 Capital cost of flood alleviation scheme.

Construct new sewer Remove existing 350 mm diameter sewer and replace it with a new 1200 mm diameter sewer. Allows for excavation, bedding material, connections, backfilling, compaction and making good.	(£)	550 000
Attenuation tank construction Excavation, installation, connecting, backfilling and making good attenuation tank arrangement including all manholes and chambers.	(£)	620 000
Decommissioning existing sewer Seal existing 350 mm diameter sewer and make good.	(£)	25 000
Preliminaries Site mobilisation, insurances etc.	(£)	165 000
Indirect costs Clients costs	(£)	220 000
Total costs	***(£)***	***1 580 000***

Table 7.15 Look-up table showing equivalent capital sum worth spending to alleviate flooding.

	Proposed flood return period in years (exceedence probability)						
Existing flood return period in years (exceedence probability)	2 (0.5)	5 (0.2)	10 (0.1)	20 (0.05)	30 (0.033)	40 (0.025)	50 (0.02)
2 (0.5)	10 458	69 746	75 750	87 941	89 837	91 440	92 110
5 (0.2)		5 229	30 638	33 640	40 413	41 740	42 977
10 (0.1)			2019	14 796	16 464	21 204	22 229
20 (0.05)				1 045	8 103	9 217	12 298
30 (0.033)					522	5 463	5 814
40 (0.025)						348	4 160
50 (0.02)							261

(Source: Green and Wilson, 2004)

discount rate of 3.5% falling to 3% in year 30. Table 7.15 shows the resulting maximum expenditures justified in increasing the return period (or reducing the exceedence probability) from one value to another.

Based on these values, the justified expenditure for decreasing the probability of a flooding event for the housing estate case study from 0.1 (14 dwellings) and 0.2 (11 dwellings) to 0.02 (all 25 dwellings) is shown in Table 7.16. The per-dwelling expenditure values from Table 7.15 have been adjusted to account for the changes in the UK consumer price index between 2004 and 2011 (17%) and shown that a total expenditure of £917 250 (2011 prices) is justified. In other words, expenditure in excess of this would yield a benefit/cost ratio less than one. Since the projected capital cost is £1.58 m, then the benefit/cost ratio of the project is 0.58 and the project should not proceed in its current form.

Exceedence probability		Value			Justified expenditure
From	To	£(2004)	£(2011)	No. properties	£(2011)
0.1	0.02	22 229	26 008	14	364 111
0.2	0.02	42 977	50 283	11	553 114
					Total = 917 225

Table 7.16 Justified expenditure to alleviate housing flooding.

7.14 Advantages and disadvantages of traditional cost–benefit analysis

Kelso (1964) believed that the final output from a traditional cost–benefit analysis should be a cardinal number that represents the dollar rate of the streams of net prime benefits of the proposal; he termed 'pure benefits'. Pure benefits measure the net benefits with the project in relation to net benefits without the project. Hill (1973) believed that this statement, one that explicitly sets out the basis for a traditional cost–benefit analysis, reveals some of the major deficiencies in the technique. Although there is some consideration of intangibles, they tend not to fully enter into the analysis. As a result, the effects of those investments that can be measured in monetary terms – whether derived directly or indirectly from the market – are implicitly treated as being more important, for the sole reason that they are measurable in this way, when, in reality, the intangible costs and benefits may have more significant consequences for the proposal.

Cost–benefit analysis is most suitable for ranking or evaluating different courses of action designed to accomplish the same ends, rather than for testing the absolute suitability of a project. This is, to an extent, because all valuations of costs and benefits are subject to error and uncertainty. Obtaining an absolute measure of suitability is an even greater limitation. Thus, the process cannot provide guidance in the allocation of investments between the various sectors such as transport, housing hospitals and so on. Given the choice between two projects from two different sectors, say a hospital and a motorway project, where the State has the resources to undertake and complete only one of them, cost–benefit analysis can give no meaningful guidance in making this selection. There exists no common scale for comparing the benefits of a new hospital against those derived from a new stretch of motorway. Therefore, the necessary conditions of traditional cost–benefit analysis of comparability and measurability of different courses of action cannot be complied with.

The advantages and disadvantages of cost–benefit analysis can thus be summarised:

Advantages

(1) The use of the common unit of measurement, money, facilitates comparisons between alternative proposals and hence aids the decision making process.
(2) Given that the focus of the method is on benefits and costs to the community as a whole, it offers a broader perspective than a financial/investment appraisal concentrating only on the effects of the project on the developers/investors/shareholders.

Disadvantages

(1)   Incorrect decisions may result from the confusion of the original primary purpose of a proposed project with its secondary consequences, simply because the less important secondary consequences are measurable in money terms.

(2)   The method is more suitable for comparing proposals designed to meet very similar objectives, rather than evaluating the absolute desirability of one proposal. This is partly because all estimates of costs and benefits are subject to errors of forecasting. A decision maker will thus feel more comfortable using it to rank a number of alternative courses of action rather than using it to assess the absolute desirability of only one option relative to the existing 'do-nothing' situation.

(3)   Although some limited recognition may be given to the importance of intangibles, they may be neglected within the main economic analysis. Those goods capable of measurement in monetary terms are usually attributed more implicit importance even though, in terms of the overall viability of the project, the intangibles may be highly significant.

The first two points can be managed effectively by employing an experienced and competent decision expert to oversee the use of the cost–benefit analysis framework. Problems arising from the third point may require use of one of the other methodologies detailed later in the text. Some efforts have been made to provide monetary valuations for intangibles to enable their inclusion in cost–benefit analysis. These techniques, in various stages of development, are introduced in the next section.

7.15   Techniques for valuing non-economic impacts

As noted in Sections 7.7 and 7.14, a number of techniques can be used to put monetary values on some non-economic effects. This is particularly relevant to the inclusion within the evaluation of environmental effects. The first set of methods comes under the classification of 'potentially applicable techniques', as care must be taken in using them. This does not mean, however, that they cannot be used. The second set is classified as 'additional methods' and is usable only in a limited number of cases. These methods still, however, remain valuable tools for making the analysis more inclusive.

7.15.1   *Techniques using surrogate market prices*

Many effects that are important within the context of the economic evaluation of a given project proposal have no established market price. Factors such as clean air, low noise levels, unobstructed views or ready access to recreational amenities are public goods for which direct market prices are rarely available. In many circumstances, however, it is feasible to estimate an implicit value for an effect such as an environmental good or service using the price paid for another good that is marketed. Surrogate market techniques provide approaches that use actual market prices to value an un-marketed effect. These methods can be useful in valuing a wide array of environmental and amenity effects.

Property values (the hedonic price approach)

The use of property values represents a leading illustration of the surrogate market methodology. The value of a building and its surrounding property is dependent on many variables, including size, construction quality, location and the quality of the surrounding environment and amenities. It does not consist solely of the construction cost of the building plus a given mark-up. Its value reflects a very extensive range of effects or attributes. Only some of these, for example floor area and type and quality of construction materials, are physical in nature. The market valuation of a property is related to the stream of benefits to be derived from it. To use the property value method to assess the market value of any given effect or attribute associated with it, all other effects must be kept constant. The basic supposition is that the difference in price between two properties, arrived at after all other variables except the effect in question have been kept constant, mirrors the decision maker's market valuation of the change in the effect under examination experienced at the two different locations. In the case of an environmental benefit, this approach controls all other variables, so that the subsequent positive price differential can then be assigned to the unpriced environmental good. In the same manner, environmental disbenefits can be measured using this technique, for instance a decrease in property value arising from increased noise pollution or visual intrusion.

The property value technique needs extensive data on the market values of individual units and on a wide range of physical characteristics associated with it. For example, a housing estate that previously enjoyed excellent air quality, finds itself subject to significant levels of air pollution due to the construction of a chemical plant nearby. The monetary valuation of the air pollution to the neighbourhood in question could then be estimated using this technique by calculating the extent to which property values had fallen in the immediate area affected by the increases in air pollution from the nearby factory compared to the house prices of a comparable neighbourhood nearby which had suffered no deterioration in air quality. The number of properties in the affected district multiplied by the price fall due to the pollution problem yields the economic cost to the community of the change in the local air quality levels. Also, a housing estate that previously enjoyed very low levels of ambient noise could find itself, due to the development of a new runway in the city airport, in the path of low flying aircraft. The monetary valuation of the noise pollution to the neighbourhood in question could then be estimated using this technique by calculating the extent to which property values had fallen in the immediate area affected by the noise increases from the over-flying aircraft compared to the house prices of a comparable neighbourhood nearby which had suffered no increases in noise levels.

Rose (1974) defined hedonic values as the implicit prices of the characteristics of a property. They are established from the observed prices of differentiated properties and the specific amounts of characteristics associated with them.

The travel cost method

This approach has been used extensively for valuing recreational and environmental facilities in cases where this cannot be achieved using entrance charges that either do not exist or do not accurately reflect the market value of the good or service in question. It is based on the proposition that the market value for a non-priced good can be obtained by treating travel costs as a 'surrogate' for admission prices. Users of or visitors to the site reveal the economic value they place on the facility by the amount of time and money they are willing to spend in travelling to and from it.

Most examples of the approach use outdoor recreational facilities such as parks, wildlife sanctuaries or general amenity areas. Since visitors to such facilities frequently incur no charge, or one that does not reflect its true economic worth, revenue generated by such amenities is not a good pointer to its inherent value. The real value of the amenity includes not only any user charge but also the total 'consumer surplus' enjoyed by the users. (The amount by which the total benefit from a good or service enjoyed actually exceeds the price paid for it by a consumer is termed the consumer's surplus, and indicates the total demand for the good as shown by each visitor's 'willingness to pay' for the amenity.)

To find the maximum implicit valuation that a consumer places on the facility we firstly assume the site charges no entrance fee and that the facility is thought to be desirable and valuable by the local population. Users therefore come from various places to spend time at the site. Although no entrance fee is charged, demand for visiting the site is not infinite as there is a cost involved in getting to it. The further away visitors live from the site, the less they are expected to visit it. Those living closer would have more demand for the facility because their outlay for each visit, as measured by their travel costs, is lower. In terms of the consumer surplus, the user most distant with the highest travel cost is assumed to have the lowest (in effect no) consumer's surplus. The closer someone lives to the site, however, the less their travel costs and the bigger their consumer's surplus will be, reflecting the high levels of benefit they derive from the site.

Information on visitation rates for different zones within the study area combined with information on how these visitation rates change with increasing admission price (or cost of travel) enable a demand curve for the site to be derived. At existing costs of travel, the initial information gathered above enables estimates of the total visitors to the facility. This is one boundary point on the demand curve. As the cost of travel increases above its present value, the visitation rates will gradually drop until a point is reached where the incremental increase in costs results in zero visits. Here lies the second boundary point. Using these two estimates, the curve can be constructed, allowing the total consumer surplus or benefits derived from the site to be estimated.

The strength of the method lies in its reliance on peoples' actions rather than their stated opinions. However, caution must be exercised when using it. It measures use values only. Travellers to the site may be getting enjoyment from other aspects of this journey that the method does not take into consideration, or the journey itself may be multipurpose in nature and not purely for the purpose of visiting the facility. The method seems to perform best for valuing remote amenity locations.

Example 7.1   Use of the travel cost approach to calculate the economic worth of a marshland area (Adapted from Knetsch and Davis, 1966)

The construction of an urban motorway would require the destruction of a marshland area that is a habitat for a number of rare species of birds. The marsh has many visitors each year. A local expert group wishes to estimate the economic value of this facility and use the travel cost approach to do this.

People visiting the site are interviewed and information is obtained regarding distances travelled and travel costs incurred, together with other details. For the purposes of the evaluation, the catchment area for the park is divided into three zones. Census information helps determine the population for each zone.

The information obtained is shown in Table 7.17. Using statistical procedures, a formula relating the visits per 1000 of population to the average cost of travel per visit, C, is derived as follows:

$$V(\text{£ per 1000 pop.}) = 500 - 100C \tag{7.1}$$

The differences in travel cost between zones can be used to simulate an entrance fee on the assumption that people's reaction to entrance fees is equivalent to their reaction to higher travel costs.

The data in Table 7.17 illustrate that, at present, with no entrance fee to the facility, the total number of visits per year is 120 000. This will be one of the boundary points on the demand curve.

To calculate the other points on the demand curve, initially assume an entrance fee of £1. This is added to the travel cost so that the cost per visit from Zone I is now £2, from Zone II it is £4 and from Zone III it is £5. For a revised travel cost of £2, and using Equation 7.1, the visits per 1000 population for the three zones are 300, 100 and 0, respectively. Multiplying each of these visitation rates by their population (in thousands), the new total number of visits is reduced to 50 000. When the entrance fee is incrementally increased by £1 further, the total trips reduce to 20 000, then 10 000 and, finally, zero when the entrance fee reaches £4. Details of these decreasing visitation rates are shown in Table 7.18.

The figures given in Table 7.18 are plotted against each other to give the demand curve for the recreational area (Figure 7.3).

Contd

Example 7.1 Contd

The total willingness of the community to pay for the amenity area in question is estimated on an annual basis through the following calculation of the area under the demand curve:

$$((120\,000 - 50\,000) \div 2) \times £1 = £35\,000$$
$$((50\,000 - 20\,000) \div 2) \times £1 = £15\,000$$
$$(50\,000 - 20\,000) \times £1 = £30\,000$$
$$((20\,000 - 10\,000) \div 2) \times £1 = £5\,000$$
$$(20\,000 - 10\,000) \times £2 = £20\,000$$
$$((100\,000 - 0) \div 2) \times £1 = £5\,000$$
$$(100\,000 - 0) \times £3 = £30\,000$$

$$\text{Total} \qquad\qquad = £140\,000 \text{ per annum}$$

Table 7.17 Visit rates to marsh by zone.

Zone	Population	Average cost of visit (£)	Visits made per year	Visits/1000 population
I	100 000	1	40 000	400
II	200 000	3	40 000	200
III	400 000	4	40 000	100
Total	—	—	120 000	—

Table 7.18 Entrance fee vs. total visits.

Average cost of entrance fee (£)	Total number of visits made per year
0	120 000
1	50 000
2	20 000
3	10 000
4	0

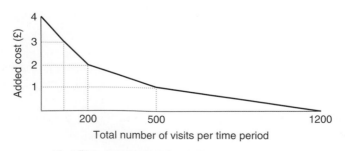

Figure 7.3 Demand curve for the marsh area.

7.15.2   Contingent valuation methods

The contingent valuation method is used where it is not possible to value the effect within either a real or 'surrogate' market. It consists of making a survey of a designated number of people using a carefully designed questionnaire and deducing the relevant economic valuations of the effects under examination from the responses of those questioned. It is based on asking the public the amount they would be willing to pay to receive or forego the benefit obtained from a particular facility for which a market does not exist, or, in the case of a negative benefit, how much you would be willing to pay to have it removed. An application of this method exists where the construction of a highway will necessitate the loss of a public park to the residents of the area. The local community do not pay to gain access to the park, so it has no inherent economic value. However, once the number of visitors over a given period is estimated, contingent valuation can be used to calculate the amount each of these visitors would be willing to pay to use the facility. Hence, an economic value for it, based on a notional market for its services indicated by the answers to the questionnaire, can be built on.

These methods thus rely on direct surveys of consumer willingness to pay or to choose certain quantities of goods and services, and seek to measure consumer preferences in hypothetical situations, rather than consumer behaviour in actual situations (the travel cost method). They are most often applied to effects such as air or water pollution and to amenity resources with cultural, ecological or sociological features. The estimates are derived from a direct bidding game approach or are deduced from choices made among different packages of goods.

Let us examine a number of techniques in this category to obtain values that, in most cases, are for environmentally related goods/effects.

Bidding games

Within this technique, each individual surveyed is requested to evaluate a hypothetical situation and to express a willingness to pay or to accept compensation for a change in the level of provision of the good in question. Public goods such as access to recreational and amenity facilities, unobstructed views and clean water, whose essential feature is that one person's consumption does not affect the quantity available for consumption by others, are readily valued using this technique. Each individual respondent to the survey is asked to state his/her willingness to pay rather than to do without the good under examination. The values obtained from all responses are summed to obtain an aggregate willingness to pay.

The simplest form of the procedure is the single bidding game. Within it, an interviewer gives a brief description of the effect to the respondent who is then requested to state the maximum monetary value they would be willing to pay for the good. Conversely, they might be requested to state the minimum level of compensation that would be acceptable to them in exchange for loss of the option to buy the good in question.

This method is quite versatile. Provided that the questions are worded carefully and that the surveys are conducted in a proper manner, it can yield a great deal of useful information. It can be carried out in 'live' face-to-face interviews or using mail-based surveys. It may, however, suffer from 'hypothetical bias', with respondents not necessarily giving 'correct' answers reflecting their truly held valuations. Values given by respondents with a vested interest in the results of the survey may result in 'strategic bias'.

Take-it-or-leave-it experiments

Trade-off games allow individual preferences among various different outcomes to be determined. It is a survey-based technique within which respondents must choose between different bundles of the effect in question. In the simple case, each outcome has two components, a certain quantity of money and a certain quantity of the effect under examination. This is termed the single trade-off game (Sinden and Worrell, 1979).

Firstly, a full description of the effect is detailed. The base outcome, No.1, will combine no money with some designated quantity of the effect, be it social, environmental or recreational. The next possible choice, Outcome No.2, will have some quantity of money to be paid by the respondent and a changed amount of the effect. The individual respondent is then asked to specify a preference between the two outcomes. The quantity of money in Outcome No.2 is varied systematically until the respondent is indifferent between it and Outcome No.1. This value constitutes the trade-off he would make for the increase in the quantity of the effect between the two outcomes. It constitutes the respondents willingness-to-pay for this increase in the effect.

For example, if one is attempting to place an economic value on a recreational park, the first stated outcome may be to pay no money and keep the original park, or pay a set amount for an additional area of parkland, say 10 hectares. The question is repeated with increasing amounts of money in the second outcome until the point is reached where the respondent is indifferent between the two outcomes; that is, paying no money while keeping the original facility and paying a certain fee and receiving additional park space. The amount specifies a price for 10 hectares of park space.

By interviewing an adequately sized, representative population sample, a value for the aggregate willingness to pay for an increased quantity of the effect can be obtained.

Delphi technique

Within this method, originally developed by Dalkey and Helmer (1963), experts rather than respondents chosen at random from the relevant community are questioned. It can aid in finding the balance between economic and environmental/social issues directly relevant to the proposal in question.

A group of experts is selected and each is asked to place a monetary value on the effect in question. Each expert's value, along with a written commentary explaining his/her basis for the price, is circulated to every other member of the group who are then asked if, in light of the other members' scores, they wish to revise their estimate. If worked properly, it becomes an iterative procedure within which successive rounds bring initial diverse values closer together.

The group does not normally meet face-to-face in one location. Individual valuations are communicated to the other members in writing rather than orally. It prevents confrontation developing within the group and the domination of any one member over the rest.

It can be a very effective method of economic evaluation, its principal strengths being its non-confrontational nature and its systematic re-interviewing of the group of experts. Its success, however, ultimately depends on how well the experts' views reflect values in society at large.

Costless choice

Costless choice, proposed initially by Romm (1969), involves giving respondents two or more options, each of which is attractive and has zero cost, and then questioning them directly to find out which one is favoured by them. For example, a respondent is asked to choose between a certain sum of money and some beneficial effect, such as less noise pollution or reduced community severance. If the effect is chosen, it can then be assigned an economic value in excess of the sum of money offered. If the money is preferred, the effect has less equivalent monetary value than the sum in question. If the amount of money is varied and the quantity of the effect is held constant, it becomes a type of bidding game. The principal difference in this case, however, is that the respondent will not have to pay anything to receive the benefit from the effect in question, nor will there be any deterioration in the existing quality of the effect if the money is taken. In both cases the respondent gains. Costless choice allows a minimum valuation of the willingness to pay for an environmental/social good to be obtained. It has the advantage of minimising the biases associated with bidding games where, in each case, a desirable and undesirable outcome is combined.

Potential problems with survey-based monetary valuation techniques

The four above contingent valuation methods depend on the individual being surveyed placing a hypothetical monetary value on various changes in the effect under examination. The value obtained constitutes the payment the respondent is willing to make to prevent a worsening of the effect in question or the compensation required for the respondent to accept deterioration in the effect. The main difficulty with the technique is its artificial nature, and, consequently, its susceptibility to various forms of bias. The four main types encountered are:

(1) *Strategic bias*. Respondents attempt to influence the outcome of the survey by not responding truthfully. This arises where they believe their responses can influence decisions by either helping to decrease costs or increase benefits in relation to the expected result in a normal unbiased market situation.

(2)   *Information bias*. This results from incomplete information being given to the respondents regarding the proposed changes in the effect whose value is being estimated. It can be reduced by the use of visual or other aids to explain the alternatives being assessed.

(3)   *Instrument bias*. This may arise from the choice of method used to collect payment for changes in the effect under examination. Use of taxation devices may result in a particularly negative reaction from some sections of a given community. For the proper working of the methods, however, above all the payment system proposed must be realistic.

(4)   *Hypothetical bias*. The error lies in the technique because of its hypothetical and artificial nature. It is unavoidable within a process where actual market behaviour is not gauged. It is possible for the respondent to make realistic choices provided the various alternatives available are clearly set out.

The existence of these biases does not imply that they are unusable. What it does imply is that care must be taken in presenting the information on the various options under consideration within the interview phase. In addition, they are both labour intensive and time consuming to carry out, and the costs involved can be prohibitively high.

Example 7.2   Use of a survey-based technique

An existing domestic landfill site near an urban centre is in need of renovation. At present it has minimal soil cover and is the source of noxious fumes. The local waste authority wishes to carry out environmental improvements to the site and will use the costless choice technique to estimate the benefit to the local community of the reduction in noxious emissions from the landfill.

The geographical area of the city affected by the noxious odours was identified and, at random, fifty residents in the area were chosen for the survey. A simplified two-option costless choice methodology is employed. The environmental emissions coming from the landfill are described to the interviewees in detail. Each respondent is offered a choice between a yearly sum of money and a 90% reduction in the level of noxious emissions emanating from the landfill. The group is divided up into five groups of ten. Each group is offered a different sum of money as an alternative to the reduced effect. If the individual chooses the reduced effect, it implies that it is valued at least as much as the cash alternative. If, at a higher cash offer, no one selects the reduced effect, it implies that a ceiling can be placed on the level of the environmental improvement.

The first group of ten interviewees were offered £5, and all select the reduced effect, thus placing a minimum value on it. The final group of ten were offered £25 and all accepted the sum of money, thus setting an upper ceiling on the value of the reduced effect. Within the group offered £15, the split was 50:50 between those accepting the cash and those opting for the reduced effect. The full results of the survey are shown in Table 7.19.

Contd

Example 7.2   Contd

While the method is not detailed enough to give a precise average value for the 90% reduction in noxious emissions, it can be assumed to be in the region of £12–15 per annum, say £13.50.

If the community affected by the noxious emissions consists of approximately 20 000 residents, the monetary benefit to the community from the environmental improvements can be estimated as follows:

$$\text{Annual benefit to community} = £13.50 \times 20\,000$$
$$= £270\,000 \text{ per annum over life of landfill site}$$

Assuming a life of 30 years for the facility, and a discount rate of 8%, the total monetary benefit of the environmental improvements in present worth terms can be calculated:

$$\text{Present value } 90\%_{\text{reduction in emissions}} = 270\,000 \times ((1.08)^{30} - 1) \div (0.08(1.08))^{30}$$
$$= 270\,000 \times 11.258$$
$$= £3.04 \text{ million}$$

Table 7.19 Costless choice decisions.

Number of individuals interviewed	Choices made		
	Cash Amount Offered (£)	90% reduction in noxious effects from landfill	Number of individuals accepting cash amount
10	5	10	0
10	10	7	3
10	15	5	5
10	20	2	8
10	25	0	10

7.15.3   *Integrating sustainability into cost–benefit analysis*

Many economists see the inclusion of environmental/social effects within the conventional cost–benefit analysis framework as essential to the future development of economic appraisal (Pearce *et al.*, 1990). Pearce and colleagues believed that proper evaluation of environmental gains and losses requires the introduction of the concept of sustainability into the cost–benefit analysis framework, and believed this could be achieved by setting a limit on the 'depletion and degradation of the stock of natural capital'.

For a given set of infrastructure projects, which we can refer to as a programme, this approach modifies the economic efficiency objective of cost–benefit analysis by requiring that the sum of the environmental damage caused by the full set of projects within the programme be zero or negative. Certain projects will tend, by their nature, to depreciate natural resources. However, this requirement states that these projects should be part of a wider programme containing 'shadow projects' which compensate for the environmental damage of others within the portfolio.

In conclusion, therefore, the concept of sustainability dictates not only that, for a given development project, the environmental costs and benefits be valued to allow their inclusion alongside the economic factors within the cost–benefit analysis framework, but also dictates that the project in question forms part of a programme where, in overall terms, the valuations of the environmental benefits must at least equal those of the environmental disbenefits. Identifying and valuing the environmental effects of a given programme of development projects and adjusting the composition of the programme if necessary with environmentally compensating projects, so that the valuation of the overall net environmental damage of the programme is zero or less, can help achieve sustainability.

7.16   Using cost–benefit analysis within different areas of engineering

This section strives to illustrate the applicability of the cost–benefit analysis methodology to the assessment of different types of engineering development projects. Most differ in terms of the type of benefit relevant to the project in question and the techniques used to estimate them. The following four sectors of development work are discussed:

(1)   Flood alleviation
(2)   Health
(3)   Agriculture/irrigation
(4)   Transport

In each case, the relevant benefits are identified and a brief description of the techniques used to value them is given.

7.16.1   Flood alleviation

The main purpose of this type of project is the reduction of risk of flooding or damage to an area from storms. Such projects result in environmental impacts with no corresponding financial flows, yet these are real gains and losses from the local community's point of view that must be accounted for within an economic analysis. This can be achieved by assigning economic values to these environmental impacts

As detailed in the case study within section 7.13, the basis for the valuation of benefits for this type of project is quite simple. It will result in a reduction both in the frequency and extent of damage loss due to flooding relative to the existing situation. The economic analysis requires that an annualised value be put on this change in probability. This is accomplished using the formula:

$$Cost \ of \ risk = Probabilitity \ of \ event \times damage - loss \ resulting \ from \ the \ event$$

(7.2)

For a given project option, a wide range of events are possible, ranging from a mild flood likely to occur once every year, with low damage loss consequences, to a major flood with a very small likelihood of occurrence but with a potential for extensive damage to property. This calculation must be undertaken for the full probability range.

For a proposal aimed at reducing the damage caused by the flooding of a local river, the sequence of calculation of the benefits of reduced damage could be:

(1)   Evaluate the probability that the river will reach certain flood discharges.
(2)   Estimate the water levels associated with each of these discharge values.
(3)   Calculate the damage/loss resulting from each of the flood levels identified above within (2).
(4)   Link all this data together using Equation 7.2.

These annualised costs due to river flooding are calculated for the proposed project and the 'do-nothing' option. They are both discounted over the life of the proposal and subtracted from each other. Once the cost of both the construction and maintenance outlays for the flood relief scheme have been estimated, the cost–benefit analysis for the proposal relative to the 'without project' situation can be arrived at.

7.16.2   Health

For developments in this sector, such as the construction of a new hospital, a general description of the benefits to the community arising from the facility can be expressed in terms of the following (Torrance, 1986):

- A reduced level of sickness (termed 'morbidity') among productive members of the community, and hence a reduced number of working days lost.
- A reduction in the number of early deaths, assuming that such occurrences deny the community as a whole the productive benefits accruing from these fatalities (termed 'mortality').

The direct economic benefits from the proposal can thus be placed in terms of the reduced healthcare costs to the community health service because the proposal under scrutiny will result in a greater number of healthier people, and in greater worker productivity within the community generated by people who work more and are sick less.

The quantification of many of the parameters within the healthcare proposal requires the analysis of vast amounts of data. Its use should be limited to the comparison of health projects with similar overall objectives because only one consequence measure, either morbidity or mortality, can be used. The technique cannot deal with changes in both at the same time (Snell, 1997).

7.16.3   Agriculture/irrigation

For a development project involving the construction of infrastructure such as canals and land drains to bring water to tracts of land previously dependant on natural irrigation by rainwater, the construction and operating costs are readily estimated. The estimation of benefits is, however, less straightforward. The benefits of the transported water can only be calculated once it is established to what direct use it is to be put. In other words, the analysis must examine the effect of the transported water on the ability of the land to produce crops economically and effectively.

For a given hectare of land, the annual benefit derived from the irrigation scheme is estimated as the economic valuation of the farmer's yearly gross return from the crops grown minus the costs for that period, termed the annual net return. This calculation is done for each hectare within the catchment area of the irrigation scheme. Multiplying each crop's net return per hectare by the number of hectares of that crop grown gives the annual economic net return for the overall project.

The final figure derived, which can be called the net benefit for the 'with project' situation, is compared with the net benefit estimated for the 'do-nothing' option, involving calculation of the net economic return from crop production achieved with irrigation by natural rainfall only. The difference yields the incremental net economic worth of the development project. (It should be noted that the production costs for the farmer are counted as negative benefits or disbenefits, rather than being counted as costs alongside the construction and operating expenses incurred by the developer of the irrigation project. This convention would not effect a net present value calculation, but would affect a benefit/cost ratio result.)

7.16.4   Transport

In the highways case study detailed earlier in the chapter, the three main user benefits associated with the economic evaluation of a transport project were estimated in detail. For other types of transport projects, such as the assessment of rail project options or the evaluation of a proposed airport facility, however, the estimation of non-user benefits is of great significance.

For a rail-based project, such as a proposed underground metro-type system, one of the main non-user benefits of the scheme is the reduction in congestion in the local road network resulting from the mode shift of a proportion of commuters from car to public transport. The lower congestion results in savings both in time and in vehicle operating costs for commuters remaining in their cars.

In other transport related cases, environmental changes (sometimes negative), such as changes in noise, air or water pollution, can affect people near a proposed facility. For example, with an airport facility, a significant non-user benefit is, in fact, a negative one, namely the social cost to people situated nearby the proposed site in houses, hospitals, industrial units and leisure outlets as a result of the increased noise levels. The Roskill Commission (Roskill Commission, 1971; Flowerdew, 1972) used a property price approach to assess this negative benefit associated with the different sites proposed for a third London Airport.

Despite the wide range of potential benefits accruing from a transport facility, the time savings associated with the project under scrutiny remain the dominant benefit

Figure 7.4 M50 Motorway, UK (Source: W.S. Atkins).

within the evaluation. Their estimation is central to the accuracy of the relative assessment of competing options. The discovery of these values and their conversion into monetary values using the correct earnings rate for both working and non-working commuters is thus crucial to obtaining a realistic economic appraisal for any transport-related evaluation.

7.17 Summary

This chapter has introduced cost–benefit analysis as central to the appraisal of major development projects. The historical background and theoretical basis for cost-benefit analysis in its traditional form is outlined, and case studies are used to illustrate how the net social benefit of a proposal can be generated, together with the numerous forms in which it can be presented. While the rigour with which the final economic valuation can be derived is seen as an inherent advantage, the traditional cost–benefit analysis method is seen as limited by the necessity that benefits can only be included if they are quantifiable in monetary terms. To overcome this disadvantage of the method, a number of techniques are detailed which allow money values to be assigned to potentially relevant effects, thus enabling their inclusion in the main evaluation. These advances reinforce cost–benefit analysis as a technique for assessing projects from many different sectors of engineering development. Three case studies plus brief description of the application of the technique to a number of these sectors illustrates the wide applicability of the cost–benefit analysis methodology.

References

Adler, H.A. (1987) *Economic Appraisal of Transport Projects: A Manual with Case Studies.* EDI Series in Economic Development, John Hopkins University Press, London (Published for the World Bank).

COBA (1996) Economic assessment of roads and bridges. In: *Design Manual for Roads and Bridges*, Vol. **1**, section 1. HMSO, London.

Dalkey, N.C. and Helmer, O. (1963) An Experimental Application of the Delphi Method to the Use of Experts. *Management Science*, **9**, 458–467.

Dorfman, R. (1965) *Measuring benefits of government investments.* Brookings Institute, Washington, DC.

Dupuit, J. (1844) On the Measurement of Utility of Public Works. *International Economic Papers*, Volume 2.

Eckstein, O. (1958) *Water Resources Development: The Economics of Project Evaluation.* Harvard University Press, Cambridge, MA.

Flowerdew, A.D.J. (1972) Choosing a Site for the Third London Airport: The Roskill Commission Approach. In: R. Layard (ed.) *Cost–benefit Analysis*, Penguin, London.

Green, C. and Wilson, T. (2004) Assessing the benefits of reducing the risk of flooding from sewers. OFWAT, Birmingham. http://www.ofwat.gov.uk/publications/commissioned/ (Accessed 8 February 2012).

Groom, B., Hepburn, C., Koundouri, P. and Pearse, D. (2005) Declining discount rates: the long and the short of it. *Environmental & Resource Economics*, **32**, 445–493.

Hill, M. (1973) *Planning for Multiple Objectives: An Approach to the Evaluation of Transportation Plans.* Technion, Philadelphia, PA.

HM Treasury (2003) The Green Book – Appraisal and Evaluation in Central Government. Treasury Guidance, TSO, London.

Kelso, M.M. (1964) Economic Analysis in the Allocation of the Federal Budget to Resource Development. In: S.C. Smith and E.N Castle (eds) *Economics and Public Policy in Water Resource Development*, pp. 56–82. Iowa State University Press, Ames, IA.

Knetsch, J.L. and Davis, R.K. (1966) Comparison of Methods for Recreation Evaluation. In: A.V. Kneese and S.C. Smith (eds) *Water Research*. Baltimore: John Hopkins University Press, Baltimore, MD.

Krutilla, J. and Eckstein, O. (1958) *Multiple Purpose River Development: Studies in Applied Economic Analysis:* John Hopkins University Press, Baltimore, MD.

Little, I.M.D. (1957) *A Critique of Welfare Economics*, 2nd edn. Oxford University Press.

McKean, R. (1958) *Efficiency in Government through Systems Analysis.* John Wiley & Sons, Inc., New York.

Pearce, D., Markandya, A. and Barbier, E.B. (1990) *Blueprint for a Green Economy.* Earthscan Publications Limited, London in association with the London Environmental Economics Centre.

Prest, A.R. and Turvey, R. (1966) Cost–benefit Analysis: A Survey. In: *Surveys of Economic Theory*, Vol. 3. McMillan, London.

Romm, J. (1969) *The Value of Reservoir Recreation.* Water Resources and Marine Sciences Centre Technical Report 19; Agricultural Experiment Station Report AE Research 296. Cornell University, Ithaca, New York.

Rose, S. (1974) Hedonic Prices and Implicit Markets: Product Differentiation in Perfect Competition. *Journal of Political Economy*, **82** (1), 34–55.

Roskill Commission (1971) *Commission on the Third London Airport*. Her Majesty's Stationary Office, London.

Sinden, J.A. and Worrell, A.C. (1979) *Unpriced values: Decisions Without Market Prices*. John Wiley & Sons, Inc., New Yok.

Snell, M. (1997) *Cost–benefit Analysis for Engineers and Planners*. Thomas Telford Publications, London.

Torrance, G.W. (1986) Measurement of health utilities for economic appraisal: a review. *Journal of Health Economics*, **5**, 1–30. (Reprinted in Culyer, A.J. (ed.) *The Economics of Health*, Vol. I and II. Edward Elgar Publishers, London.)

UKWIR (2007) The Role and Application of Cost Benefit Analysis. Vol. 1: Generic Guidance. Report Ref. No. 07/RG/07/09. UK Water Industry Research, London.

US Federal Interagency River Basin Committee, Subcommittee on Benefits and Costs (1950) *Proposed Practices for Economic Analysis of River Basin Projects*. Washington, DC.

US Flood Control Act of June 22, 1936, Section 1 (49 United States. Stat. [1936]).

Chapter 8

Economic Analysis of Renewable Energy Supply and Energy Efficient Projects

8.1 Introduction

Renewable energy and energy efficiency are topics that have become increasingly important in almost all fields of engineering over the last few decades. The design and maintenance of wind farms, electric vehicles, light emitting diode lighting systems and low-energy buildings involve inputs from electrical, mechanical, building services and civil engineers, as well as from many other disciplines. In addition to the growing importance of this area to engineers, the economic analysis of such renewable energy and energy efficient projects warrants special attention for a number of reasons, including:

- investment structures differ from conventional energy projects because they typically involve relatively high up-front capital costs and low operational costs;
- they mitigate externalities often associated with conventional energy projects, such as global warming and adverse health effects;
- frequently they are not tried and tested technologies and, as such, involve greater risks than conventional technologies;
- important input parameters to economic models can be highly sensitive due to market characteristics (e.g. volatile energy prices) and technological advances;
- the intermittent nature of many renewable energy supply systems means that system boundaries must, in certain instances, include backup or storage costs.

For these reasons, particular attention must be given to: the choice of discount rate; the avoided costs to society of externalities such as carbon dioxide emissions; risk; methods for dealing with uncertainty, such as sensitivity analysis or Monte Carlo analysis; and designing and costing renewable projects which provide the same level of services as the conventional alternatives.

8.2 Policy context

It must be recognised that the market for energy supply and energy efficient systems is largely driven by international agreements and the policies of national

Engineering Project Appraisal: The Evaluation of Alternative Development Schemes, Second Edition. Martin Rogers and Aidan Duffy.

governments. Therefore, it is important to understand the context of energy policies and drivers.

It is widely accepted that greenhouse gases (GHGs) are almost certainly linked to rising global temperatures, which, in turn, have undesirable effects on ecosystems and human societies due to changing weather patterns and rises in sea level. GHGs comprise a variety of gases, the most important of which include carbon dioxide (CO_2), methane (CH_4), nitrous oxide (N_2O) and ozone (O_3). These are emitted from all sectors of modern economies, mainly due to their direct and indirect reliance on the use of fossil fuels for energy conversion processes, but also due to agricultural activities such as livestock farming (resulting in significant methane production) and tillage (nitrous oxides). Direct and indirect global GHG emissions in 2008 were primarily accounted for by: transport (13.8%); buildings (15.3%); industry (21.2%); energy production (9.6%); land use change (18.2%); waste management (3.6%); and agriculture (14.9%) (Baumert *et al.*, 2005).

A second problem associated with the world's heavy reliance on fossil fuels relates to security of supply. Significant proportions of the world's oil and, to a lesser extent, gas production occur in relatively politically unstable countries in the Middle East and Eastern Europe. The world's largest oil producers are members of a cartel – OPEC – which has the capacity to set global energy prices. It is worth mentioning that coal, although more geographically dispersed, is often not a good substitute for oil or gas due to its lower energy density and greater environmental impacts and, therefore, does not effectively offset these risks. The result is that the economies and societies of many importing countries – including most European countries – are exposed to the threat of oil price shocks and supply interruptions. For example, oil prices rose fourfold during the oil embargo of 1973, while as recently as 2006 the European Union was threatened with a gas supply interruption.

The third main concern to energy policymakers in most industrialised countries is the recent sustained rise of global energy prices due to rising demand in emerging economies and its impact on competitiveness and growth. This trend is likely to be compounded in the future due to a decrease in the availability of the more easily accessible oil resources and the need to exploit more expensive sources, such as deep-sea reserves and tar sand deposits. It is argued that 'peak oil' – the maximum rate of global oil production - has already been reached, or soon will be, although this prediction is widely contested. Figure 8.1 shows nominal and real oil prices over the 25 years up to 2010. It can be seen that real prices increased fivefold between 1998 and 2010.

Energy policies in industrialised countries tend to reflect these concerns of environmental sustainability, cost competitiveness and security of supply. For example, in November 2010, the European Commission (EC) issued *Energy 2020 A strategy for competitive, sustainable and secure energy* (COM/2010/0639 final) which outlined its energy strategy up to 2020. The European Union (EU) is highly dependent on imported energy – importing 82% of oil (2010), for example. Under its '20-20-20' targets, the EU plans to cut 2020 GHG emissions by at least 20% compared to 1990

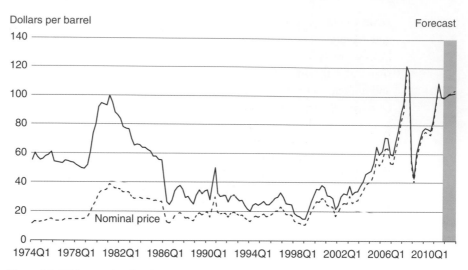

Figure 8.1 US imported real and nominal crude oil prices (Source: EIA, 2011).

levels, generate 20% of its energy needs from renewable resources and reduce primary energy needs by 20% through energy efficiency improvements. 2050 targets include an ambitious 80–95% reduction in energy-related GHG emissions compared to 1990. Meeting these targets will involve considerable investment in energy efficient and renewable energy supply projects.

A variety of EU directives is aimed at achieving these targets. The Energy Performance in Buildings Directive (EPBD) promotes energy efficiency in buildings by providing information to consumers in the form of building energy rating certificates. A similar energy labelling initiative exists for consumer goods such as white goods, light bulbs and cars (an emissions label in the latter case). The Biofuels Directive sets targets for fossil fuel replacement and new regulations have been introduced governing carbon dioxide emissions from car engines.

A central plank of EU energy policy is the European Union Emissions Trading Scheme (EU ETS), a cap-and-trade scheme whose members include the largest GHG emitters (more than 10 000 members) in the industrial sector, which accounts for approximately 40% of EU-27 emissions. Historical EU ETS carbon prices are shown in Figure 8.2; these are expected to rise to between 31 and 35 €/tCO$_2$by 2020 (Point Carbon 2010). Carbon costs have become an important factor in energy project appraisal.

Individual member states employ a variety of market-based and regulatory policies aimed at achieving targets allocated under the EU's 20-20-20 initiative. Such national policies include:

- regulation of products and services by setting minimum energy efficiency or emissions standards (e.g. building regulations);
- carbon pricing (e.g. carbon taxes);
- subsidies and tax incentives that favour efficient and renewable technologies, such as capital subsidies and preferential feed in tariffs;

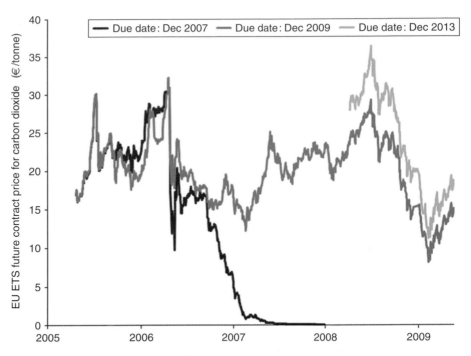

Figure 8.2 EU ETS future contract prices for carbon dioxide from 2005 to 2009 (Data source: European Climate Exchange, 2012).

- research and development support to encourage the development, commercialisation and deployment of energy efficient and renewable energy supply technologies and systems.

8.3   Renewable energy supply and energy efficient technologies

Renewable energy supply (RES) and energy efficient (EE) technologies can be broadly divided into:

- energy efficient technologies, including both passive (e.g. insulation and natural ventilation in buildings) and active (e.g. high-efficiency boilers and regenerative braking systems for cars) systems; and
- renewable energy supply technologies which harness renewable energy such as the wind, waves and biomass.

Energy efficient technologies aim to reduce the overall use of energy by a system, thus reducing resource depletion and the emission of harmful pollutants. For example, dwelling insulation and draught proofing will reduce heat energy losses to the environment, the quantities of heat which must be generated, the amount of fossil fuels consumed and, ultimately, associated GHG emissions. Similarly, energy supply using fossil fuel as the primary energy source can also involve energy efficient technologies. For example, combined heat and power (CHP) involves the recovery of waste heat from the electricity generation process, thus doubling efficiencies. Heat pumps and condensing gas boiler may also be regarded as energy efficient technologies, when compared to their alternatives.

Renewable energy supply involves the conversion of energy from renewable sources, such as the wind, sun, tides, waves, rivers, deep rock and biomass, into forms of energy which can be used for thermal, electrical and mechanical power. These energy sources are replenished over a very short time frame and, therefore, do not involve the use of fossil fuels, which are regenerated over millions of years. This is not to say that renewable energy is completely environmentally benign. For example, the fabrication, construction and maintenance of such projects typically involve the use of fossil fuels and, therefore, the emissions of GHGs. Nevertheless, such emissions tend to be much lower per unit of energy produced than for efficient fossil-fuelled technologies.

There is a very wide variety of energy efficient and renewable energy supply technologies which can be deployed across projects in all engineering disciplines. It is therefore not practical to give a comprehensive summary of these technologies in this text. However, a brief overview of some of the most important technologies is given below, so that the reader will have an appreciation of the factors relevant to their economic performance and assessment. Examples relate mainly to the built environment and energy supply sectors; they highlight:

- the current status of the technology and important factors to consider when modelling its performance;
- the extent to which particular technologies are favoured by governments and the types of incentives typically available for their promotion; and
- technology trends and their impacts on fixed and variable costs, such as primary energy inputs and capital costs.

8.3.1 Fossil-fuelled power generation

Fossil-fuelled thermal power plants use oil, gas and coal (and, to a lesser extent, peat and biomass) to produce steam which drive turbines and electrical generators. They represent the most common form of electricity generation worldwide and this situation is unlikely to change in the foreseeable future. For example, in China, coal-fired electricity generation capacity is expected to rise from 87 GW in 1990 to 1250 GW by 2030 (IEA, 2008). Globally, about 20% of electricity is produced using gas and 40% from coal. Therefore, technological advances in increasing their efficiencies and mitigating emissions are of central importance for achieving global GHG targets. It is expected that significant advances will be made in coal-fired power generation technology, with 'clean coal' and advanced steam cycle technologies increasing efficiencies from approximately 35% today to as much as 50% by 2050 (IEA, 2008).

Despite these projected improvements in the energy efficiency of fossil-fuelled power generation plant, the technology will continue to produce very large quantities of carbon dioxide. Carbon capture and storage (CCS) is an emissions mitigation technology for the power generation sector that involves the removal of carbon dioxide from combustion gases. The carbon dioxide is then transported using pipelines, ships or trucks and subsequently injected into suitable onshore or offshore geological formations. The technology is not yet commercialised; technologies for capturing gases are expensive (25–50 $/tonne CO_2 avoided), transport can be expensive and the permanence of storage over long time-scales is not yet proven (IEA, 2008).

Combined heat and power (CHP) involves the recovery and use of waste heat from the power generation process, resulting in an increase in efficiency from approximately 40% to over 80%, depending on the extent to which the heat can be gainfully used. CHP is a relatively mature technology, but high deployment costs at the relatively small scales required to allow heat to be produced close to its point of use have resulted in poor penetration of this technology. The technology has been most successfully deployed in the chemicals, refining and pulp/paper sectors but has significant potential in the commercial and even domestic sectors. Depending on the technology used, CHP costs between 600 and 1800 $/kW installed (UNEP, 2011).

8.3.2   Biomass and bioenergy

Bioenergy refers to the extraction of energy from organic material, or biomass. Forms of biomass are varied and include:

- energy crops (trees, grasses, crops used for fuel production such as oil seed and maize);
- organic residues (from forestry, agriculture, landfill gas).

Biomass is converted into energy carriers such as heat, power and fuel using a wide variety of processes. For example, wood pellets are made using compressed saw dust for combustion and heat production in domestic boilers. Landfill gas must be collected using wells and pipework, cleaned and can be used to fuel CHP plant to produce electricity and heat. Energy crops such as rapeseed are processed and blended with additives to produce biodiesel for use as a transport fuel.

Because biomass can provide energy for a variety of economic sectors using relatively mature technologies, it has significant potential to reduce global GHG emissions. However, costs are still high, although these are likely to fall as bioenergy becomes more widely deployed. Future bioenergy costs are difficult to predict due to competing land uses for food production and medium-term uncertainty about global dietary trends. In 2008, electricity produced from biomass cost between 0.04 €/MWh and 0.12 €/MWh (gas-fired power plant costs are in the region of 0.08 €/MWh). Costs in 2050 are projected to be in the region of 0.03 €/MWh and 0.08 €/MWh. The cost of producing biofuels for transport, currently at between 0.07 and 0.20 €/GJ could drop to between 0.05 and 0.08 €/GJ (IEA, 2008). Table 8.1 gives typical plant size, efficiencies and capital costs for a range of bioenergy technologies.

Table 8.1 Typical plant size, efficiencies and capital costs for a range of bioenergy technologies (Source: IEA, 2008).

Conversion type	Typical capacity	Net efficiency (%)		Investment cost (US$/KW)
Anaerobic digestion	<10 MW	10–15	(electrical)	
		60–70	(heat)	
Landfill gas	<200 kW–2 MW	10–15	(electrical)	
Combustion for heat	5–50 kW$_{th}$ residential	10–20	(open fires)	
		40–50	(stoves)	~23 (/kW$_{th}$)
		70–90	(furnaces)	370–990 (/kW$_{th}$)
Combustion for power	10–100 MW	20–40		1975–3085
Combustion for CHP	0.1–1 MW	60–90	(overall)	3333–4320
	1–50 MW	80–100	(overall)	3085–3700
Co-firing with coal	5–100 MW existing	30–40		123–1235
	>100 MW new plant			+ power station costs
Gasification for heat	50–500 kW$_{th}$	80–90		864–980 (/kW$_{th}$)
Gasification for CHP using gas engines	0.1–1 MW	60–80	(overall)	1235–3700

Supports for bioenergy include regulations which impose minimum thresholds for biofuel use (e.g. EU 'Biofuel Directive' requiring replacement of 5.75% of petrol and diesel with biofuels by 2010) and heat tariff supports such as the UK Renewable Heat Incentive (RHI), which provides long-term guaranteed financial support for renewable heat technologies.

8.3.3   Wind power

Wind power is one of the most mature renewable energy technologies, yet accounted for just 2–2.5% of worldwide electricity consumption in 2010 (REN21, 2011). However, it is a rapidly growing market, with global installed capacity growing 100-fold to almost 200 GW between 1990 and 2010. Optimistic projections are for 450 MW of capacity to be installed by 2015. The International Energy Agency (IEA) estimates that wind will supply between 9 and 12% of global electricity by 2050 (IEA 2008).

Like many renewable energy supply technologies, wind turbines have relatively high up-front capital requirements and low operational costs when compared to conventional fossil-fuelled power plant. Lower operational costs are largely due to the avoided cost of fuel. Capital costs are falling due to learning-by-doing, economies of scale and through the development of larger plant. In 2007 the average turbine capacity was approximately 1.5 MW while the largest turbines in 2011 were in the order of 6 MW, although in the last several years capital costs have increased due to rapid market growth and supply bottlenecks.

A significant disadvantage with large-scale wind penetration in a large electricity network is intermittency, since power is only generated when the wind blows.

Large-scale penetration therefore needs storage or backup, which must be adequately costed into the project. Storage technologies can be very expensive and include pumped hydro storage plant and/or compressed air storage. Backup typically involves dispatching a fossil-fuelled power plant to make up any shortfall in wind power.

The costs and technical challenges associated with onshore and offshore wind power are significantly different. Onshore wind turbines are a relative mature technology; the turbines are easily accessible and subject to less aggressive environments than offshore installations. Although wind resources are greater offshore, foundations, construction, durability and maintenance are all complicated by the maritime environment. Onshore wind capital costs are in the region of 1200 €/kW (2010) and may fall as low as 700 €/kW by 2030 (EWEA, 2011). Offshore wind energy costs in 2010 were in the region of 3000 €/kW with significant scope for cost reduction over time. The cost per unit of electricity produced is dependent on a variety of factors, including capital cost, the wind resource available and discount rate. Unit costs are in the region of 94 £/MWh for onshore installations and between 157 and 186 £/MWh for offshore plant (Mott McDonald, 2010). Wind turbine capital cost projections made by the European Commission (EC) and the European Wind Energy Association (EWEA) are shown in Figure 8.3.

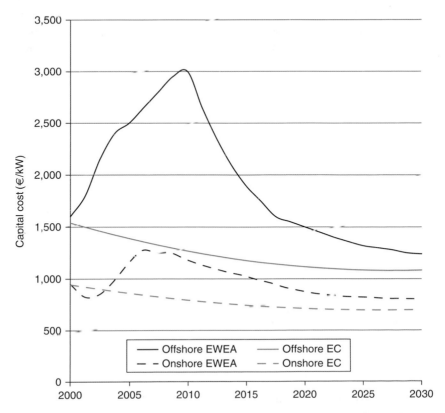

Figure 8.3 Projected capital costs for onshore and offshore wind turbines (Source: EWEA, 2011).

8.3.4   Solar energy

There is a range of solar energy technologies which convert energy from the sun into electricity or heat. Photovoltaic (PV) systems convert energy from the sun directly into electricity using arrays of semiconductor cells known as PV modules. Solar water heaters (SWH) use a 'collector' where solar energy is used to heat water either directly or indirectly using a heat transfer fluid; the technology is widely deployed for generating low temperature hot water for domestic dwellings, but can also be used for commercial and industrial applications at higher temperatures. Concentrated Solar Power (CSP) uses mirrors to concentrate sunlight to produce high pressure steam and drive turbines to generate electricity; it is an emerging technology with applications in regions with very high levels of solar energy.

The deployment of PV installations has been growing in recent years, with average annual growth rates of 49% between 2005 and 2010. This has been driven by a variety of factors, including subsidised feed-in-tariffs (FITs) in some countries and falling capital costs. Module prices almost halved between 2008 and 2010 and cost in the region of 1.30–1.80 \$/Wpeak (REN21, 2011). These costs exclude 'balance of system' costs for electrical wiring, inverters, frames and so on that are required to condition and transport the electrical energy generated; these can equal or exceed module costs for small domestic systems. Further cost reductions are envisaged and technology learning rates are in the order of 15–22% (IEA 2010).

Heat-producing technologies such as SWH are well placed for significant growth if the right installation costs, energy prices and policy environments exist. Domestic SWH are typically used for the production of hot water and European system costs (including collector and balance of system) are in the region of 800–1200 €/kWth for small, forced-circulation (pump-assisted) systems and are projected to fall to between 300–650 €/kWth by 2030; unit production costs are in the region of 0.05–0.16 €/kWh depending on location (ESTTP, 2007). This compares to heating costs of approximately 0.04 0.20 €/kWh for other sources, such as gas, oil and electricity.

8.3.5   Buildings

Achieving significant energy efficiency and GHG mitigation in the built environment over the coming decades will require both efficient new construction and upgrading of existing structures. Energy efficient interventions for both centre on:

- passive measures which minimise thermal losses and maximise microenvironmental benefits, such as sunshine, daylight and shading to enhance thermal gains, internal lux levels and cooling at appropriate times;
- active measures which use plant and equipment to deliver, recover and manage energy in the most efficient way possible.

A very wide range of active and passive interventions for buildings are possible, including: highly efficient glazing systems; improved wall and roof insulation; solar thermal devices; ground source heat pumps; PV; efficient heat pumps for heating and cooling; efficient white and brown goods; highly efficient lighting; bioenergy; and combined heat and power. Typically, a number of different measures are considered in combination when designing or upgrading new and existing buildings. It is important to understand that such measures cannot be assessed in isolation and must be assessed together. For example, insulation upgrades will reduce boiler capacity or more efficient lighting will reduce the optimum size of a PV system. In general, passive energy efficient measures should be considered before renewable energy supply solutions.

Energy policies to promote the uptake of technologies in houses include:

- capital subsidies for energy efficient upgrades, such as insulation, and renewable technologies including solar water heaters;
- regulations setting minimum emissions or energy use thresholds for new dwellings, or substantial alterations to existing dwellings;
- information on the energy and emissions performances of new dwellings or dwellings which are being sold; and
- feed-in-tariffs for renewable energy supply systems producing electricity, such as PV.

8.4   Economic measures for renewable energy and energy efficient projects

Many of the conventional economic measures described in this book can be applied in the assessment of renewable energy supply and energy efficient projects. Due to the particular characteristics of energy projects described above, a number of additional measures specific to energy projects are described to assist in this process. Typically, no one measure will provide a definitive answer to the question being addressed. Therefore, it is best practice to assess a project using a variety of appropriate economic measures and to use this information to decide whether and how to rank, exclude or proceed with a project.

Some of the aspects which should be considered before undertaking an investment appraisal of an energy efficient or renewable energy supply project are described first; general and specific economic measures are then introduced.

8.4.1   Aspects of renewable energy supply and energy efficient analysis

8.4.1.1   Investor perspective

It is important to be clear about the investor perspective for a particular renewable energy supply or energy efficient project because a project may be attractive to one investor, but not to another. For example, a polluting coal-fired power plant may make a good economic return to a private investor but, when the health and environmental costs of the pollution are considered, it might give a negative return to society.

For large energy projects as well as energy policies it may be important to assess the project or technology from a number of different perspectives. For example, it may be most appropriate to base government calculations of capital subsidies for renewable technologies on the capitalised lifetime value of carbon avoided. However, these subsidies may not incentivise investors, who may use higher discount rates, consider taxation effects and ignore negative externalities. Therefore, to avoid costly policy failures, governments should consider the project from the private investor's perspective as well as from a public policy standpoint.

8.4.1.2   System boundaries

The boundary considered when assessing energy projects should include the entire lifecycle, typically including planning, procurement, design, construction, operation, maintenance, major refurbishment and decommissioning. When comparing different technologies, the same level of service should be provided; for example, storage and backup may need to be included when intermittent generation (such as wind) is being compared to dispatchable electrical generation, such as gas-fired power plant. Similarly, revenues and costs should be comparable. Where emissions' costs such as carbon taxes apply (or will apply over the investment's lifetime), then these should be accounted for in private investments; where emissions are external costs, then they should be included when evaluating from a societal perspective.

The boundary of the energy projects being analysed must be appropriate in terms of the time horizon considered, the system scope and, where projects are being compared, these must be consistent. For example, comparing offshore wind turbines with a 20-year lifespan to conventional fossil-fuelled power plant that will last for 40 years might involve accounting for the reconstruction of the turbines after year 20 to provide a further 20-year service life. Alternatively, the production costs of energy can be 'levelised' over the investment lifespan for the purpose of comparison. A fair comparison of these technologies might also involve extending the scope of the fossil fuel system to include Carbon Capture and Storage (CCS) and the wind turbine system to include storage or backup, so that they can be compared on a like-for-like basis.

Consideration may need to be given to the effects across system boundaries. For example, large-scale electrical grid storage systems, such as compressed air energy storage or pumped storage hydropower, have the effect of increasing the use of variable power supply sources such as wind, since this is often curtailed during off-peak periods. Where wind power is considered without storage or backup, the marginal increase in conventional electricity costs associated with the lower electricity demand asset utilisation rates for dispatchable plant may need to be considered.

In the case of retrofitting existing buildings, vehicles or energy infrastructure, the boundary should exclude historic costs, energy uses and emissions (such as 'embodied' energy and emissions) because these can be regarded as 'sunk'(already incurred and unrecoverable) and, therefore, irrelevant to the project being considered.

8.4.1.3 System size

Engineering analysis and system optimisation are required to determine the most appropriate size to be used in any analysis. The choice of optimisation criteria, such as net present value, simple payback, the cost of carbon abated or other parameters, is dependent on both the project being considered and the investment perspective. For example, a householder could choose the optimal size for a domestic PV system by maximising NPV. A government may choose the most appropriate system size based on minimising the cost of carbon abated, and favour this size with appropriate subsidies. A very risky project (deep geothermal energy extraction, for example) may be optimised by minimising simple payback periods.

Competing systems must be sized on a comparable basis. For example, sizing competing renewable energy supply technologies, such as PV and wind turbines, on the basis of total units of grid electricity avoided will be misleading when maximum demand and capacity tariffs apply. Since PV output is more predictable, it may meet more daytime maximum demands than the intermittent wind option, thus reducing capacity and demand charges.

8.4.1.4 Capital costs

Capital costs typically represent a significant proportion of renewable energy supply and energy efficiency investment costs, since operating and maintenance costs are usually low or negative (in the case of energy efficient investments). For some technologies, such as photovoltaics and some solar water heaters, recurrent costs are insignificant. It is important, therefore, to obtain accurate capital cost data before a reliable economic model can be developed. In many instances, these data should be sourced locally, as local market conditions, construction techniques, regulations and capital subsidies can be significantly different to those prevailing where else. Some typical capital costs for renewable technologies are shown in Figure 8.4.

Because many renewable energy supply and energy efficient technologies are relatively new, learning rates tend to be high. 'Learning rates', 'learning-by-doing' or 'experience curves' describe how the cost of a product tends to decrease as deployment increases. A learning rate is the rate at which the unit cost of a product decreases with every doubling of the market size. Figure 8.5 shows experience curves for PV and wind systems on a log-log scale, indicating that PV costs per kW installed in 2010 were one fiftieth of those in 1976. It can also be observed that onshore and offshore wind costs dropped significantly up to about 2006 before rising again due to high market demand. Therefore, the timing of investments in renewables is important, as a project may be significantly more attractive in the future due to lower investment costs. This has important implications where projects have long lead times due to planning or other factors. Such projects include onshore wind turbines and many policy initiatives. Furthermore, policymakers must be aware of the endogenous nature of policies favouring preferred energy technologies. For example, generous PV FITs in Germany and Renewable Obligation Certificates (ROCs) allowances

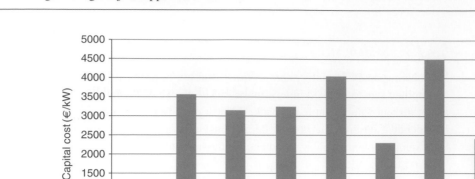

Figure 8.4 Typical capital costs for renewable energy supply systems (Source: JRC, 2009).

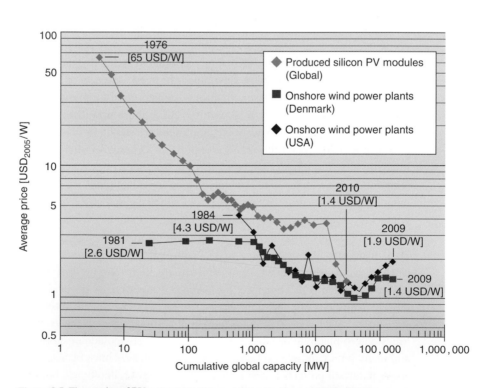

Figure 8.5 Time series of PV and wind turbine capital costs (Source: IPCC, 2011).

for offshore wind in the United Kingdom contributed to surges in demand for these technologies and resulted in increased capital costs.

When estimating capital costs, account should be taken of any capital subsidies available as well as avoided capital costs. For example, building-integrated PV cladding is currently expensive, but the avoided cost of conventional cladding will significantly reduce this cost.

8.4.1.5   Operating and maintenance costs

Operating and maintenance (O&M) costs comprise both fixed and variable costs. The former include costs which do not vary with production levels, such as insurances and personnel, whereas the latter may include fuel and maintenance. O&M costs vary greatly between technologies or even within technology categories. For example, it has already been mentioned that many renewables, such as those based on solar technologies, have very low operating costs as they have no fuel requirements and low staffing needs. Within a particular technology group, O&M costs can depend on the technology used (e.g. turbine versus Otto cycle CHP). Costs are also affected by the degree of deployment within a geographic area, with low penetration normally associated with relatively higher prices. Table 8.2 shows typical operating and maintenance costs for some energy efficient and renewable energy supply technologies.

Technology	Cost (€/MWh)	
	Lower	Upper
Passive technologies (insulation, solar gain, natural ventilation)	0.00	0.00
Combined heat and power (CHP)[*]	5.08	10.15
Large-scale electricity generation from coal or gas with CCS[†]	120.75	132.25
Photovoltaics (PV) – large scale[§]	1.74	3.48
Photovoltaics (PV) – domestic[§]	2.61	5.22
Wind – onshore[**]	12.00	15.00

Table 8.2 Typical operating and maintenance costs for some energy efficient and renewable energy supply measures.

Source: *UNEP (2011); [†]Mott McDonald (2010); [§]IEA (2010); **EWEA (2005)

8.4.1.6   Externalities

An externality is a cost (or benefit) which is not accounted for by the project directly but is incurred by third parties. External costs and benefits can be referred to as 'negative externalities' and 'positive externalities' respectively. A common example of a negative externality is the emission of pollutants, which have negative impacts on society. Negative externalities associated with electricity generation using fossil fuels include respiratory illnesses due to particulate emissions; sulfate emissions leading to acid rain, slow growth of forests and freshwater acidification; and carbon dioxide emissions resulting in global warming, droughts and sea level rises.

In some jurisdictions, policymakers have responded by regulation and/or by attempting to internalise some of these costs. In many countries, nitrogen oxides (NOx) and sulfur oxides (SOx) emissions limits have been introduced for coal-fired power plant exhaust gases to curb the negative effects of these emissions on third parties. Where emissions cannot be adequately controlled, taxes and subsidies can be used to counteract the cost of a negative externality. For example, carbon taxes exist in a growing number of countries worldwide. In Europe, the Emissions Trading Scheme imposes a cost on GHG emissions from large polluters; this is supplemented with carbon taxes in a number of countries (including Denmark, Finland, Republic of Ireland and Sweden) which impose costs on pollution from smaller sources outside the scheme. Many countries subsidise energy technologies that have low impacts, which can be seen as rewarding their relatively lower external costs, which result in a lowering of externalities. For example, guaranteed feed-in-tariffs are used to subsidise various forms of renewable energy supply technologies, especially wind; and capital grants for home insulation reduce heating requirements and, consequently, emissions from home-heating fossil fuels.

Best estimates of external costs should be used when an appraisal is from the perspective of society. Table 8.3 provides estimates for the external costs of electricity generation in European countries considering a set of health and environmental effects.

Since carbon dioxide emissions contribute to global warming, which has negative impacts on the environment and society, there is a cost for every unit emitted. The quantification of this cost, however, is problematic and various estimates exist for the 'social cost of carbon' (SCC). A meta-analysis by Yohe *et al.* (2007) indicated a

Table 8.3 External costs (quantifiable impacts of global warming, public health, occupational health and material damage) of electricity production in various European countries in €cent/kWh (Source: EC, 2003).

Country	Coal & Lignite	Peat	Oil	Gas	Nuclear	Biomass	Hydro	PV	Wind
AT				1–3		2–3	0.1		
BE	4–15			1–2	0.5				
DE	3–6		5–8	1–2	0.2	3		0.6	0.05
DK	4–7			2–3		1			0.1
ES	5–8			1–2		3.5*			0.2
FI	2–4	2–5				1			
FR	7–10		8–11	2–4	0.3	1	1		
GR	5–8		3–5	1		0–0.8	1		
IE	6–8	3–4							0.25
IT			3–6	2–3			0.3		
NL	3–4			1–2	0.7	0.5			
NO				1–2		0.2	0.2		0–0.25
PT	4–7			1–2		1–2	0.03		
SE	2–4					0.3	0–0.7		
UK	4–7		3–5	1–2	0.25	1			0.15

*co-fired with lignite

mean value of 43 \$/tC (approximately 12 \$/tCO$_2$) with a standard deviation of 83 \$/tC; the large standard deviation is due to the differing assumptions and climate models employed in the various studies.

8.4.1.7   Backup and storage

Many renewable energy technologies provide either predictable or unpredictable intermittent energy supplies. Tidal energy is intermittent but predictable whereas wind energy is unpredictable (at a time horizon beyond several days); neither, therefore, is dispatchable (can be called upon to provide electricity whenever needed). Hydropower and conventional fossil-fuelled power plant, on the other hand, are predictable and dispatchable. Where renewable energy supply technologies are variable and demand is likely to exceed supply during certain periods, then a reliable and continuously available supplementary supply source is required. This may take the form of a backup energy supply technology (such as conventional generation on the grid) or an energy storage device which has sufficient capacity for the interruption period.

Figure 8.6 illustrates the need for storage or backup in renewable energy supply systems. It shows daily national electricity supply and demand for a hypothetical situation where all electricity is supplied using two-way tidal power (in reality, supply would be more diversified due to the different timings of tidal flows around a country's coastline). When supply exceeds national demand, electricity is available

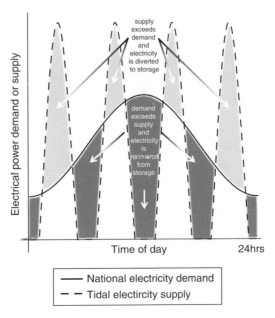

Figure 8.6 Top-up electricity needed and storage available where electricity is generated using tidal power.

for storage; this can then be recovered to supply energy when 'top-up' is required when the tide is not flowing. The cost of storage must be included in any economic appraisal of such a system.

A variety of different levels of backup and/or storage are required for different energy supply technologies and scales:

- autonomous or 'island' systems with intermittent generation which require the constant availability of energy must include the costs of a dedicated storage or backup system;
- the installation of very large intermittent renewable electricity supply systems (i.e. which constitute a significant proportion of the national supply) will require backup and/or storage costs to be included;
- the installation of small, grid-connected intermittent renewable electricity supply systems (i.e. which constitute an insignificant proportion of the national supply) do not need backup or storage as the marginal cost to the system for their provision is small;
- the installation of intermittent renewable heat supply systems (such as solar water heaters but not biomass systems) will always require backup and/or storage.

8.4.1.8   End-of-life

Cash flows at the end of an energy project can, in some cases, have a significant impact some economic measures such as NPV, IRR and CBR. Some projects have significant residual value; for example, wind farm sites with high wind resources are likely to have a high future resale value which should be accounted for. Conversely, nuclear plant has high decommissioning costs.

8.4.1.9   Energy project revenues

It is important to fully account for revenues when assessing energy projects. For energy efficient projects, the avoided cost of energy (and carbon if not included) should be considered. For example, high-efficiency lighting will reduce consumption of mains electricity, so the avoided costs may be the sum of the reduced consumption and the domestic tariff, including any carbon taxes where appropriate. For renewable energy supply projects revenues can include:

- the market value of the energy sold to third parties, such as electricity output to retailers or heat output to heat consumers;
- contracted feet-in-tariffs (FITs), for example those offered by governments to support favoured technologies such as wind or PV;
- the value of alternative energy sources avoided as a result of on-site production, for example where CHP displaces the need to buy grid electricity and boiler gas;

- avoided carbon taxes when taking a private investor perspective and the avoided social cost when taking a societal viewpoint;
- other avoided costs, such as maintenance costs.

8.4.1.10 Discount rate

Energy projects tend to be long-term investments: large fossil-fuelled electricity power plants have typical operating lifespans of 30–50 years; dams used for hydroelectric power generation have lifespans in excess of one hundred years; even investment in domestic wall insulation could have a lifespan of hundreds of years. Consequently, the choice of discount rate for present worth calculations will have a significant impact on the economic viability of the project.

The choice of discount rate varies depending on the investor. It can be influenced by a variety of factors, including expected future interest rates and the rate of return required by the investor. Some practitioners incorporate risk into the discount rate; however, it is better to deal with risk by undertaking uncertainty analysis (discussed later in this chapter). Governments, which can borrow at relatively low interest rates and take a long-term view of capital investments, tend to use lower discount rates (typically in the range 4–8%) than industry. Industry values vary greatly, lower values being used for low-risk, long-term investments in mature markets (typically 7–10%) with higher values in some industries where the opportunity cost of capital is high (>15%).

In the electrical power industry, where large-scale and long-term investments have a long history, the 'weighted average cost of capital' (WACC) is often used for discounting purposes or as a hurdle rate for investment decisions. WACC is the minimum rate that a utility must earn to fund its various capital components, including debt as well as preferred and common shares; it may be adjusted to account for corporation tax. WACC figures are typically in the range of 5 to 8%.

8.4.2 Conventional economic measures

8.4.2.1 Net present value

This is a useful tool for evaluating mutually exclusive RES and EE projects and projects which include significant social costs (as opposed to 'private costs') and benefits. Social costs are those costs which are not considered by a private investor and can include the cost to society of pollutants such as GHGs. NPV, however, is not a good measure when different energy projects of different scales are being compared, since it does not take account of the greater opportunity to invest at a favourable return for the larger project. Nonetheless, NPV is always useful, as it can be used to verify the results of other measures employed.

Example 8.1

A local authority constructing social housing wants to identify the most economical heating system to employ. The average heating requirement for each house is 10 000 kWh. Three options are identified: oil, gas and wood-pellet boilers, all of which have a life span of 15 years. Capital costs, fuel costs and maintenance costs are obtained for each option (Table 8.4) and a cost of 30 €/tCO$_2$-eq is allocated to carbon dioxide equivalent emissions. Boiler efficiencies and fuel tariffs are used to estimate fuel use and annual energy costs and the cost of emissions are estimated. An 8% discount rate is used to estimate the NPV for each option. Calculate the NPV for each.

Solution

The annual fuel consumption of each option is first calculated based on the annual heat demand and plant efficiencies. Fuel CO$_2$-eq emissions intensities are then applied to estimate annual emissions. Fuel and carbon and tariffs are used to determine the annual fuel and carbon costs for each option. These are combined with capital and maintenance costs to determine the NPV for each option based on the 8% discount rate. The results in Table 8.4 indicate that a gas boiler gives the best NPV of the options considered.

Table 8.4 Calculation of NPV for different heating options.

		Oil boiler	Gas boiler	Wood-pellet boiler
General				
Heat energy demand	(kWh/annum)	10 000	10 000	10 000
Boiler efficiency	(%)	0.85	0.9	0.8
Fuel used	(kWh/annum)	11 765	11 111	12 500
CO$_2$-eq	(kg/kWh)	0.275	0.185	0
CO$_2$-eq emitted	(kg/annum)	3235	2056	0
Unit costs				
Fuel	(€/kWh)	0.08	0.05	0.06
CO$_2$-eq	(€/kg)	0.03	0.03	0.03
Life-cycle costs				
Capital	(€)	1500	1500	8000
Energy	(€/annum)	941	556	750
Maintenance	(€/annum)	250	150	250
CO$_2$-eq	(€/annum)	97	62	0
NPV				
Discount rate		0.08	0.08	0.08
Lifespan		15	15	15
NPV		**(12 527)**	**(8067)**	**(16 559)**

8.4.2.2   Internal rate of return

Internal rate of return is useful since it can be compared easily to an energy company's hurdle rate for investment, such as the weighted cost of capital (WACC), and can be used to determine whether to accept or reject individual energy projects. However, IRR is not appropriate for assessing mutually exclusive projects since investment scale is not considered (see NPV above). Projects with multiple IRRs and unequal lives must be treated carefully as outlined in Chapter 5 (Sections 5.4.3 and 5.4.7).

Example 8.2

An energy supply company has an option to build an offshore wind farm comprising 30 turbines, each with a rated output of 3 MW. The estimated capital cost is 3000 €/kW and maintenance costs are projected to average 60 €/MWh. Calculations based on plant availability and the wind resource of the site indicate an annual electricity output of 305 GWh. A FIT of 150 €/MWh is guaranteed over the project life of 20 years. If the company's hurdle rate is 10%, should it consider investing in the project?

Solution

Table 8.5 show that total capital costs for the 90 MW wind farm will be €270m, generating annual revenues of €46m with maintenance costs of €18m. This yields an internal rate of return of approximately 8%, lower than the company's hurdle rate. The project should not proceed.

Installed capacity	(MW)	90	
Unit capital cost	(€/kW)	3000	
Total capital costs	*(€m)*	*270*	
Annual energy produced	(GWh)	305	
Feed-in-tariff	(€/MWh)	150	
Annual revenues	*(€m/annum)*	*46*	
Unit operating costs	(€/MWh)	60	
Annual operating costs	(€m/annum)	18	
IRR		**8%**	

Table 8.5 Internal rate of return for a 90 MW offshore wind farm.

8.4.2.3   Simple payback period

Simple payback is a quick and simple way of comparing competing energy projects but suffers from a number of drawbacks as outlined in Chapter 3.2.5. The technique is most useful for projects with short paybacks, since the discount rate will not have a significant effect on results over a short period. It is also useful for analysing projects with high risk, as it gives an indication of how long a capital investment will be at risk. It is not useful for ranking projects.

> ### Example 8.3
>
> A paper manufacturer uses 80 kW pumps to move process water to batching tanks. The process requires variable flow, which is achieved by manually. The performance of a variable speed drive (VSD) for the pump has been simulated and it is estimated that it will result in savings of 543 895 kWh/annum of electricity. The capital cost of the VSD estimated to be €22 250 with no additional maintenance costs. The manufacturer pays 0.08 €/kWh for electricity. What is the simple payback period (n_p) for the project?
>
> #### Solution
> Firstly, annual savings are calculated as the product of the annual energy savings and the unit electricity cost:
>
> Annual energy savings $= 543\,895$ kWh/annum $\times 0.08$ €/kWh $= 43\,512$ €/annum
>
> Then, the n_p is calculated by dividing the capital cost by this saving:
>
> $n_p = $ €22 250$/43,512$ €/annum $= 0.51$ years
>
> The use of n_p is justified given the very short payback period and indicates that the capital invested will not be at risk for a significant period.

8.4.2.4   Discounted payback period

Discounted payback period (DPP) is the time required to repay an investment using discounted cash flows. It overcomes some of the drawbacks of the simple payback technique because it takes account of the time value of money. However, it should not be used for selecting from, or ranking, multiple energy projects, as it does not take account of benefits after the payback period and, therefore, does not give a true picture of the value of a project over its lifespan.

> ### Example 8.4
>
> A homeowner would like to install a photovoltaic system on a dwelling and wants to know how long it will take to pay back. The homeowner expects it to have a long payback period, so has decided to estimate the discounted payback period. The 1.76 kWp system will cost €7040, is assumed to have no running costs and will produce 1558 kWh of electricity per annum, which can be sold to the grid at a FIT of 0.5 €/kWh. What is the discounted payback period if a discount rate of 8% is used?
>
> #### Solution
> Cash flow analysis shows outgoings of €7040 in year 0 (captial costs) and recurrent revenues of €779 thereafter (1558kWh ×0.5 €/kWh). Cumulative discounted cash flows are then calculated and show that the DPP occurs during year 17 (Table 8.6). Using interpolation (145/(145+66)), a discounted payback period of 16.7 years is estimated.

Table 8.6 Cumulative discounted cash flow for a 1.76 kW PV system with a feed-in-tariff showing the discounted payback period.

Year	0	1	2	3	...	15	16	17	18	19	20
Capital cost	(7040)	0	0	0	...	0	0	0	0	0	0
Annual revenues	0	779	779	779	...	779	779	779	779	779	779
Annual cash flow	(7040)	779	779	779	...	779	779	779	779	779	779
Annual discounted cash flow	(7040)	721	668	618	...	246	227	211	195	181	167
Cumulative discounted cash flow	*(7040)*	*(6319)*	*(5651)*	*(5032)*	*...*	*(372)*	*(145)*	***66***	*261*	*441*	*608*

8.4.2.5   Benefit/cost ratios

The benefit/cost ratio (BCR) is used to measure the extent to which the benefits of a project outweigh its costs (where a ratio >1 indicates a desirable project). Ratios provide a simple way of ranking different projects. Cost–benefit analysis is frequently used for public projects where social costs and benefits are incorporated. The technique is not suitable for mutually exclusive projects unless modified to consider the incremental benefits of alternative projects. The benefit/cost ratio is defined as the ratio of the present value of benefits divided by the present worth of costs:

$$BCR = (PV(\text{benefits})) / (PV(\text{costs}))$$

where BCR is the benefit/cost ratio, PV(benefits) is present value of the project benefits and PV(costs) is the present value of the project costs.

Example 8.5

A northern European country is considering whether to subsidise the purchase of domestic solar water heaters to help achieve emissions reductions targets agreed for 2020. Failure to meet these targets will entail the purchase of carbon credits through the EU ETS, at rates that are likely to increase over time. It is estimated that a subsidy of €1496 is required for each domestic installation in order to make it financially attractive to the homeowner (and investor); this is based on offsetting the negative net present value of the system to a private investor over a projected lifespan of 20 years. Based on an initial carbon dioxide prices of 30 €/t, rising linearly to 80 €/t after 20 years, the discounted value of the avoided emissions is estimated to be €646 per installation, yielding a BCR of:

$$BCR = €646/€1496 = 0.43$$

Since the BCR is less than one, the costs of the policy would outweigh its benefits, so the government should identify a different project. Since it is likely that the BCR will improve over time due to decreases in capital costs and increases in carbon costs and energy tariffs, it may be worthwhile to monitor the BCR of this policy periodically.

8.4.3   Special economic measures

8.4.3.1   Savings/Investment Ratio

The savings/investment ratio is a special case of the benefit/cost ratio; it is useful for assessing energy efficiency projects in particular. Unlike the benefit/cost ratio calculation, however, costs only include the main 'over and above' costs (such as capital or additional maintenance) associated with the energy efficient investment. Savings are determined by calculating the difference between the original costs and the new, reduced, costs. An SIR greater than one indicates a desirable project:

$$SIR = PV(savings) / PV(costs)$$

where SIR is the savings/investment ratio, PV(savings) is present value of the net savings due to lower energy, replacement, maintenance costs and so on, and PV(costs) is the present value of the over and above costs such as capital, replacement, and maintenance.

> **Example 8.6**
> A company is trying to decide whether to install a CHP plant on its site to produce heat and electricity for its operations. The plant will reduce the annual costs of electricity purchased from the grid and gas, which is used in its boilers. The company's calculations show that a 1 MW electrical (MWe) CHP can provide 5519 MWh of electricity, which can be used on site, and 6071 MW of heat annually. With an average electricity tariff of 10 €c/kWh, a boiler gas tariff of 4 €c/kWh and a boiler efficiency of 80% (giving a heat tariff of 5 €c/kWh), annual savings will be €855 414. The capital cost of the installation is estimated to be €1.5m and running costs (primarily gas and maintenance) are estimated to be €479 711; a major maintenance event in year 10 will cost approximately €300 000. Using a savings/investment ratio approach, is the project worthwhile?
>
> *Solution*
> Nominal and discounted cash flow are shown in Table 8.7. The resulting savings/investment ratio is:
>
> $$SIR = €5072/€4412 = 1.15$$
>
> Since the SIR is greater than one, the project is worthwhile for the company.

Table 8.7 Calculation of the savings/investment ratio (SIR) for a CHP plant (prices in €'000s).

Year	0	1	2	3	4	5	6	7	8	9	10	11	12	13	14	15	16	17	18	19	20
Savings		855	855	855	855	855	855	855	855	855	855	855	855	855	855	855	855	855	855	855	855
Costs	1500	480	480	480	480	480	480	480	480	480	780	480	480	480	480	480	480	480	480	480	480
Discounted savings		737	636	548	472	407	351	303	261	225	194	167	144	124	107	92	80	69	59	51	44
Discounted costs	1500	414	357	307	265	228	197	170	146	126	177	94	81	70	60	52	45	38	33	29	25
NPV savings *5072*																					
NPV costs *4412*																					

8.4.3.2   *Levelised cost of energy production*

The levelised cost of energy (LCOE) is the life-cycle breakeven energy generation price when considering all life-cycle costs. It allows different technologies to be compared which involve different levels of capital, operational, maintenance and decommissioning costs. For example, it allows production costs between domestic PV generation to be compared with that for combined cycle gas turbine (CCGT) power plant. It is, therefore, useful in ranking investment alternatives. The levelised cost of producing one unit of energy using technology x is given by:

$$\text{LCOE}_X = NPC_X / \sum_{i-1}^{n} (E_i / (1+r)^n) \tag{8.1}$$

where LCOE_x is the levelised cost of energy production of technology x (€/kWh), NPC_x is the net present life-cycle cost of technology x, E_i is the energy produced or saved by technology x in year i (kWh), r is the discount rate and n is the lifespan of the technology (years).

> *Example 8.7*
> Consider a renewable energy company wishing to invest in renewable technologies in an environment where the same feed-in-tariff is offered for all such investments. The options include:
>
> - a 3 MW offshore wind turbine costing €9 m, with annual operating costs of €612 000, producing 10.2 GWh of electricity each year;
> - a 3 MW onshore wind turbine costing €4.05 m, with annual operating costs of €82 000, producing 8.2 GWh of electricity each year;
> - a 2 MW PV form costing €2.2 m, with annual operating costs of €10 000, producing 0.9 GWh of electricity every year.
>
> Using Equation 8.1 and a discount rate of 8% gives the results shown in Table 8.8. It can be seen that the most attractive technology is onshore wind, assuming that all are subject to the same FITs.

Table 8.8 Levelised costs of energy calculations for hypothetical offshore and onshore wind and PV technologies.

Technology		Offshore wind	Onshore wind	PV Farm
Capacity	(MW/MWpeak)	3	3	2
Capital costs	(€m)	9.00	4.05	2.20
Operating costs	(€/annum)	612 000	82 000	10 000
Net present cost	(€m)	15.0	4.9	2.3
Energy output	(GWh/annum)	10.2	8.2	0.9
Net present energy output	(GWh)	100	81	9
LCOE	*(€/kWh)*	*0.15*	*0.06*	*0.26*

Table 8.9 Matrix showing which economic measures can be used for different decision types.

Decision	Net Present Value	Internal Rate of Return	Simple Payback Period	Discounted Payback Period	Benefit/ Cost Ratio	Savings/ Investment Ratio	Levelised Cost of Energy Production
Accept/Reject	Acceptable	Suitable	Helpful	Helpful	Suitable	Suitable	Suitable
Select from competing alternatives	Suitable	Not appropriate	Not appropriate	Not appropriate	Not appropriate	Not appropriate	Not appropriate
Ranking options	Acceptable	Not appropriate	Not appropriate	Not appropriate	Suitable	Suitable	

8.4.4   Choosing economic measures

The choice of the economic measures to use in the appraisal of an energy project depends on a variety of factors. A key consideration is the type of decision which is being made. Table 8.9 gives a summary of the different economic measures that can be used for different decision types.

Accept/reject decisions relate to individual projects and whether they reach particular thresholds. For example, if an IRR is greater than the hurdle rate (or minimum acceptable rate of return - MARR) used by a company, then this will have a significant bearing on whether to proceed. Similarly, if the levelised cost of energy is lower than the available FIT, then an important investment criterion will have been met.

In the case of competing alternatives, such as projects competing for the same site or financial resources, only one project can be selected. Many measures do not discriminate on the basis of investment size, so these are not appropriate. However, all other things being equal, NPV does give greater values to larger projects and is, therefore, the most suitable method for assessing competing alternatives.

Where an investor has a set of projects, some or all of which can be undertaken, then they may be ranked in order of preference for programme or other reasons. Simple and discounted payback techniques are not reliable because they do not take account of returns after the payback period. IRR tends to overestimate the value of projects where cash flows are front-loaded. NPV can be used, but feasible combinations of projects should be assessed together.

8.5   Estimating GHG emissions

To correct for the negative externalities associated with fossil fuels, many governments have introduced measures such as carbon taxes or tradable certificates. In a perfect market, these carbon costs would equal the social cost of fuel-related carbon dioxide emissions discussed previously. When assessing an energy-related project, it is important to include these costs. And, in order to assess such costs, the emissions or avoided emissions of CO_2-eq must be estimated.

It has been seen that energy and emissions targets are set at national levels and over long time horizons, so the impacts of policies on emissions from all economic sectors must be considered. The decision to build a power station, for example, will result in increased emissions not just from the energy sector of an economy but also from the construction and manufacturing sectors. For this reason, 'life-cycle assessment' (LCA) is regarded as best practice, where emissions from all stages of a project's life are accounted for. For a private investor, however, an LCA approach may not be appropriate but is often necessary for large projects with significant societal impacts.

Policymaking involves the efficient allocation of scarce resources. Where emissions' mitigation is the prime consideration, efficient policies will favour projects providing the greatest carbon dioxide savings at the least cost. This is measured using the marginal abatement cost (MAC), which is the change in total cost when emissions are reduced by one unit. The abatement costs of policies can be compared using MACs.

This section describes LCA and how it can be used to develop MACs.

8.5.1   *Life-cycle assessment*

We know that fossil-fuelled heat and power generation involves the combustion of hydrocarbons and the release into the atmosphere of harmful pollutants, including: carbon dioxide, which results in global warming and ocean acidification; particulate matter, such as coal fly ash from coal-fired power stations, which has adverse health impacts, contributes to short-term atmospheric cooling and contaminates soils and water with heavy metals; and radioactivity from coal combustion. However, these 'direct' emissions are not the only emissions that can be attributed to heat and power production: the maintenance of the plant requires energy use and emissions; the plant itself must be manufactured and this requires energy; the raw materials used in that manufacturing process had to be extracted; and the machinery used in their extraction had to be manufactured. In fact, there is an almost infinite set of interlinked upstream emissions-generating activities which can be attributed to the production of heat and power (or the production of any product or service, for that matter). Similarly, although operationally cleaner, energy used in the manufacture, operation, maintenance and disposal of RES technologies results in the release of pollutants.

Energy use and emissions may be 'direct', that is resulting from the activity being considered. For example, direct emissions from the manufacture of structural steel result from the energy used in the steel mill. 'Indirect' emissions emanate from 'upstream' (e.g. the production of pig iron for the structural steel manufacturing process) and 'horizontal' (e.g. emissions associated with insuring the steel mill) activities.

Life-cycle assessment (LCA) is a technique that is used to estimate the environmental aspects and impacts over the full lifespan of a product or process. ISO 14044, 'Environmental Management – Life Cycle Assessment – Requirements and Guidelines', describes the four main phases in an LCA:

(1)   goal and scope definition, where the system boundary and purpose of the study are defined;
(2)   life-cycle inventory analysis (LCI), where data on inputs (e.g. energy) and outputs (e.g. GHGs) are collected and/or calculated;
(3)   life-cycle impact assessment (LCIA), where the results of the LCI are assessed in order to clearly understand their environmental significance;
(4)   interpretation, in which the results are discussed and conclusions and recommendations are made in the context of the goal and scope definition.

LCA is useful in the cost–benefit analysis of renewable energy projects for a number of reasons:

- to quantify and price all emissions associated with a project;
- for identifying opportunities to improve the environmental performance of a system;
- to communicate with stakeholders such as government, industry and the public.

LCA addresses the environmental aspects and potential impacts throughout an investment's life cycle from raw material acquisition through production, use, end-of-life treatment, recycling and final disposal (i.e. cradle-to-grave). Energy use and emissions during the life cycle of a product or services include both direct and indirect emissions and can be characterised as:

- 'embodied' emissions associated with the production of a product or delivery of a service;
- operational emissions (or emissions' savings), such as those associated with fuel use, consumables, insurance;
- maintenance emissions due to planned and forced maintenance, including major plant overhauls;
- decommissioning emissions and accounting for recycled materials.

8.5.2   Carbon dioxide equivalent

Each GHG has a different effect on global warming, termed its Global Warming Potential (GWP). GWPs are measured using a normalised scale where a GWP of one is equivalent to the warming effect of carbon dioxide; GWPs of each of the main GHG gases are shown in Table 8.10. Where an environmental aspect involves a number of different GHG gases, their emissions-weighted GWPs can be summed to give a total carbon dioxide equivalent (CO_2-eq). Reported CO_2-eq emissions typically use a 100-year time horizon.

Greenhouse gas	Formula	Time Horizon (years)		
		20	100	500
Carbon dioxide	CO_2	1	1	1
Methane	CH_4	72	25	7.6
Nitrous oxide	N_2O	289	298	153

Table 8.10 Global warming potentials (GWPs) for the main greenhouse gases (Source: Forster *et al.*, 2007).

8.5.3   Energy and GHG accounting methods

A variety of accounting approaches exists for energy and CO_2-eq emissions associated with the production and delivery of products and services. These include process analysis, input–output analysis and a variety of combinations of these techniques collectively referred to as 'hybrid analysis'; Crawford (2008) provides a detailed overview. All techniques involve obtaining the product of an emissions intensity and activity level to give total emissions for a particular activity.

Input–output (I–O) analysis was first developed by Wassily Leontief in the 1930s and has had environmental applications since the 1970s. A national economy is divided into a number of sectors, each assumed to produce a single, uniform good or service. I–O tables, which are constructed from national accounts and trade balance data and are available for all advanced economies, show the extent to which a monetary output from one sector is an input to each other sector. These can be combined with direct sectoral energy and GHG emissions data from national energy and environmental accounts to estimate the total sectoral emissions. The technique includes all upstream inputs, but assumes sectoral homogeneity, which can result in significant error at the level of an individual product or service. Sectoral emissions intensities for a selection of European countries are shown in Table 8.11; these can be multiplied by sectoral expenditure on a good or service to estimate total emissions using Equation 8.2:

$$E_{tot} = \sum_{i=1}^{n} EI(IO)_i \times P_i \qquad (8.2)$$

where E_{tot} is total emissions for the product or service ($kgCO_2$-eq), $EI(IO)_i$ is the I–O energy intensity of sector i ($kgCO_2$-eq/€), P_i is the expenditure on the product or service which is categorised under sector i and n is the number of sectors in the economy.

Process analysis involves identifying the supply chain upstream of a product or service and measuring and summing material quantities and/or activity levels. These data are combined with process emissions intensities that are representative of the material or activity being studied to give total emissions for a particular product or service. Process emissions intensities are available from a number of publically and commercially available databases (for example, the Inventory of Carbon and Energy – http://www.bath.ac.uk/mech-eng/sert/embodied/ – and Ecoinvent – http://www. ecoinvent.ch/). Although process-based emissions intensities are accurate for the

Table 8.11 Sectoral emissions coefficients of consumption for selected European countries (gCO$_2$-eq/€).

Sector	AT	DK	FR	DE	IT	NL	PT	ES	SE	UK
Agriculture, Fishing, Forestry	1206	1370	1266	1039	829	947	1137	991	1062	1534
Food, Tobacco, Beverages	401	697	512	462	500	463	650	573	364	435
Textile, footwear	155	224	135	274	307	202	358	295	129	191
Furnitures	773	675	791	860	889	645	1557	1089	569	871
Books, printing	347	184	182	273	354	190	528	341	276	232
Gas, water, electricity	1503	2647	1028	3767	2516	2619	3682	2956	870	2956
Household Equipment	2135	1656	1791	2558	3043	1859	3679	3112	1460	2605
Household appliances	442	434	374	497	726	527	822	789	265	606
Vehicle purchase, maintenance	290	342	280	424	538	307	536	543	205	459
Construction and maintenance of the house	208	221	150	213	262	167	520	303	193	176
Transport	444	1213	377	578	548	689	951	712	444	712
Restaurants and Hotels	140	283	197	217	249	247	350	196	128	159
Telecommunications	75	120	51	118	133	73	132	171	74	111
Insurances	53	38	47	63	55	45	74	58	28	78
Real estate, rents	76	44	19	35	31	43	73	60	83	35
Public and miscellaneous services	625	686	762	583	1077	1256	1718	956	464	741
Social services	71	84	81	107	37	82	82	66	54	88
Health care	86	86	64	118	105	103	308	119	39	115

extent of the supply chain analysed, it is ultimately, and often unknowingly, necessary to truncate the supply chain and disregard a very large number of activities which individually have a relatively small effect but collectively can represent a significant error.

Emissions can be estimated for a product (or service) using Equation 8.3:

$$E_{tot} = \sum_{j=1}^{m} PEI_j \times Q_j \tag{8.3}$$

where E_{tot} is total emissions for the product (kgCO$_2$-eq), PEI_j is the process energy intensity of material j (kgCO$_2$-eq/kg), Q_j is the quantity of product j (kg) and m is the number of different materials in the product.

Process and I–O emissions accounting techniques can be combined into 'hybrid' techniques to benefit from their respective strengths while minimising the effects of their weaknesses and include tiered, I–O-based and integrated. The choice of technique depends on factors including data availability, available analytical tools, expertise, time constraints, system boundaries and required accuracy. Detailed descriptions are not covered here and can be found in Crawford (2011).

In summary, I–O analysis is complete when applied to sectors where the product or service can be treated homogenously and is often suitable for policy analysis where policies are aimed at single or multiple industrial sectors. However, when individual products/services are analysed, process analysis can be applied where system boundaries are limited and data applicable. For more complete boundaries, an appropriate hybrid approach must be adopted.

8.5.4   *Marginal abatement cost of carbon*

In the study of energy and emissions, the marginal abatement cost (MAC) can be defined as the cost of avoiding the last unit of GHG using a particular RES or EE technology. In other words, it is the long-run average cost of avoiding one unit of GHG using a particular technology or system. The MAC of CO_2-eq emissions using technology x is the additional cost of energy production, expressed per tonne of CO_2-eq saved, of technology x over the displaced (or alternative) technology. It can be expressed as:

$$MAC_x = \left(LCOE_x - LCOE_{DT} \right) / \left(EI_{DT} - EI_x \right) \tag{8.4}$$

where MAC_x is the marginal CO_2-eq abatement cost of energy production of technology x compared to production using the best alternative technology (€/tCO_2-eq), $LCOE_x$ is the levelised cost of producing one unit of energy using technology x (€/kWh), $LCOE_{DT}$ is the levelised cost of producing one unit of energy using the displaced technology (€/kWh), EI_{DT} is the emissions intensity of the displaced technology (tCO_2-eq/kWh) and EI_x is the emissions intensity of technology x (tCO_2-eq/kWh).

A number of studies of marginal abatement costs have been undertaken which rank the relative cost effectiveness of different technologies in reducing GHG emissions. The broad range of technologies considered in these studies makes them suitable for policy makers who can use the results to inform decision making about targeting subsidies, feed-in-tariffs, carbon taxes, national retrofit programmes and other policy measures. Figure 8.7 shows indicative marginal abatement costs for a variety of electricity generation technologies in OECD countries compared to a reference, 'business-as usual' case. An indicative emissions mitigation potential for each technology is also shown on the horizontal axis.

The choice of the displaced technology is important because it provides the baseline emissions against which the energy efficient or renewable energy technology will be judged. Furthermore, the emissions analysis of the displaced technology must follow the same methodology and boundaries adopted for the comparator project. For example, additional insulation to a dwelling with a gas-fired boiler may displace the need for additional heat provided by a gas boiler. For wind power, grid electricity is displaced; for large-scale electrical energy storage technologies such as pumped hydro or compressed air storage, gas-fired peaking plant may be displaced.

8.6   Uncertainty

The future of renewable energy and energy efficiency is dependent on price and policy developments both at local and global levels. The renewal of global agreements, such as the Kyoto Protocol limiting GHG emissions, may significantly influence

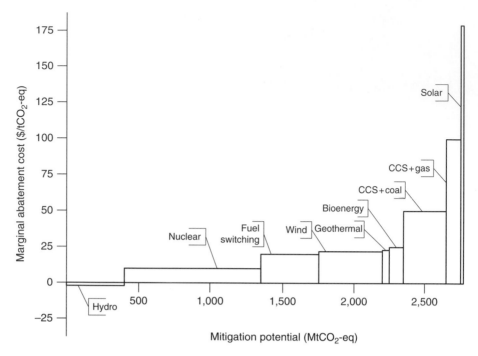

Figure 8.7 Indicative marginal abatement costs and potentials for a variety of electricity generation technologies in OECD countries (various data sources).

future carbon prices, FITs and subsidies. In addition, the future prices of fossil fuels will set the benchmark against which investment in renewables and energy efficiency will be judged. A rapid market uptake of renewable energy supply and energy efficient systems will reduce the time required to achieve further technology learning and make it more attractive to invest in the technology.

The lifecycle economic appraisal of renewable energy supply and energy efficient technologies is complicated by the unpredictability and endogenous nature of key input variables such as:

- renewable energy prices supports, such as feed in tariffs, which are subject to political risk, public opinion and economic circumstances;
- fossil fuel prices which are dependent on global growth, the discovery of new reserves and security of supply;
- capital cost projections which are subject to the growth of the renewables market and technology learning;
- the energy and emissions intensities of many energy carriers (e.g. heat, electricity) which are dependent on the uptake of renewable energy supply and energy efficient technologies;
- discount rates which are subject to national and global economic cycles.

Give the volatility of many of the key input variables, it is important to incorporate uncertainty into any long-term energy investment appraisal. There are many ways of doing this, including sensitivity analysis, scenarios and Monte Carlo analysis.

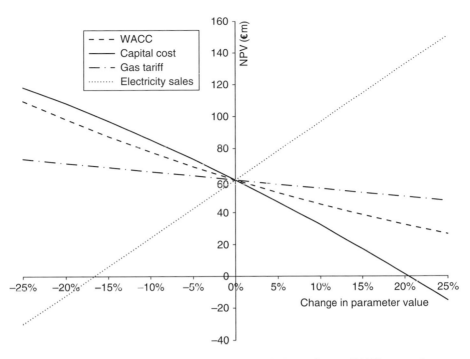

Figure 8.8 Sensitivity analysis results for the Compressed Air Energy Storage (CAES) case study.

8.6.1 *Sensitivity analysis*

Sensitivity analysis involves allowing the values of key input variables (such as those listed above) to vary over their expected range and examining the resulting impact on the dependent variable, such as NPV or IRR. For example, an evaluator may take the view that a ± 25 % variation in fossil fuel prices should be assessed over the project lifespan to identify the corresponding change in NPV. Figure 8.8 shows the results of a sensitivity analysis for a large-scale electrical energy storage project case study showing the NPV (vertical axis) for different percentage changes in key input variables (horizontal axis). It can be seen that the project is most sensitive to electricity sales (the greatest change in NPV per unit change in the variable).

8.6.2 *Scenarios*

Scenario analysis involves the development of future contexts based on past trends and current understanding. Typically, scenario development involves a 'business-as-usual' scenario where current trends are assumed to continue into the future; these can have a high degree of accuracy over a short-to-medium time horizon. In addition, a number of exploratory scenarios may be developed which assess possible future social, technological, environmental, economic and political contexts. The likely impact of each scenario on key input variables is then assessed and the resulting values of the economic measure are determined for each scenario.

8.6.3   Monte Carlo analysis

Monte Carlo analysis involves repeated sampling of input probability distributions and the computation of an output distribution for the chosen economic measure. It has the advantage of providing information on the range and probability of an outcome, rather than just a range, as in the cases of sensitivity and scenario analysis. However, it is dependent on the availability of input distributions for key variables, such as capital costs, energy tariffs and carbon prices. Such information can be based on historic distributions, samples of expert opinions (e.g. using the Delphi method) or on the judgement of the assessors.

8.7   Case studies

8.7.1   Compressed air energy storage

A utility company proposes to develop a compressed air energy storage (CAES) plant in a salt cavern. It will be capable of delivering 140 MW of dispatchable electricity to the national grid over an eight-hour period covering weekday peak times (typically 16:00–19:00) when electricity prices are highest (the plant does not operate at weekends due to low prices during the period). The technology involves the creation of a subterranean cavern in a salt deposit using solution mining and the construction of a gas-fired power plant; capital costs are shown in Table 8.12. Compressors pump air into the cavern during off-peak times, favouring the use of surplus wind capacity on the network. During peak periods, the air is recovered from the cavern, mixed with natural gas and combusted to drive gas turbines and produce electricity. Compression and storage occurs during the cheapest off-peak hours; generation during the most valuable peak hours.

The construction period is estimated to be three years with the following phasing:

- years 1 and 2: costs are equally divided between the following: land costs; pre-financial close costs; subsurface costs; installation costs; other costs;
- year 3: major surface equipment; balance of plant; interconnection costs.

Site procurement	8.7
Pre-financial costs	4.0
Cavern construction	22.8
Main plant	72.3
Interconnection	10.6
Balance of plant	8.0
Installation	7.8
Other	6.5
Total (inc.contingency at10%)	154.9

Table 8.12 CAES case study capital cost breakdown.

All values are millions of Euros.

The plant will have a heat rate of 4200 kJ/kWh (i.e., 4200 kJ, or 1.17 kWh of gas will be required to produce 1 kWh electricity) and will operate for a 30-year period, with annual maintenance costs spread evenly over that time. After this time, the residual value of the investment includes scrap value of the power plant (assumed to be negligible) and the value of the site and cavern (assumed to be the sum of the site and cavern construction costs).

The tariff for gas used in the regeneration process is 0.016 €/kWh while the revenues are the hourly electricity system marginal prices (SMP) as well as a capacity payment of 80 000 €/MW. Inflation-corrected half-hourly system marginal prices from the previous three years were used to estimate annual revenues; the sum of the SMP and use of system (UOS) charge of € 2.5/MWh was used to estimate compression costs. Maintenance costs are estimated to be 11.58 €/MWh. An analysis of the company's capital structure indicates a weighted average cost of capital (WACC) of 7.5%.

Is the project a worthwhile investment from the perspective of the utility company?

Solution

Firstly, total annual electricity output from the plant is calculated:

Total electrical output = 140 MW × 8 h/d × (365 − 104) d/annum = 292.32 GWh/annum

Annual costs and revenues are then calculated:

Gas cost = 292.32 GWh × 1.17 × 8 h × 16 €/MWh = 5.47 m€/annum

Compression cost = $\left(\sum_{i=1}^{n} (SMP_i + UOS) \times P_{in} \right)$ = sum of the products of the SMP + UOS for each operational period and the electricity used by the compressor over that period (P_{in}) = 9.34 m€/annum

Operational cost = 292.32 GWh × 11.58 €/MWh = 3.39 m€/annum

Electricity revenue = $\left(\sum_{i=1}^{n} (SMP_i) \times P_{out} \right)$ = sum of the products of the SMP for each operational period and the plant's electrical output over that period $(Pout)$ = 26.79 m€/annum

Capacity revenue = capacity tariff × plant's power output = 80 €/kW × 140 MW = 11.2 m€/annum

The NPV, IRR, LCOE and DPP are then calculated to establish respectively whether the investment: gives a positive cash position in today's money; exceeds the WACC hurdle rate; compares favourably with other peaking plant unit costs; and the duration for which investor's money will be at risk. Table 8.13 shows the annual nominal and discounted costs, revenues, profits and cumulative position for the CAES project. It can be seen that after a 30 years of operation the project has a positive NPV of 60.4 m€ and the DPP is 16.2 years. The IRR is calculated from the undiscounted cash flows to be 11.53%. The LCOE is given by:

LCOE = sum of discounted annual costs / sum of discounted annual electricity output = 303.6 m€ / 2779.1 GWh = 10.92 c/kWh.

A summary of results is given in Table 8.14. The project is an attractive investment as it yields a positive NPV, its IRR exceeds the hurdle rate of 7.5% and it produces a discounted positive cumulative cash flow after 16.2 years for a further period of 18.8 years. The LCOE falls within the range of comparable values; for example, LCOE for combined cycle gas turbine (CCGT) peaking plant values can vary from 7–12 c/kWh. However, with only two operational plants worldwide and no recent history of

Table 8.13 Annual nominal and discounted costs, revenues, profits, energy and cumulative cash position for the CAES project.

Year	Undiscounted			Discounted			
	Costs (m€/annum)	Revenues (m€/annum)	Cash flow (m€/annum)	Costs (m€/annum)	Cash flow (m€/annum)	Cumulative Cash (m€)	Energy (GWh)
1	32.0		−32.0	29.8	−29.8	−29.8	0.0
2	32.0		−32.0	27.7	−27.7	−57.4	0.0
3	90.9		−90.9	73.2	−73.2	−130.6	0.0
4	18.2	38.0	19.8	13.6	14.8	−115.8	218.9
5	18.2	38.0	19.8	12.7	13.8	−102.0	203.6
6	18.2	38.0	19.8	11.8	12.8	−89.2	189.4
7	18.2	38.0	19.8	11.0	11.9	−77.2	176.2
8	18.2	38.0	19.8	10.2	11.1	−66.2	163.9
9	18.2	38.0	19.8	9.5	10.3	−55.8	152.5
10	18.2	38.0	19.8	8.8	9.6	−46.2	141.8
11	18.2	38.0	19.8	8.2	8.9	−37.3	131.9
12	18.2	38.0	19.8	7.6	8.3	−29.0	122.7
13	18.2	38.0	19.8	7.1	7.7	−21.3	114.2
14	18.2	38.0	19.8	6.6	7.2	−14.1	106.2
15	18.2	38.0	19.8	6.2	6.7	−7.4	98.8
16	18.2	38.0	19.8	5.7	6.2	−1.2	91.9
17	18.2	38.0	19.8	5.3	5.8	4.6	85.5
18	18.2	38.0	19.8	5.0	5.4	10.0	79.5
19	18.2	38.0	19.8	4.6	5.0	15.0	74.0
20	18.2	38.0	19.8	4.3	4.7	19.7	68.8
21	18.2	38.0	19.8	4.0	4.3	24.0	64.0
22	18.2	38.0	19.8	3.7	4.0	28.0	59.5
23	18.2	38.0	19.8	3.4	3.7	31.8	55.4
24	18.2	38.0	19.8	3.2	3.5	35.3	51.5
25	18.2	38.0	19.8	3.0	3.2	38.5	47.9
26	18.2	38.0	19.8	2.8	3.0	41.5	44.6
27	18.2	38.0	19.8	2.6	2.8	44.4	41.5
28	18.2	38.0	19.8	2.4	2.6	47.0	38.6
29	18.2	38.0	19.8	2.2	2.4	49.4	35.9
30	18.2	38.0	19.8	2.1	2.3	51.7	33.4
31	18.2	38.0	19.8	1.9	2.1	53.8	31.1
32	18.2	38.0	19.8	1.8	2.0	55.7	28.9
33	18.2	69.5	51.3	1.7	4.7	60.4	26.9
				303.6	60.4		2779.1

Measure	Unit	Value
NPV	m€	60.4
IRR	%	11.5
LCOE	c/kWh	10.9
DPP	years	16.2
n_p	years	7.8

Table 8.14 Results for CASE analysis.

construction, CAES is a relatively immature technology with a high risk profile. Investors must decide if the investment exposure period (of 7.8 years in nominal terms) is acceptable.

Figure 8.8 shows a NPV sensitivity analysis of the project. It can be seen from the slopes of the lines that the project is most sensitive to the value of electricity sales: a fall of 17% in expected revenues results in a negative NPV. The project is also relatively sensitive to changes in capital cost where an increase of 20% results in a negative NPV. It may, therefore, be prudent to reduce capital cost risk given the relative uncertainty of this parameter.

8.7.2   Policy assessment of domestic PV

A government is considering promoting domestic PV systems in order to reduce GHG emissions. Specifically, it wishes to identify what feed-in-tariff would be appropriate to encourage the uptake of domestic PV, the associated cost of carbon abated and whether this would represent value for money for the exchequer. Although the PV systems will be located on domestic dwellings, they will feed electricity directly onto the local distribution network, thus any FIT will apply to all energy produced.

The approach adopted involves estimating the LCOE for commonly-available PV system sizes and using this as the minimum tariff required to make it worthwhile for a rational investor to adopt the technology. The life-cycle emissions savings for PV-derived electricity over that from the grid are estimated and used to determine the MAC for the technology. A comparative analysis with the LCOEs and MACs of alternatives is then undertaken. Finally, a CBA is carried out.

A number of different PV sizes are available on the market which suit different house sizes; their outputs under representative conditions have been determined. It is assumed that systems have a lifespan of 20 years but that inverters (an electrical component) must be replaced at year 10. The system sizes, costs, outputs and LCOEs are shown in Table 8.15. An investor discount rate of 10% is used and a net €100 end-of-life disposal cost is assumed for all systems.

Firstly, the LCOE is calculated for the different domestic PV sizes. The net present costs of the systems, including initial capital costs, inverter replacement costs and disposal costs are calculated. These are divided by the sum of the discounted energy flows produced by each system to give LCOEs. Results are shown in Table 8.16 and range from 0.70 to 0.96 €/kWh. This is expensive when compared to the average wholesale price of electricity of society approximately 0.09 €/kWh. Therefore, FITs in the range of 0.70 to 0.96 €/kWh are required to incentivise investment; the cost to would be approximately 0.09 €/kWh less than this (the value on the

Table 8.15 PV system input parameters.

Size	(kWp)	0.47	1.41	1.72	2.82	4.23	5.64
System cost	(€)	3103	7651	9151	14 474	21 297	28 120
Output	(kWh/annum)	414	1243	1517	2487	3730	4974
Inverter cost	(€)	666	1412	1541	1782	2270	4024

Table 8.16 NPVs, discounted energy flows and resulting LCOEs for each PV system.

	PV system size (kWp)					
	0.47	1.41	1.72	2.82	4.23	5.64
Net present costs (€)						
Year 0	3103	7651	9151	14 474	21 297	28 120
Year 10	257	544	594	687	875	1551
Year 20	39	39	39	39	39	39
NPC	*3398*	*8234*	*9784*	*15 200*	*22 211*	*29 710*
Sum of discounted energy flows (kWh)						
Years 1–20	414	1243	1517	2487	3730	4974
Sum	*3529*	*10 586*	*12 914*	*21 173*	*31 759*	*42 346*
Energy costs (€/kWh)						
LCOE	*0.96*	*0.78*	*0.76*	*0.72*	*0.70*	*0.70*
Subsidy	*0.87*	*0.69*	*0.67*	*0.63*	*0.61*	*0.61*

Table 8.17 Marginal abatement cost (MAC) calculations for the PV systems.

		PV system size (kWp)					
		0.47	1.41	1.72	2.82	4.23	5.64
Embodied emissions	$(kgCO_2$-eq)	1118	3353	4090	6706	10 059	13 411
Annual avoided emissions	(kg/annum)	196	588	717	1175	1763	2350
Life-cycle emissions	$(kgCO_2$-eq)	−2800	−8399	−10 245	−16 797	−25 196	−33 595
Emissions intensity	$(kgCO_2$-eq/kWh)	−0.34	−0.34	−0.34	−0.34	−0.34	−0.34
Subsidy (FIT-wholesale price)	(€/kWh)	0.87	0.69	0.67	0.63	0.61	0.61
MAC	$(€/tCO_2$-eq)	*2559*	*2029*	*1970*	*1853*	*1794*	*1794*

wholesale market), resulting in subsidies in the range 0.61–0.87 €/kWh. It should be noted that the price is highly dependent on the discount rate. If, say, low-cost government-backed finance was available to individual investors at a 6% discount rate, then the FIT required would drop to between 0.53 and 0.74 €/kWh.

To estimate MACs, life-cycle emissions are first calculated as the sum of embodied emissions (manufacture and installation of PV cells, wiring, inverter, frame etc.), avoided operational emissions (from grid electricity), operational and maintenance emissions (inverter manufacture and replacement) and any end-of-life emissions (here assumed to be negligible). These are determined using process analysis and electricity emissions inventory data. Life-cycle emissions are divided by total energy produced to give emissions intensities per unit of electricity produced. The required subsidies are then divided by these emissions intensities to give MACs; Table 8.17 gives the results of these calculations. The MACs, which are between 1794 and 2559 €/tCO$_2$-eq are significantly greater that the maximum cost of carbon traded on the EU ETS of approximately 35 €/tCO$_2$ (Figure 8.2), which in this case is taken to represent the opportunity cost to society.

A benefit/cost ratio for each PV system can be simply calculated by dividing the required subsidy by the positive externalities (health and environmental) associated

		PV system size (kWp)					
		0.47	1.41	1.72	2.82	4.23	5.64
Costs	(€/kWh)	0.87	0.69	0.67	0.63	0.61	0.61
Benefits (health, emissions, etc.)	(€/kWh)	0.04	0.04	0.04	0.04	0.04	0.04
Benefit/Cost Ratio		*0.04*	*0.05*	*0.06*	*0.06*	*0.06*	*0.06*

Table 8.18 Benefit/cost ratios for the different PV systems.

with avoiding the production of electricity using fossil-fuels, here taken to be 3.75 c/kWh. It can be seen (Table 8.18) that all of the ratios are very much less than one, indicating that the costs to society greatly outweigh the benefits.

In summary, the PV option analysed results in an LCOE almost 8–11 times greater than that produced using a mix of conventional technologies, has a MAC over 50 times greater than historic maximum carbon prices and has BCR considerably less than one. On these measures, a policy promoting the technology cannot be justified. It is highly unlikely that a sensitivity analysis would alter this conclusion.

References

Baumert, K., Herzog, T. and Pershing, J. (2005) Navigating the Numbers – Greenhouse Gas Data and International Climate Policy. World Resources Institute, Washington, DC.

Crawford, R. (2008) Validation of a hybrid life-cycle inventory analysis method. *Journal of Environmental Management*, **88**, 496–506.

Crawford, R. (2011) *Life Cycle Assessment in the Built Environment*. Spon Press, New York.

EC (2003) External Costs – Research results on socio-environmental damages due to electricity and transport (EUR 20198). European Commission, Brussels. [Online] http://www.externe.info/externpr.pdf (accessed 28 February 2012).

EIA (2011) Short-Term Energy Outlook. [Online] http://205.254.135.24/steo/archives/jul11.pdf (accessed 4 October 2011).

ESTTP (2007) Solar Heating and Cooling for a Sustainable Energy Future in Europe. ESTTP, Brussels(www.esttp.org).

European Climate Exchange (2012) EU ETS future contract prices 2005–2009. [Online] http://www.eea.europa.eu/data-and-maps/figures/eu-ets-future-contract-prices-200520132009 (accessed 28 February 2012).

EWEA (2005) *Wind Energy – The Facts, Volume 2, Costs and Prices*. European Wind Energy Association, Brussels.

EWEA (2011) Pure Power Wind energy targets for 2020 and 2030. European Wind Energy Association, Brussels (http://www.ewea.org).

IEA (2008) Energy Technology Perspectives – Strategies and Scenarios to 2050. OECD/IEA, Paris.

Forster, P., Ramaswamy, V.,Artaxo, P. *et al.* (2007) Changes in Atmospheric Constituents and in Radiative Forcing, in Climate Change 2007: The Physical Science Basis. Contribution of Working Group I to the Fourth Assessment Report of the Intergovernmental Panel on Climate Change (eds S. Solomon, D. Qin, M. Manning *et al.*). Cambridge University Press, Cambridge, UK, and New York, NY.

IEA (2010) Technology Roadmap – Solar photovoltaic energy. OECD/IEA, Paris.

IPCC (2011) Summary for Policymakers. In: *IPCC Special Report on Renewable Energy Sources and Climate Change Mitigation (SRREN)*. Intergovernmental Panel on Climate Change (IPCC), Geneva.

JRC (2009) 2009 Technology Map of the European Strategic Energy Technology Plan (SET-Plan); Part – I: Technology Descriptions. JRC-SETIS Work Group, JRC Scientific and Technical Reports, European Commission.

Mott McDonald (2010) UK Electricity Generation Costs Update, June 2010.Mott MacDonald, Brighton, UK. [online] http://www.decc.gov.uk/assets/decc/statistics/projections/71-uk-electricity-generation-costs-update-.pdf (accessed 28 February 2012).

Point Carbon (2010) Carbon 2010.Carbon Market Insights 2010 conference, 2–4 March 2010, Amsterdam (www.pointcarbon.com).

REN21 (2011) Renewables 2011 Global Status Report. REN21 Secretariat, Paris (http://www.ren21.net).

UNEP (2011) Energy Technology Factsheet – Cogeneration. United Nations Environment Programme, Paris.

Yohe, G.W. *et al.* (2007) Perspectives on climate change and sustainability. In: M.L. Parry *et al.* (eds) *Climate Change 2007: Impacts, Adaptation and Vulnerability. Contribution of Working Group II to the Fourth Assessment Report of the Intergovernmental Panel on Climate Change*. Cambridge University Press, UK.

Chapter 9

Value for Money in Construction

9.1 Definition of Value for Money

Value for Money is a term used to assess whether or not an organisation has obtained the maximum benefit from the goods and services it both acquires and provides within the resources available to it. Some elements of the assessment may be subjective and, therefore, difficult to measure. As a result, judgement is required when considering whether Value for Money has been acceptably achieved or not. It measures not only the cost of goods and services, but also takes into account a mix of quality, cost, fitness for purpose and timeliness. Taken together, they enable stakeholders to judge whether good value has been achieved.

Obtaining Value for Money can be described in terms of the 'three Es' – economy, efficiency and effectiveness. These can be defined as follows:

- Economy relates to the assiduous use of resources so as to save on time/cost/effort.
- Efficiency relates to the delivery of the same level of resources for less time/cost/effort.
- Effectiveness relates to the delivery of a better service for the same cost/time/effort.

The achievement of Value for Money within an organisation requires the existence of a culture or attitude within it which seeks continuous improvement. There is a commonly held misconception that Value for Money within infrastructure projects can be achieved simply by building cost effectively. However, if the project fails to provide the necessary level of performance, the anticipated benefits will not materialise. Value for Money requires that the needs of the end users be taken into consideration as much as the requirements of those providing the finance for project. Clearly defining the means by which the project will meet the requirements of the end users can be a difficult process. Investors and those third parties with a legitimate

Engineering Project Appraisal: The Evaluation of Alternative Development Schemes, Second Edition.
Martin Rogers and Aidan Duffy.
© 2012 John Wiley & Sons, Ltd. Published 2012 by John Wiley & Sons, Ltd.

interest in the project may have quite a different perspective on the purpose of the project to those who may ultimately use the facility. A clear and unambiguous set of project objectives will help minimise the prospect of any confusion in this regard.

The main benefits of promoting the principles of Value for Money include the clarification of project objectives within the organisation promoting and financing the project, minimising the risk of an activity failing to deliver the intended outcome, on time and within budget, increased openness and transparency within the organisation by publicly demonstrating a commitment to Value for Money, and being seen to comply with statutes and regulations, thus reducing the risk of failure to identify and comply with such requirements.

9.2   Defining Value for Money in the context of a construction project

The prime objective of Value for Money in this context is that the completed construction project achieves the optimum combination of whole-life cost and quality so as to meet the promoter's requirements, with quality referring to a number of relevant factors including functionality, durability, long-term adaptability and maintenance. It is important to the construction process as it allows project promoters and managers to have confidence in addressing subsequent questioning of the decision making processes that formed part of the design and construction processes. Thus, when internal or external audit is performed, Value for Money will allow it to be readily demonstrated that resources were used in an optimum way, that best advice available was obtained and implemented and that risks were managed properly and that informed judgements were made.

9.3   Achieving Value for Money during construction

The Value for Money framework, as detailed within this chapter and published by the Treasury Public Enquiry Unit (HM Treasury, 1997), incorporates a number of management tools that provide a structured approach which should be used in order to plan and manage the project from the start. The framework ensures that Value for Money is incorporated within the project by:

- Defining the project carefully in order to meet the user needs and to ensure that sufficient time and resources are allocated to enable the project to be properly planned.
- Managing and assessing all possible risks associated with different procurement methods, and, where appropriate, making recommendations to the project promoter.
- Integrating value, risk and cost management methods within the project management processes.
- Taking long-term sustainability and whole-life costing issues into consideration, recognising the need to maintain, repair and replace/dispose of components within the construction process.
- Using teamwork and partnership to minimise waste and the potential for conflict.

- Seeking to promote the integration of design and construction.
- Appointing consultants and contractors on the basis of Value for Money rather than solely on the basis of lowest tender.

It should be noted that the above framework may require modification to suit the needs and requirements of different developers/promoters and procurement strategies.

This framework is described in greater detail within Section 9.6. Before we describe it in detail, however, two concepts, 'whole-life costing' and 'milestones', are introduced. Both are of central importance to the Value for Money framework.

9.4   Whole-life costing

The concept of whole-life costing is central to the concept of Value for Money. The whole-life costs of an infrastructure project comprise the costs of acquiring/constructing it, together with the costs of operating and maintaining it throughout its useful life until its disposal/decommissioning. Long-term costs over the useful life of a project are more reliable indicators of Value for Money than the initial construction costs, as money spent on good design can be recouped many times over within the construction and maintenance outlays. Higher costs at the design and construction stage may be necessitated so as to achieve significant savings over the life of the project, with an integrated approach to design, construction, operation and maintenance improving its design quality, sustainability and buildability, driving out waste, reducing cost and addressing the sustainability of the project. Addressing these issues in a meaningful way will positively impact on issues such as the direct costs of energy usage and will ensure that the chosen design results in the convenient, cost effective and safe operation and maintenance of the facility.

9.4.1   *Relationship between Value for Money framework and whole-life decision making*

Parties within the supply chain for a construction project need to possess reliable data on the operational and maintenance costs of their products. The Value for Money framework seeks to integrate the design and construction processes, with the design team taking responsibility for the cost and quality implications of their design, with input from those who will be in charge of the operation and maintenance of the facility seen as essential. The framework should also involve the design team in the process of advising how the chosen design will affect cost, health and safety during both the construction and operational phases and of detailing how the design will support the speed of construction and the operational efficiency of the completed project. Designers operating within the framework must take account of the needs of the end users of the facility to avoid costly changes at a later stage in the delivery of the project, must make the sustainability of the completed facility a priority, taking into consideration its whole-life costs, and must operate a regime where continuous improvement can be demonstrated.

9.4.2   Baseline costs vs. actual performance over 'whole-life'

It is imperative that the expected operational running costs of the proposed project be established over its entire life and compared with those incurred by similar infrastructure projects. This requires that quantified estimates of running, maintenance and other associated costs, such as disposal/decommissioning, be produced. Net present value should be used as a basis for deriving a worth for such estimates. Operational running costs should then be compared with those incurred at other comparable facilities presently in use. If the proposed costs are found to be higher, justifications must be put forward. This process of comparing the baseline capital and operational costs of the project with benchmarked costs for other similar facilities should help designers deliver better value, should allow higher initial capital costs to be incurred in situations where subsequent reduced running costs will result in net savings – thus helping to reduce predicted whole-life costs without reducing quality of value – by using value engineering during the design process. (Value engineering works by enhancing whole-life value rather than by squeezing profit margins and/or initial construction costs.)

9.5   The concept of 'milestones'

The Value for Money framework introduces the concept of 'approval milestones'. At each gateway or milestone, the chosen Value for Money approach to the delivery of the construction project in question can be confirmed and validated independently of those directly responsible for its project management.

They occur at key planning stages in order to ensure that risks are being managed appropriately and that the project remains affordable. The staged approval process helps ensure that Value for Money is achieved at each of these milestones. Each gateway approval requires a Value for Money review, a financial review and a review of the project delivery management systems.

9.5.1   Value for Money review

This assures the promoter that the project provides Value for Money by ensuring that both the project's objectives and the project brief meet the promoter's needs, that the risks associated with the project have been properly identified, evaluated, allocated in a realistic fashion and are being actively and effectively managed, that all design, construction, procurement and funding options have been properly evaluated and the recommended options fully justified, and that the chosen design takes fully into account the maintenance, operating and disposal costs of the project, in order to provide a Value for Money whole-life solution.

9.5.2   Financial review

This review ensures that the latest estimate for the project is compared with the previously approved budget, with a fully reasoned justification accompanying any reported budget overrun, that the latest estimate comprises a baseline estimate plus an allowance for risk, that the risk allowance itself relates to identified risks only and is not an assumed contingency provision, and that funds are available to meet the planned expenditures up to the point of the next milestone.

This review must ascertain that the project remains affordable and that the appropriate capital, life cycle and whole-life cost management and reporting procedures are in place for the project and are being adhered to by the project management staff.

This review should, in the normal course of events, be carried out by the project promoter's financial team who should report directly to the promoter.

9.5.3   Project management delivery systems review

This review ensures that the appropriate management structure has been put in place and that individuals have been appointed to manage the project having received the recommended training and having been adjudged suitably competent for their respective roles within the project management framework.

The review must also ensure that the appropriate quality, cost, time and change controls are in place.

These processes should normally be carried out by the project management team which may wish to involve the internal audit team from within the sponsoring organisation. Their involvement at such approved milestone points within the project will provide the promoter and his senior management team with an assurance that the project is progressing satisfactorily. If this is not the case, then necessary corrective action can be identified. The involvement of internal audit also allows such assurances to be obtained independently of the project management team.

The project advances to the next stage once approval is obtained from the project promoter.

9.6   Detailed description of the Value for Money framework

The Value for Money framework can be described as a set of milestones at or around which reviews must be undertaken. These milestones normally occur when the project options are assessed and a preferred one chosen, immediately prior to the competitive procurement process, immediately prior to the award of the contract, once the asset is ready for delivery and after service delivery as part of an operational review. The Value for Money framework involves execution of the following activities:

- Value Management Studies
- Risk Management Studies
- Project Execution Plans
- Control Procedures
- Project Reports

These are the main tools which permit project managers to navigate their way from the initial conceptualisation of the project to its final delivery.

A diagrammatic representation of the Value for Money Framework is detailed in Figure 9.1, with details of the milestones within it given in Figure 9.2.

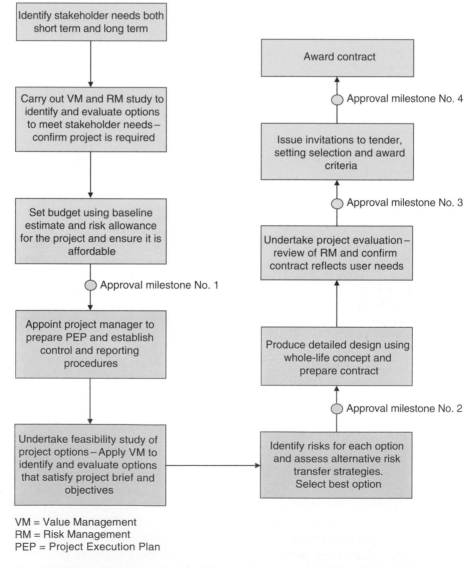

VM = Value Management
RM = Risk Management
PEP = Project Execution Plan

Figure 9.1 Diagrammatic representation of Value for Money framework with approval milestones.

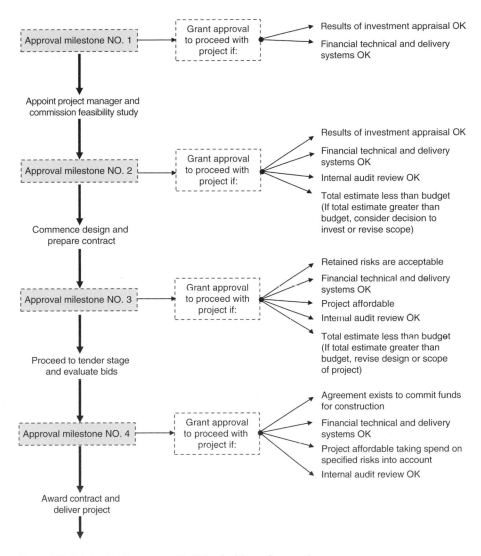

Figure 9.2 Details of milestones within Value for Money framework.

9.6.1 Value Management

Value Management is a strategic approach to the achievement of maximum value for money in a project which is consistent with the objectives of the sponsoring individual/organisation. It requires a structured team-based approach to problem solving which can readily be applied to the setting of objectives, concept, design and construction stages as well as the continuing management of the project. Value Management helps developers ensure that their investment in construction produces valuable assets which are cost effective to construct, use and maintain. It helps both to define what 'value' means to a developer when meeting a perceived need and to deliver that value via the design and construction process.

The focus of Value Management is on obtaining value for money, defined as the optimum combination of whole-life cost and quality to meet the requirements of the customer. Whole-life costing assesses the cost of an asset over its useful life, taking account of initial capital costs, finance costs, operational costs, maintenance costs and replacement/disposal costs at the end of its useful life. Discounting is employed to bring all future costs and benefits to present day values. Value Management is a process used to review all aspects of the project against consumer requirements on a continuing basis.

Value Management involves project participants working to achieve the optimum value for money solution, with value being appraised through careful analysis of function, with the objective of identifying the most cost effective way of achieving key functions within the delivery of the project.

Value Management can achieve a better understanding of client needs, can help define the objectives and specific needs of the client in simple clear and easily understandable terms, can affect the full consideration of all feasible project options, can provide optimum value for money solutions and can provide an effective mechanism for team building.

The decision to construct an infrastructure project is a significant one, involving substantial investment of time, money and effort, producing a valuable asset positively affecting the life of those who own and use it. It is of central importance to attaining value for money through bringing about improvements in the overall value and performance of infrastructure projects. Value Management achieves this by identifying and evaluating the need for the project prior to making a financial commitment, by identifying and prioritising the most important project objectives and by ensuring that the key features of the design maximise the effectiveness of the project in terms of cost, quality and meeting the developer's ultimate needs. Eliminating unnecessary costs is vital to Value Management's success in this regard.

As stated in CIRIA's 1996 document *Value Management in Construction: A Client's Guide*, some construction professionals argue that Value Management 'is what they do anyway'. In some situations this may well be true. While competent construction professionals may well achieve value for money some of the time, the objective must be to achieve it all of the time. Value Management, with its planned and rigorous approach, can do this.

Using Value Management, developers can have confidence that value for money can be achieved. There are a number of key points, termed 'milestones' or 'gateways', within every infrastructure project, of the type described in Section 9.5, at which the developer or his appointed staff must take important decisions. Value Management ensures that these decisions are taken in a manner that is:

- rational;
- explicit; and
- accountable.

Rational decisions require that the nature of the problem which the project is addressing is understood, that the decisions are made in the light of agreed objectives, that different well-structured project options for achieving the agreed objectives are considered, that the decisions are made on the basis of the best available data and that the decision makers comprise experts in each of the relevant fields.

In addition to decisions being rational, they must also be made in such a manner that is both clear and understandable. The structure provided by Value Management helps ensure that decision process is explicit, allowing key stakeholders to participate in a meaningful way in the making of decisions, and resulting in a heightened level of commitment to the project on their part.

Accountable decision making helps ensure that all decisions taken are seen to be rational, providing stakeholders with the ability to provide explanations subsequently for any of their pronouncements. A properly planned and documented Value Management process will provide evidence of rational and explicit decision making when judged against value for money criteria relevant to the infrastructure project in question.

The appointment of a value manager by the project's promoter is seen as vital in assisting the development and implementation of a value management plan. A value manager should have thorough knowledge of, and experience in, value management and its associated activities, and should show competence in a range of management skills, including the ability to analyse complex problems, to seek innovative solutions, to challenge assumptions regarding needs and approaches, to motivate the project participants towards the achievement of project objectives, to communicate effectively with fellow team members and to lead in a respectful and confident manner.

To conclude, in order to maximise the effectiveness of Value Management, senior management must be entirely committed to its introduction and implementation, and the developer must ensure it addresses key objectives, identifies and appoints experienced and effective staff and ensures that design team members are fully conversant with their role in the Value Management process.

Many of the economic and non-economic tools of project appraisal detailed within this text are vitally important to an effective value manager.

9.6.2   *Risk Management*

The objective of Risk Management is to help ensure that risks are identified at the outset of the project, their potential impacts are allowed for, and, where possible, these risks are minimised.

Risk Management is thus a systematic process involving an *identification* process to determine what the risks are, an *assessment* process to determine the likelihood of the risks occurring and their potential impacts, a *response* process where action is taken to manage the risks and a monitoring and control process which identifies options for dealing with risks and their impacts, and monitors the implementation of the preferred course of action.

Risk identification

Risk identification is the key to successful Risk Management. Engineering techniques and management practice should be applied in order to determine where things might go wrong. It is imperative when identifying risks to distinguish between the source of a given risk and its impact.

Risk assessment

Risk assessment enables the likelihood of occurrence and the potential impacts on project overrun to be both understood and quantified. Techniques for risk assessment can be both qualitative and quantitative.

A qualitative risk assessment comprises a descriptive written assessment, which must include details of the stages within the project when it is likely to occur, the elements of the project that could be affected, the factors that could cause it to occur, any relationship or interdependency on other risks, the likelihood of it occurring and how it is likely to affect the project.

A quantitative risk assessment assigns a numerical probability to the likelihood of a risk occurring. A typical scale is as follows:

- 0   = impossible for risk to occur
- 0.5 = 50% chance of risk occurring
- 1.0 = risk will definitely occur

The possible consequences of a risk occurring can be quantified in terms of cost, time or performance; that is, the additional cost arising relative to the baseline estimate for the project, the additional time beyond the baseline estimate of completion date for the project and the extent to which the consequent under-performance results in the project failing to meet the user requirements for standards and performance.

Risk response

Risk response must be considered only after the possible causes and effects of the risk in question has been considered and fully understood. The response will consist of one or more of the following project management activities:

- Avoidance of the risk in situations – where risks have such serious consequences on the outcome of the project that they become totally unacceptable in terms of the attainment of the project's objectives, measures might include a review of the project's objectives, in some cases resulting in a replacement project or even its cancellation.
- Reduction (up to complete elimination of) the risk – actions can include the re-design of the project, the use of different equipment or materials, the use of different methods of construction to avoid inherently risky construction

techniques, changing the project execution plan to package the work content in a different manner and changing the contract strategy to allocate the risk between participants within the projects differently.

- Transferring of the risk elsewhere – in cases where acceptance of a risk would not achieve Value for Money, it might be possible to transfer it to another party who would then be responsible for the consequences of the risk if it occurs. The objective of doing this is thus to transfer risk to another party in a better position to control it. Possible transfers are from client to design consultant, from client to contractor, from contractor to subcontractor, from client to an insurer by means of an insurance policy or from contractor or subcontractor to a bank by means of a warranty, bond or guarantee. Risk transfer usually involves the payment of a premium, resulting in an increase in baseline cost and a reduction in risk allowance. The provision of Value for Money requires that the total cost of the risk reduces by more than the cost of the premium.
- Retention of the risk – those not avoided or transferred are retained by the client and must be managed in order to minimise their potential negative impact on the project.

It is best practice to allocate risks to those best placed to manage and cope with them.

9.6.3   Project execution plan

This is the central management document which relates to the project strategy, organisation, control procedures, responsibilities and the working relationship between the client and the project manager. It comprises a formal statement of the users' needs, project brief and the strategy agreed with the project manager for attaining them. The scope of the plan will depend on the size and type of the project being considered. It is, in essence, a document which will be regularly updated in real-time and used by all parties both as a means of communicating information and as a monitoring, control and performance monitoring tool.

The project sponsor is primarily responsible for the preparation of the project execution plan. (The project sponsor is the client representative who acts as a single focal point of contact with the project manager, representing the interests of the client organisation/project developer and, in particular, the end users of the built asset.) The document is divided into two general areas, one containing matters relating to organisation and responsibilities within the client body and one relating to the project execution team. The project sponsor will develop those elements relating to the client organisation and determine and define the roles and responsibilities of the key personnel involved. The project manager, in contrast, will be responsible for developing those elements relating to the project team's activities and the project execution strategy.

The project sponsor must be convinced that the project execution plan represents a viable and realistic plan for executing the project and realising its core objectives. The sponsor is required to review the document in detail with all parties within the project in order to be sure that the plan as a whole and the responsibilities that flow from it are structured in such a way that all parties have the capability and resources to discharge their responsibilities.

A checklist for a construction project execution plan can be a useful tool in ensuring its effective implementation. It can comprise the following headings:

- General description – description and location of project along with progress to date.
- Project objectives – described in terms of user needs.
- Project brief – purpose of project, schedule of accommodation, relevant standards, operational requirements, equipment and special services, maintenance requirements, environmental needs, disposal criteria and statutory requirements.
- Constraints – external factors limiting the design and construction of the plan, including planning and site conditions together with the availability of utilities, and internal factors, arising from decisions or policies of the client organisation, such as confidentiality, safety standards or procurement policies.
- Cost control – Current capital and whole-life cost estimates, budgetary control and risk allowance.
- Programme – Overall timescale for project, together with pre-construction, construction and occupation programmes and milestone activities.
- Change control – Proposed and approved changes.
- Prioritisation – Cost or time, cost or quality, time or quality.
- Internal management – Personnel, responsibilities, authority, delegation.
- Procurement – Procurement route, form of contract and performance assurance mechanism.
- Roles and responsibilities – Statement detailing the procurement route, the roles and responsibilities of external parties, to be incorporated into their contract. It will cover the project manager, designers, the cost consultant and other specialist consultants, contractors and suppliers.
- Co-ordination – Communications, cost, change and quality controls, health and safety procedures, commissioning procedures and reports.
- Occupation – Facilities management, maintenance, commissioning, staff recruitment and training, programme, costs.

9.6.4   Control procedures

One of the main causes of cost overruns and the non-achievement of Value for Money on construction projects are changes to design, particularly after the award of the contract. Such changes invariably arise from unclear or ambiguous project definition, inadequate time spent in project planning, risk analysis and risk management, or due to a change in circumstances within which the project is taking place. In many cases, the consequences of changes during the construction stage can be considerably greater than the perceived direct impacts of these changes.

Changed can best be handled through sensible project planning and review. Where a possibility for change is identified, it should be regarded as a project risk and dealt with within the risk management plan. A robust change control procedure, including Value for Money criteria, should be in place in order to evaluate and manage change when it occurs within the project.

The necessity for change within a construction project can be minimised by ensuring that the project brief is comprehensive and agreed by all major parties, through taking all proposed legislation into account and having early talks with external authorities to anticipate their requirements, by undertaking comprehensive site investigations and ensuring that all designs are as fully developed and coordinated as required by the construction project before the contract is signed, and through good project management, forward planning and the identification and management of risks.

Before approval is given for a change, a change control procedure should consider factors such as the reason for the change, its source, the whole-life cost, time and procurement consequences of the change, the risks associated with the change and its impacts, properly evaluated alternatives to the proposed change, proposals for avoiding or mitigating time overruns, the source of funding of any cost overrun and the impact of planning constraints, building warrant and other statutory approvals.

Lack of time control can also result in the non-achievement of Value for Money on construction projects. The programming of activities necessary to complete a construction project is normally carried out using computer planning software. The client or those acting on the client's behalf must insist that the final programme on which decisions are made should be simple, straightforward and easily deciphered. It must identify clearly those tasks which lie on the critical path. This process is closely related to cost control and should include: a time budget, indicating the overall project duration as fixed either by specific constraints or by the contract strategy and identified as a key parameter for the management of the project; a time plan indicating the division of time into interlinked time allowances for readily identifiable activities with definable starting and finishing points; and a time checking procedure which will monitor closely the actual time spent on each activity against the allowance within the time plan, reporting any divergence as soon as it is identified.

Quality control is also an integral part of Value for Money. It determines the final quality of the project being constructed. It requires a clear specification of the testing and verification regime required to provide assurance of compliance with the specification. Confidence in the quality control activities carried out by the contractor is essential.

9.6.5   Project reports

Project reports are recurring statements issued by the project manager to those backing the project financially. It contains narrative from the project manager backed up by reports from the design team and the contractor. Its aim is to summarize the current status of the project, detail the significant problems that need to be resolved and the measures that will be required to do so.

The project report should comprise the following:

- An executive summary, not exceeding one page, detailing the current status of the project, list the key outstanding issues associated with it needing resolution.
- The main text of the report giving detail to the issues listed within the summary.
- Appendices containing data backing up the information within the main text.

The report provides a continuing review of the project to date and provides a trigger for those involved with the management of the project to resolve any problems or issues outstanding at that juncture.

The main text of the report addresses quality, time and cost issues pertaining at the time of its writing. In terms of quality, it states how the project, as designed and constructed, meets the requirements of the project execution plan; it lists potential areas of non-compliance and the different options available to rectify them. In terms of time, it compares actual progress against programme, listing the areas where delays are occurring and detailing the steps being taken to remedy them. In terms of cost, it compares total actual costs incurred with those budgeted for, highlighting the areas of divergence from budget, the reasons for this and measures being taken to get it back on track. The section on cost should also state the expenditure on risk allowance, a comparison between actual and predicted cash flow, again explaining the reasons for any divergence and actions being taken to correct them, and a statement of contract claims submitted up to that point in the project, with an estimate of the potential financial implications of these claims. All supporting data relating to quality, time and cost issues associated with the project should be contained within the report's appendices.

9.7   Value for Money and design

Whole-life Value for Money requires that the performance of the constructed project once in service should be of central importance to the designer when the project is being planned and devised. Hence, the design with the lowest capital cost may not necessarily be the preferred option to proceed with. Value for Money involves more than ensuring that a project is delivered on time and within budget. Good design should result in the delivery of a project which contributes to the environment, delivers a range of social and economic benefits and is adaptable to accommodate any one of a number of future possible uses. Good design aims to deliver a constructed project which meets the requirements of all stakeholders, especially the end users.

High-quality design requires observing a set of objectively set principles that establish if a construction project works for all potential end users. A well designed infrastructure project functions well in use, is built to last, is designed in a way that will permit its completion to specification, on budget and within the permitted time, and is respectful of its context.

Good quality design is of fundamental importance to the construction of a high-quality infrastructure project, generating value to the economy. Delivering value involves maximising the requirements of different stakeholders while also keeping the use of resources to a minimum. Value thus becomes an exercise in maximising the benefits flowing from the project while minimising the resources needed to complete it.

The UK National Audit Office (2004) listed the following six value drivers which can form the basis for ensuring good design within an infrastructure project:

- Maximise business effectiveness
- Ensure effective project management and delivery
- Endeavour to achieve required financial performance
- Maximise positive impacts on locality
- Minimise operation and maintenance costs
- Ensure compliance with third party requirements

These are applicable to all stages of the project from initial conception to final delivery. It is very important for good design that each value driver has a metric which makes it quantifiable. Some will be monetary, others non-monetary. Taken together, they attempt to balance the financial and functional requirements of an infrastructure project in order to deliver value for money.

9.7.1   Maximising business effectiveness

Infrastructure projects, once completed, allow certain activities to be conducted within them. An office building allows people to conduct commercial activities related to their business; a toll road allows cars to progress speedily towards their destination; while a sports stadium allows fans of the particular sport to support their local team. Good design should allow easy accessibility and encourage efficient use. The finished project should provide a comfortable, safe and healthy environment and should be capable of accommodating any changes in activity likely to occur over its economic life. Facility management techniques should be in place to monitor the finished project's effectiveness to allow for continuous improvement to meet changing needs.

Metrics for measuring business effectiveness will vary depending on the nature of the infrastructure project. Examples of possible metrics include:

- Revenue generated
- Footfall (number of visitors)
- Operating costs
- Space utilisation ratio
- Customer satisfaction surveys

9.7.2   Delivering effective project management

To deliver a project which maximises value and minimises waste, a project team should be selected not just on a cost basis but on the basis that it possesses the required technical competence to complete an infrastructure project that is well designed and constructed. The team should display significant levels of integration, coordination and communication skills covering all facets of a complex project, becoming involved at the earliest possible stage of its inception. It is also important that the team communicates effectively with all the relevant stakeholders, involves all contractors/subcontractors and consultants/subconsultants at the appropriate times during the design and construction phases and should adhere to recommended best practice within the construction industry. The project team must develop the project execution plan detailing the requirements of the finished project (given within the output specification), the organisation of the project team and the plan for executing the project.

The effectiveness of the different processes used within the construction process should be assessed using industry standard key performance indicators, with the project reviewed at certain milestone points during the design and construction process.

Possible metrics for measuring effective project management other than standard key performance indicators include some quantitative measure of the degree or level of success of any milestone review carried out during the design/construction process.

9.7.3   *Achieving the required financial performance*

This value driver addresses the affordability of the project and the optimisation of the net benefits of it to those using the facility. The budgeted cost agreed for the project must constitute an optimum balance between its required functionality and the cost of delivering the finished product. Given the financial outlay required to complete the project, the business case for it must compute how much the organisation can afford to invest in its construction, operation, maintenance and ultimate decommissioning, expressing this figure in present value terms. Regardless of cost, the requirement that the stated design deliver the required functionality is central to providing value for money. A project that costs more to construct or maintain than originally budgeted will adversely impact on its fundamental economic viability. Both the capital and whole-life budgets as stated within the business case for the project must be both affordable and sustainable over the entire economic life of the project.

The developer should undertake an investment appraisal at the outset of the project, identifying the different project options and comparing them on the basis of their capital and whole-life costs and benefits. The lowest cost option will not necessarily provide best value for money. Good design of a particular option may generate a wider range of whole-life benefits which may more than compensate for higher capital and whole-life costs.

Metrics for measuring the achievement of the required financial performance include the overall capital cost of the project, the total net present value (net present value of benefits minus net present value of costs) taking account of both capital and whole-life benefits/costs, and the payback period. Numerous worked examples of these techniques are given in earlier chapters of this text.

9.7.4   Achieving a positive impact on the local environment

The impact of the completed project on the surrounding area and those that use or visit it is of significant importance in maximising its value. Good design creates space that is easily accessible and possesses a distinctive character and creates the potential to have a positive economic and social impact on the locality. The mere act of renewing or regenerating an outdated asset may, in itself, have a positive impact on the local community, with the perceived willingness to invest giving a vote of confidence in the local economy.

A number of design measures can be taken to ensure that an infrastructure project impacts positively on the local environment and delivers benefits to the community. Design should reflect the patterns of development that are distinctive locally; where appropriate it should promote/incorporate public spaces that are safe and uncluttered. It should enable ease of movement, creating places that are both accessible and connected to each other, and should promote adaptability, creating development that can change with changing social, economic and technological conditions.

In summary, the project should give a positive impression, demonstrating that there is a clear vision at the basis of its design.

Metrics for measuring the achievement of a positive impact on the local environment include the results of public/private surveys, the winning of design awards from recognised professional bodies and the recorded views of the planning authorities.

9.7.5   Minimising operational maintenance costs and environmental impact

This important value driver not only measures the degree to which the design has eased the day-to-day maintenance and operation of the finished project but also measures impact on the natural environment and the extent to which sustainability has been incorporated within it, particularly in the use of sustainable construction materials.

Ease of day-to-day maintenance and operation results from a design which delivers good layout, structure and engineering systems and minimises both the energy consumed by the lighting, heating and ventilation systems and the carbon footprint of the facility. The requirement for continuing maintenance should be minimised through use of high-quality finishes, maximising the durability of all components, ensuring that they can be easily and efficiently replaced when necessary.

The methods of construction and the materials used should be carefully thought through by the design team, with the minimisation of both waste and energy during the construction process a high priority. The ability to recycle all construction materials used, once the end of the project's useful life is reached, should also be considered. The maintenance manager should be encouraged by the design team to use monitoring techniques which will actively manage the facility's performance during its life, thus ensuring waste is reduced, sustainability is improved and functional effectiveness monitored at all times.

Metrics for measuring the minimisation of operational and maintenance costs and environmental impact include annual heating and ventilation costs, annual cleaning costs, results of independent environmental sustainability assessments and level of integration of maintenance management team with the initial briefing process for the project.

9.7.6   Ensuring compliance with third-party requirements

Stakeholders comprise not only those responsible for initiating the project and bringing it to fruition, but encompass all those with an interest in the project or having any interest in influencing it. Good design requires that all such parties be consulted at the appropriate time. The views and concerns of Statutory Authorities with a direct influence in the planning, design, construction and operation of the facility must also be allowed for. To achieve this, the project team must develop a strong working relationship with the Statutory Authorities, while allowing consultation with other third parties within the community in order to gauge the project's impact on the locality.

Metrics for measuring compliance with third party requirements include the ease with which planning consent was achieved and public surveys that assess local support for the facility.

9.7.7   Assessing the relative importance of the different value drivers

Not all the above value drivers are of equal importance to a specific project. It is important for those managing a project to be able to assess the relative importance of the different value drivers on a case-by-case basis.

To do this, the project team should establish the project specific value drivers, deriving the relative importance weightings of the different value drivers in consultation with the client and end users, using these limits or boundaries to derive a 1 to 10 scale. The team can then agree a performance level for each value driver on this scale from 1 to 10. These individual scores are each multiplied by their importance weighting to compute an overall value index for the project, allowing the performance of the project against all value drivers to be assessed.

Table 9.1 details a worked example illustrating how the value for money index is computed for a hypothetical toll road project. A maximum score of 1000 is achievable, with 850 considered excellent and less than 500 indicating the need for improvements into the future.

9.7.8   Concluding comments on achieving Value for Money through good design

The above six value drivers form the basis for ensuring good design within an infrastructure project. They will aid in maximising the impact of the project on the local

Table 9.1 Value for money index for a sample toll road project.

Value driver	Importance weighting	Metric	1	2	3	4	5	6	7	8	9	10	Weighted value score
Business effectiveness	30	Revenue per annum						X					30×6 Revenue = £6 m per annum
Score 1 = £1 m per annum, score 10 = £10 m per annum													
Effective project management delivery	15	Program review					X						15×5 50% satisfaction rating in review
Financial performance	25	Capital cost						X					25×6 Capital cost = £20 m NPV
Score 1 = £12 m per annum, score 10 = £30 m per annum													
Impact on local environment	5	Survey						X					5×6 60% satisfaction rating in survey
Maintenance operation and impact on environment	15	O and M costs								X			15×8 M and O costs = £1 m per annum
Score 1 = £0.5 m per annum, score 10 = £2 m per annum													
Third party requirements	10	Survey							X				10×7 70% satisfaction rating in survey
													625 TOTAL

environment, enhancing its engineering performance and ensuring that the finished product is of use to the maximum number of people. Design, if done well will have a disproportionately large impact on how the facility performs during its useful life.

9.8   Is there a conflict between Sustainability and Value for Money

In the context of the delivery of an infrastructure project, sustainability can be defined as simultaneously delivering economic, social and environmental outcomes for it. In practice it can involve a greater emphasis on the environmental implications of the design, construction and operation of the facility in addition to the more traditional focus on economic and social objectives. The defining feature of the concept of a sustainable project is a significant reduction in its environmental impacts. These can include measures to reduce energy consumption and emissions of carbon dioxide, minimise the use of resources such as water and construction materials, reduce the release of pollutants, promote sustainable travel choices such as public transport, cycling and walking, and maximise the use of sustainably sourced and recycled materials.

There is a perception that sustainable development cannot be justified on the basis of Value for Money, because of the following perceptions:

- the relatively high capital costs of sustainable options,
- the long payback periods involved,
- the difficulty of spending now to save later, given the budgetary pressures and the separation of capital and operational budgets.

In 2005, the UK National Audit Office investigated how sustainable project options were appraised, and found that Value for Money was not considered when establishing the business case for the project. Once projects were under way, however, in most cases both contractors and clients proposed suggestions for sustainable options in relation to individual aspects of the project. In general, the report noted that the extent to which Value for Money assessments were undertaken varied for the projects examined, with no explicit appraisals of sustainable options on a whole-life basis observed. The perception therefore persists that sustainable buildings do not deliver Value for Money because they cost more.

However, case studies undertaken by the National Audit Office (2007) indicated that the additional capital costs of sustainable building options were recouped during its life through savings in the cost of energy and water. However, the research indicated that, in certain situations, the extra capital cost of some sustainable building options could prevent their adoption, regardless of the potential benefits it would deliver during the project's operational phase. Furthermore, where sustainable options are not integral to the design process, it can be particularly easy to remove them when there is downward pressure on capital budgets.

If Sustainability and Value for Money are to exist side by side, the evaluation of different project options at the planning stage must demonstrate explicitly that a sustainably designed infrastructure project is capable of delivering tangible benefits. To do so, the evaluation process may need to consider wider impacts that are often ignored despite their potential to deliver large savings. For example, the benefits of reducing a project's greenhouse gas emissions tend not to be included in a standard cost–benefit analysis, even though the benefits to society of doing so can be readily estimated using the social cost of carbon. However, there are many social and environmental benefits for which there is no standard approach to their quantification; hence they tend to be excluded from tools of evaluation such as cost–benefit analysis. Reconciling Sustainability and Value for Money may require use of a wider range of decision making tools, such as some of those detailed within Chapter 11–15. Decision models which allow an option's attributes to be measured on different measurement scales will permit the social and environmental benefits of sustainable design to be assessed alongside the more readily measured economic benefits.

9.9   The role of better managed construction in delivering projects on time and within budget

Construction forms an important part of any national economy. Within the United Kingdom, for example, construction normally accounts for greater than 8% of the national GDP. Its end product is the enhancement of the built environment which impacts directly on society through the delivery of better public services in the health, education and transport sector, the promotion of social cohesion and the maintenance and enhancement of the natural environment. Well managed construction activities, implementing value for money practices as detailed within the preceding sections of this chapter, provide opportunities for public and private sector clients to deliver infrastructure projects efficiently with consequent social and economic benefits to all citizens. However, in the absence of such practices, the risks to value for money from inefficiencies, lack of safety, waste and the failure to deliver the project on time and within budget are substantial.

Working on the premise that well managed construction is integral to improved efficiency and delivery, the UK's National Audit Office (2001) identified the need to tackle inefficient working practices within the construction industry and emphasised the need for improvements in construction performance leading to value for money gains. Since 2001, therefore, the UK Office of Government Commerce has implemented a range of construction improvement initiatives aimed at improving construction delivery capability for Government sponsored projects.

The National Audit Office (2005) detailed the progress made between 2001 and late 2004 in improving construction performance on a number of fronts, including delivery of projects on time and to budget, and on value for money savings arising from improved cost predictability and reduction in costs.

The ability to deliver a construction project to budget and on time is one of the critical factors in bringing an infrastructure project to successful completion. The 2005 National Audit Office Report examined the cost and time predictability of construction projects across Central Government over the 1999 to 2004 time frame. In 1999, their analysis indicated 25% of projects delivered within budget and 34% of projects delivered on time. By late 2004, these statistics had improved to the point where 55% of projects were delivered within budget, with 63% delivered on time.

In relation to Value for Money savings arising from improved cost predictability and reduction in costs, the 2005 National Audit Office Report analysed 142 projects completed between April 2003 and December 2004 and concluded that, since 1999, the average level of overspending on projects had decreased from 6.5% to 4.1%.

While drawing robust conclusions from such statistics must always be done with care and caution, the above figures would indicate that establishing Value for Money procedures does enhance the probability that a project will be delivered on time and within budget. The 2005 National Audit Office Report notes that continued improvement in construction delivery performance can be achieved by:

- establishing effective construction programmes;
- developing and supporting well-focused and capable clients;

- basing both design and decision making on whole-life costing;
- using appropriate procurement and contracting strategies;
- working collaboratively through fully integrated teams;
- evaluating the project's performance.

9.9.1   Establishing effective construction programmes

The establishment of an effective construction programme necessitates that plans and programmes be streamlined using a coherent project management approach, so that an understanding exists of both the current and planned construction requirements. This exercise will allow an organisation to package projects in an efficient manner, with similar projects grouped together into one contract to reduce overall setup costs.

9.9.2   Developing and supporting well-focused and capable clients

This may involve supplying appropriate intelligent systems for good clients who may lack experience in the delivery of infrastructure projects, with providing professional expertise and helping exert management control over projects in order to minimise the risks of cost and time overruns.

9.9.3   Basing both design and decision making on whole-life costing

This requires the investment of more time and resources in the design of an infrastructure project prior to key decisions being made, ensuring all relevant stakeholders are involved, with proposals being subjected to independent challenge. The affordability of the project's running costs over its whole life should be ascertained as well as a wider assessment of its economic, social and environmental impacts.

9.9.4   Using appropriate procurement and contracting strategies

Clients must clearly understand the procurement strategy which is most appropriate to their circumstances and capabilities, only using suppliers who have an established track record in developing the skills of their workforce, and are committed to health and safety and sustainable development. Supply tenders must be awarded on the basis of clearly communicated evaluation criteria, using only contracts promoting collaborative working within the project team.

9.9.5   Working collaboratively through fully integrated teams

A shift in attitude is required to move from the hitherto traditional 'adversarial' method of managing an infrastructure project to a collaborative framework. Joint training involving both the client and contractors in areas such as project management techniques and procurement approaches can facilitate this cultural change. Clients working within an integrated team from the earliest possible stages of a contract are better able to identify, articulate and share the objectives of the infrastructure project. Once these objectives are identified and shared, the process of deciding and identifying the most cost effective design solutions can be expedited.

9.9.6   Evaluating the project's performance

This can be achieved through the creation of appropriate measures and targets reflecting progress towards the delivery of the project on time, within cost and to the required quality. Regular evaluations of the achievement of all key targets including those for cost and time predictability, whole-life costs and economic, social and environmental benefits must be performed. An assessment of the level of performance of all parties involved within the completion of the project should also be carried out.

References

CIRIA (1996) *Value Management in Construction: A Client's Guide*. Special Publication 129, CIRIA, London.

HM Treasury (1997) *Guidance Document No. 2, Value for money in construction projects*. Treasury Public Enquiry Unit, The Stationary Office, London, UK.

National Audit Office (2001) *Modernising construction*. Report by the Comptroller and Auditor General, House of Commons 87 Session 2000–2001, 11th January. The Stationary Office, London, UK.

National Audit Office (2004) *Getting value for money from construction projects through design: How auditors can help*. Prepared by Davis Langdon and Everest, Holborn, The Stationary Office, London, UK.

National Audit Office (2005) *Improving public services through better construction*. Report by the Comptroller and Auditor General, House of Commons 364-1 Session 2004–2005, 15th March. The Stationary Office, London, UK.

National Audit Office (2007) *Building for the future: Sustainable construction and refurbishment on the government estate*. Report by the Comptroller and Auditor General, House of Commons 324 Session 2006–2007, 20th April. The Stationary Office, London, UK.

Chapter 10

Other Economic Analysis Techniques

10.1 Introduction

A major shortcoming of the cost–benefit analysis framework is its inability to include within its framework all relevant benefits and disbenefits that cannot be quantified in monetary terms. Even in situations where it is feasible to make a realistic estimation of a proportion of the benefits and disbenefits, there will always be a significant number of attributes that cannot be expressed in monetary values. A number of methodologies exist which allow the non-monetary valuations of attributes to be assessed rigorously within a formal, structured framework. Many of these will be addressed in detail in Part II of the book. A number of these techniques go beyond the pure monetary evaluation of decision criteria while still having the framework of cost–benefit analysis (CBA) at their basis. Three of these are discussed in this chapter:

(1) Cost Effectiveness Analysis
(2) Planned Balance Sheet
(3) Goal Achievement Matrix

All three methods retain, albeit in the case of the last two to a very limited extent, the very narrow based structure of CBA, which enables only a limited range of impacts to be valued. They are, therefore, seen as techniques that seek to reform CBA rather than completely reject it. They are usually termed 'reformist' or 'cost–utility' techniques.

10.2 Cost effectiveness

Within this technique the expected benefits/consequences of a proposed project are quantified. Their estimation is not, however, done in terms of money values. Some measure of effectiveness is used instead of the monetary valuations of benefits and is set against its discounted cost.

Engineering Project Appraisal: The Evaluation of Alternative Development Schemes, Second Edition.
Martin Rogers and Aidan Duffy.
© 2012 John Wiley & Sons, Ltd. Published 2012 by John Wiley & Sons, Ltd.

Take, for example, a major urban centre where the transport planners wish to introduce a major new public transport system to the city centre district. Details are given in the Example 10.1. Before final approval is given, a number of alternative proposals must be considered. These include a 'do-nothing' option but also with a do-minimum' management option that involves relatively low cost improvements to the existing rather limited urban rail system (minor extensions, efficiency improvements etc.). This will be the lowest cost option and is used as the option against which all others are assessed. The benefit is measured in some physical unit, while the costs are expressed in monetary terms. In this case an appropriate measure of benefit for each alternative proposal would be the average cost per trip for every additional rail user over and above the level of the 'do-minimum' option. The cost effectiveness (C/E) of a given alternative, i, can thus be expressed as:

$$C/E = \frac{C_i - C_{do\text{-}minimum}}{P_i - P_{do\text{-}minimum}}$$ (10.1)

where C_i is the present value of the cost of option i, $C_{do\text{-}minimum}$ is present value of the cost of do-minimum option, P_i = is the number of passengers for option i, $P_{do\text{-}minimum}$ is the number of passengers for the do-minimum option.

Other measures of effectiveness for a given option might be the reduction in accident levels, savings in travel time or improvement in air quality resulting from it.

Example 10.1   Choosing an urban rail system using cost effectiveness analysis

Eight options are considered within the cost effectiveness evaluation of a proposed urban rail system. One is the 'do-nothing' option (Base), termed the baseline case, with another being the 'do-minimum' or 'management' option (Min), which involves minor upgrading and extension of the existing heavy rail lines in the city area. The remaining six 'live' options are defined as:

GB$_1$:   A limited guided bus option involving the construction of two new lines.
GB$_2$:   An extensive guided bus option, involving the construction of four new lines.
LRT$_1$:   A limited light rail option involving the construction of three suburban lines.
LRT$_2$:   An extensive light rail option involving the construction of three suburban lines and a major city centre ring line.
HR: A heavy rail option involving one extra line to the airport and one to the western suburbs of the city.
Met: A 'metro' option involving the construction of three suburban lines and one city centre line linking the east and west sides of the city

Contd

Example 10.1 Contd

Assume a discount rate of 8% and a 25-year design life. Table 10.1 indicates the total capital/operating cost for each option. These capital/operating costs are converted into equivalent annual costs using the capital recovery factor as follows:

$$
\begin{aligned}
\text{CRF} &= r(1+r)^N \div ((1+r)^N - 1) \\
&= 0.08(1.08)^{25} \div ((1.08)^{25} - 1) \\
&= 0.0937
\end{aligned}
\tag{10.2}
$$

Table 10.2 indicates the total annual cost of each option together with its average annual usage. The total annual costs are estimated by multiplying each of the values in Table 10.1 by the capital recovery factor calculated above.

The differences in cost and passenger usage for each project option are compared with those of the 'do-nothing' and 'do-minimum' options in Tables 10.3 and 10.4. It can be seen that the two light rail options have a ratio of change in annual cost to change in passenger usage that is less than unity in both Tables 10.3 and 10.4. This indicates that both are more cost effective than either the 'do-nothing' or 'do-minimum' options. Table 10.4 shows that the heavy rail option, with a ratio of less than unity, is more cost effective than the 'do-minimum' option, but marginally less cost effective than 'do-nothing'. None of the Metro or Guided Bus options can be justified compared with these two baseline alternatives. Thus, only three options would appear feasible on the basis of this analysis, with the second light rail option performing best, followed by the first light rail and heavy rail options. Of the non-feasible options, GB_2 performs worst, with GB_1 and Metro placed above it.

Table 10.1 Capital/operating costs of project options.

Project option	Total capital/operating cost (£)
Base	12 700 000
Min	37 500 000
GB_1	70 000 000
GB_2	80 000 000
LRT_1	96 000 000
LRT_2	110 000 000
HR	155 000 000
Met	208 000 000

Table 10.2 Cost per passenger for each rail option.

Project option	Total annual cost (£)	Total annual passenger use	Cost per passenger (£)
Base	1 189 721	1 090 000	1.09
Min	3 512 954	3 000 000	1.17
GB_1	6 557 514	5 650 000	1.16
GB_2	7 494 302	6 000 000	1.25
LRT_1	8 993 163	9 000 000	1.00
LRT_2	10 304 666	10 500 000	0.98
HR	14 520 211	14 250 000	1.02
Met	19 485 186	17 500 000	1.11

Project option	Change in annual cost (£)	Change in annual passenger use	Ratio (£/passenger)
Base	—	—	—
Min	2 323 233	1 910 000	1.216
GB$_1$	5 367 794	4 560 000	1.177
GB$_2$	6 304 582	4 910 000	1.284
LRT$_1$	7 803 442	7 910 000	0.987
LRT$_2$	9 114 945	9 410 000	0.969
HR	13 330 490	13 160 000	1.013
Met	18 295 466	16 410 000	1.115

Table 10.3 Comparison of each option with 'do-nothing'.

Project option	Change in annual cost (£)	Change in annual passenger use	Ratio (£/passenger)
Min	—	—	—
GB$_1$	3 044 560	2 650 000	1.149
GB$_2$	3 981 348	3 000 000	1.327
LRT$_1$	5 480 209	6 000 000	0.913
LRT$_2$	6 791 711	7 500 000	0.906
HR	11 007 257	11 250 000	0.978
Met	15 972 232	14 500 000	1.10

Table 10.4 Comparison of each option with 'do-minimum'.

10.2.1 Final comment

In the previous example, cost effectiveness was estimated using one sole measure – passenger usage. It is possible to use more than one measure of effectiveness and this can be achieved using a linear scoring function, where weightings must be assigned to the different measures in order to arrive at an overall effectiveness score. In such a system, the total score, S, for a given option, i, over j measures is derived as follows:

$$S_i = \sum_j w_j x_{ij} \qquad (10.3)$$

where w_j is the importance weighting placed on the jth measure of effectiveness, x_{ij} is the score for measure j on option i (this assumes that all measures can be estimated on the same scale).

The cost effectiveness method, in effect, retains half the basic structure of CBA. The costs, both construction and maintenance, are measured in monetary terms No attempt is made to measure the benefits in monetary terms. The project output is described in physical terms and expressed per unit cost. This expression is then compared with some baseline case that may be the 'do-nothing' and/or the 'do-minimum' scenario.

The most important step in the process is establishing which measure or measures of effectiveness to use within the analysis. In the above transport example, the measure employed was level of passenger usage; that is, the number of actual public transport users each option will attract. This criterion is fundamental to the viability of any transit system. If, on the other hand, the analysis had an essentially environmental perspective, measures such as ambient air quality or resulting increase in noise level may be employed.

The method employs a standard economic principle in assessing whether, relative to the baseline case used, the marginal benefit as indicated by the measure of effectiveness is justified by the additional cost. Within Tables 10.3 and 10.4, those options for which the ratio dips below one indicate that the increased effectiveness of the option in question over the base case is achieved at an extra cost which is economically justifiable. A ratio equal to one means the option is exactly as economically justifiable as the base case, while a value in excess of one indicates non-viability relative to base.

Although cost effectiveness analysis may seem to be a straightforward economic and engineering approach, in practice certain restrictions may make the decision less immediate. Say, for the above example, the Government indicates that only transit systems capable of catering for annual passenger levels of 10 000 000, only two of the options, heavy rail and metro, are viable. Neither are economically justifiable relative to the 'do-nothing' option, with the heavy rail proposal only justified relative to 'do-minimum'. In such a situation an analysis concentrating on a wider range of measures of effectiveness, or one which encompasses a broader set of project options may have to be considered.

Finally, it should be noted that the analysis does not take account of the developer's ability to fund the scheme. It is possible that the most cost effective solution may be too expensive, and therefore some upper boundary may have to be placed on option costings. Cost effectiveness is a powerful tool but one which should be handled carefully. Particular attention must be paid to selection of relevant and quantifiable measures of effectiveness.

10.3   The Planned Balance Sheet

Lichfield's Planning Balance Sheet (PBS) (Lichfield, 1969, Lichfield *et al.*, 1975) was a method devised for the analysis of different project options within regional and urban planning. It recognised what it saw as fundamental deficiencies in CBA and overcame these not by totally rejecting the method but by modifying it to suit the needs of the problem under examination. Unlike cost effectiveness analysis which widens the scope of the decision process by retaining monetary valuation for only half the analysis, namely the costs, but choosing a different process for evaluating benefits, the Planned Balance Sheet diverges more fundamentally away from CBA.

While CBA is seen to be a powerful instrument of project evaluation, it identifies a basic problem inherent in the use of this methodology by practitioners, namely that the evaluation of a project must relate to an unambiguous monetary uni-dimensional criterion, since a comprehensive cost–benefit analysis approach requires a transformation of all project option effects into one single monetary dimension. This severe restriction is responsible for many difficulties in the practical application of CBA, as attributes/criteria that cannot be readily transformed into monetary units are omitted from its framework. Given these limitations in the application of CBA, several adaptations have been developed, one of which is the Planned Balance Sheet (PBS). Another is the Goal Achievement Matrix (GAM), described further in Section 10.4.

10.3.1   The relevance of pure CBA to planning problems

Before the Planned Balance Sheet is examined in detail, it is important to highlight in some detail the perceived deficiencies in CBA in its treatment of project appraisal for physical planning problems that led to the emergence of the Planned Balance Sheet. The utility of CBA for tackling option choice problems in physical planning can be expressed in terms of a set of criticisms focussing on the following areas:

- CBA has difficulty in defining who exactly is affected by the proposed development project. It also has difficulty dealing with variations both in attributes that occur over the entire spatial area of the study zone in question and with individual groups and institutions on which different costs/benefits are incident.
- CBA encounters difficulty in assigning meaningful monetary values to costs and benefits, particularly in the context of a physical planning problem, when trying to accommodate the variations over space and among individual groups and institutions referred to in the preceding point.
- Again, in the context of urban and regional planning, there is difficulty in summarising costs and benefits to single discrete values when they may vary across different locations in the study area and for the different actors upon whom the costs/benefits are incident.
- There is a general problem of assigning a meaningful discount rate to the problem under examination to allow costs and benefits occurring within different time frames to be directly compared.

Each of these points can be validly directed at both CBA and cost effectiveness analysis.

Despite these obvious problems with CBA in tackling physical planning problems, it does provide a method for structuring information on different ways of tackling a physical planning problem in terms of its costs and benefits, and gives the decision makers some basis for the support of one option over another. In addition, it is of some assistance in assessing the monetary compensation due to those suffering financial loss as a result of the proposal. There is also merit in allowing a single net benefit figure to be assigned to each option, making choice of the most economically viable one straightforward.

For these reasons, planners believed that any revised method should not completely discard CBA. Rather, the chosen method should modify CBA to take account of the particular evaluation problems associated with a physical planning problem.

There are four major weaknesses of CBA in its ability to analyse a physical planning problem:

(1)   The omission from the analysis of impacts/attributes that cannot readily be transformed into monetary units, even though such impacts tend to involve environmental and social attributes, such as aesthetics, air and noise pollution, community severance and cultural values, which are of vital importance to the quality of life of communities, and thus may be important factors in the projects' desirability. Lichfield felt that CBA under-represented these 'intangible' non-quantifiable criteria within the decision making process and believed that the Planned Balance Sheet had the ability to incorporate non-monetary/non-quantified impacts within the same visual framework as the monetary valuations, thus helping decision makers give relatively intangible impacts their due regard in choosing a preferred option.

(2)   The greater suitability of the framework for ranking/comparing physically similar projects or projects designed to attain the same ends rather than for assessing the absolute desirability of a project. CBA is particularly limited in this regard, given its solely monetary basis. It can be used in the comparison of two proposed landfill sites, but cannot meaningfully compare a landfill project with a school project, where the basic nature of the impacts are so different that direct comparison lacks reliability. At present, no common scale exists for comparing the monetary benefits of two such diverse proposals. Comparability and measurability are necessary conditions for the application of CBA to alternative proposals (Hill, 1973).

(3)   Lichfield (1971) felt that CBA ignored questions of distributional equity. Within PBS, he sets out the incidence of costs and benefits against various affected groupings within society. PBS identifies 'producers' responsible for establishing and running the project. Each producer is paired with a group of 'consumers' which consumes the goods / services of the 'producers' in real or notional transactions. Costs, too, are represented as passing from producer to consumer in these transactions.

(4)   CBA is, in the main, applied to projects in a single sector of activity, whereas many planning problems are multisectoral, with many divergent groups affected in many differing ways, with these impacts often not open to being valued in monetary terms or even quantified. Such proposals are much more amenable to analysis using the Planned Balance Sheet.

10.3.2   Lichfield's Planned Balance Sheet

The Planned Balance Sheet was developed by Lichfield (1969) and applied to practical problem solving in the physical planning area. It involves the comparison of different solutions to a particular planning-based problem. It permits the different actors on whom the costs and benefits are likely to be incident to be broken down or 'disaggregated' into two distinct groups of actors, firstly producers and operators and, secondly, consumers. As a result of transactions of goods and services between producers and consumers, these two groupings incur costs and are the recipients of benefits that are classified as being either fixed or operating items. In addition, another major feature of the Planned Balance Sheet is its classification of costs and benefits into three major categories; firstly, items are assessed in monetary terms, secondly, items are assessed in quantitative but non-monetary terms and, lastly, non-quantitative non-monetary based items are usually assessed on some agreed qualitative scale. While CBA is criticised for ignoring societal goals, overvaluing efficiency, overemphasising money at the expense of subjective aspects of the quality of life and for ignoring social and environmental costs and benefits, subsequent approaches, including Lichfield's Planned Balance Sheet, incorporate many concepts of CBA into their own methodology.

The Planned Balance Sheet (PBS) attempts to reform the CBA framework through the promotion of these 'disaggregating' techniques. It was devised in order to introduce the rigour of CBA into the urban and regional planning process. It attempts to highlight the consequences of certain planning proposals and to provide decision makers with information. Thus, planning needs a rigorous decision making tool such as CBA to arrive at planning decisions that would be in the public interest. Lichfield believed that CBA techniques could be adapted for this purpose, without any loss of the objectives of the planners themselves. The method presents a useful summary of the impacts relevant to the planning problem. It does not pretend, as Massam (1980) notes, to give a single 'best' solution to the particular problem. The final decision involves some subjective evaluation by the decision maker that is not specifically allowed for within the method.

PBS is thus a set of social accounts covering the costs and benefits of resources. Measurements in the balance sheet are based on opportunity costs and consumer willingness either to forego alternative goods and services for these benefits, or to receive goods and services in compensation for costs. This information is gathered, where possible, through observation and analysis of revealed preferences. Where actual prices or 'shadow' prices are not available, other quantified units of measurement are used, including notional point systems or qualitative/graphical scales.

The PBS methodology goes some way towards meeting the criticisms of CBA regarding identification of how income is distributed within a proposed project. It avoids problems of weighting the relative importance of impacts/effects by leaving the ultimate weighting to the decision makers after analysis. The Planned Balance Sheet avoids the problems inherent in trying to aggregate impacts by taking the opposite tack of promoting disaggregation (Carley, 1979). The PBS methodology presents data in a clearly disaggregated format, that is all costs and all benefits are identified with respect to all relevant community groupings. Such a layout does, however, impose on the decision maker the need to both digest large amounts of information and to make explicit trade-offs between monetary and intangible impacts on the project.

The reasoning behind the PBS methodology lies in its demonstration of the incidence of costs and benefits of a project on various sectors of society. It measures all the 'good' and 'bad' consequences of alternative courses of action, tracing the effects of planned projects on clearly defined community incidence groups.

10.3.3   Layout of the Planned Balance Sheet

Tables 10.5 and 10.6 illustrate the broad outline of a Planned Balance Sheet. The various sectors affected by the proposed plans are considered both as producers (expressed as X1, Y1, and Z1) and as consumers (expressed as X2, Y2 and Z2). The following notation is used:

- If a money symbol (e.g. '£') precedes a letter, then the cost or benefit can be expressed in monetary terms,
- If 'M' followed by a numerical subscript is recorded in the cell, then the cost can be expressed in quantitative but non-monetary form,
- An 'i' represents an intangible item, with the subscripts indicating the nature of the intangible, and
- A dash '—' in a cell indicates that no cost or benefit of that type would accrue to that sector if that particular plan under consideration was implemented.

Once the balance sheet is prepared, the costs and benefits that would accrue to each sector are reduced, aggregated and compared for each plan under consideration.

Table 10.5 Benefits and costs for producers.

	Plan A				Plan B			
	Benefits		Costs		Benefits		Costs	
	Cap.	Ann.	Cap.	Ann.	Cap.	Ann.	Cap.	Ann.
X1	£a	£b	—	£d	—	—	£b	£c
Y1	i_1	i_2	—	—	i_3	i_4	—	—
Z1	M_1	—	M_2	—	M_3	—	M_4	—

Cap.: capitalised cost
Ann.: annual cost

Table 10.6 Benefits and costs for consumers.

	Plan A				Plan B			
	Benefits		Costs		Benefits		Costs	
	Cap.	Ann.	Cap.	Ann.	Cap.	Ann.	Cap.	Ann.
X2	—	£e	—	£f	—	£g	—	£h
Y2	i_3	i_4	—	—	i_7	i_8	—	—
Z2	M_1	—	M_3	—	M_2	—	M_4	—

Cap.: capitalised cost
Ann.: annual cost

Example 10.2   An illustration of the Planned Balance Sheet (Lichfield, 1969).
One case study of note, the cost–benefit analysis of an urban expansion scheme in Peterborough, UK, entailed Lichfield (1969) using the Planned Balance Sheet methodology to evaluate five planning alternatives.

The Planning Balance Sheet method groups the community into various homogeneous sectors, distinguished by the kind of operations they wish to perform. It then evaluates and compares the alternatives from the point of view of the advantages (benefits) and disadvantages (costs) accruing to every sector from each alternative, in order to see which would provide the maximum net advantage (benefit).

Benefits/costs are measurable in monetary terms, in physical or time units, or may not be measurable at all.

A strict conclusion cannot be drawn from PBS – it does not aim to provide a conclusion in terms of net profit or rate of return. Its value lies in its ability to expose/highlight the implications of each set of proposals to the whole community, thus leading to the selection of a plan which will best serve the total interests of the entire community.

The technique used in the Peterborough study involved two basic steps:

(1)   Firstly, a summary of available data on all aspects of the development from the perspective of the five alternatives is compiled. These aspects include:

- Land
- Engineering services
- Road system
- Residential development
- Industrial development
- Open space/recreation

Data can be expressed in terms of numbers of physical units or approximate money costs for items thought to be critical, and for which sufficient information is available for an estimate to be attempted. An extract from the data table is given in Table 10.7.

(2)   The second step involves the preparation of the balance sheet itself. The community groupings/sectors are divided into 'producers' and 'consumers', those who will be involved in creating/operating the services resulting from the project, and those likely to use/pay for the services. The following is a list of the subgroups into which the two groups are divided.

- Producers
  - The developing agency
  - Current landowners
  - Local authorities

- Consumers
  - The public
  - Current landowners/house owners
  - Ratepayers

Contd

Example 10.2 Contd

The various sectors of the community were defined in relation to the major operation(s) they would be undertaking within the proposed town (shopping, working etc.) and then grouped with reference to major land uses (residential, industrial etc.)

A broad outline of the Peterborough Planned Balance Sheet example is shown in Tables 10.8 and 10.9.

Both costs and benefits are treated together in discussion, (costs as negatives, benefits as positives). The five project alternatives are ranked according to the objective or subjective measurement described in the available data, with '1' indicating greater benefit/less cost, and '5' indicating the opposite.

The points registered on the ranking scale are totalled for producers/operators and consumers separately, and combined into a grand total for each alternative. The alternative with the lowest total score is deemed the most preferable course of action.

Table 10.7 Extract from data table.

	Options				
Item	1	2	3	4	5
Land					
(a) Developed	N++	N+	N	N	N+
(b) Undeveloped (acres)	6888	6888	6888	6888	6888
(c) Regional Open Space	N	N	N	N	N
Main services					
(a) Storm Water	N	N	N	N	N
(b) Gas	N	N	N+	N++	N
(c) Foul Drainage (£ million)	4.4	4.0	2.4	5.5	4.0

N indicates amount unknown, N+ indicates a greater amount than just N,
N++ indicates a greater amount than just N+ etc.

Table 10.8 Extract from balance sheet (producers/operators).

Sector	Units	Options					Order				
		1	2	3	4	5	1	2	3	4	5
Developing agency											
Land acquisition	(£) million	n	N	n	n	n	3	2	1	1	2
Mains services	(£) million	n+	n+	n	n++	n+	3	2	1	4	2
Building const'n	(£) million	n+	n+	n+	n+++	n	4	1	3	5	2
Site engineering	(£) million	n+	N	n++	n++	n	2	2	2	1	2
Returns	(£) million	n	N	n	n	n	0	0	0	0	0
Local authority											
Mun. costs	(£) million	n	n+	n++	n+++	n+	1	2	4	5	3
Mun. revenues	(£) million	n	N	n	n	n	0	0	0	0	0

'n' indicates an unknown amount, 'n+' indicates a greater amount than just 'n' etc.

Table 10.9 Extract from balance sheet (consumers).

Sector	Units	Options 1	2	3	4	5	Order 1	2	3	4	5
The public											
In central area	£ million/ int.	n++	n+	n	n	n+	1	2	4	4	3
Residential area	£ million/ int.	n+	N	n	n	n	1	2	3	3	3
Working public	£ million/ int.	n	N	n	n	n	1	2	3	4	2
Current occupiers											
Central area	£ million	n+	n++	n+++	n+++	n++	1	1	3	3	2
Countryside	£ million	n	N	n	n	n	2	1	2	2	3
Ratepayers	£ million	n	N	n	n	n	1	2	4	5	3

'int.' denotes intangible; 'n' indicates an unknown amount, 'n+' indicates a greater amount than just 'n' etc.

10.3.4   Conclusions

The Planned Balance Sheet has proved a convenient aid for tackling the evaluation of plans, particularly in its ability to categorise its relevant effects. Rather than proceeding item-by-item in the usual manner, the balance sheet is structured around the affected groups of individuals. The nature of the balance sheet dictates that an evaluation cannot take place without explicitly considering both the items of cost and benefit and the groupings on which those items will be incident. The nature of many multisectoral proposals, such as those found in the urban and regional planning area, is such that the decision maker will find it more straightforward to identify those parties likely to be affected and then proceed to identify the manner in which these are likely to be affected. Unlike the more clear-cut single-sector projects, such as the comparison of a number of different highway project options, the identification of relevant costs and benefits may not be straightforward. It may prove easier to identify the actual relevant groupings on whom the costs and benefits are incident.

10.4   Hill's Goal Achievement Matrix

Hill (1966, 1968) developed the Goal Achievement Matrix in an attempt to overcome what he viewed as basic weaknesses in the Planned Balance Sheet. He believed that Lichfield's technique was limited in scope and not the optimal approach, since costs and benefits could only be compared if it was possible to relate them to a common objective, and this was not always achieved within the Planned Balance Sheet. Hill sought to rectify this shortcoming by incorporating objectives and non-monetary based criteria into the cost–benefit analysis framework. While recognising that the balance sheet had a broader perspective than the strictly monetary-based CBA system, he believed it failed to recognise that costs and benefits could only be compared if it was possible to relate them to a common objective. Hill proposed the view that costs and benefits should always be defined and identified in terms of *goal achievement*. Benefits signify progress in the direction of the desired objectives while costs signify a move in the opposite direction. Costs, benefits and the goal should be defined in the same units where possible. A traditional CBA would dictate that monetary measurement is used throughout.

Thus, along with the Planned Balance Sheet, the Goal Achievement Matrix is the best-known technique that aims to 'reform' CBA by providing an alternative framework within which both costs and benefits can be aggregated. Both seek to develop the basic procedure of CBA to suit a multisector problem.

Hill developed the Matrix for assessing options within single-functional sectors, such as transportation or waste management. He believed it could not be used to compare options involving multisectoral operations, as the method was not capable of taking into consideration the interaction and interdependence of objectives related to different sectors. Some have argued, however, that unless it is required to identify an optimal project proposal, application of the method does not require an understanding of the manner in which the achievement of one objective impinges on the level of achievement of others. Thus, if the process was to shortlist a number multisectoral proposals, arguments in favour of using the matrix could be put forward.

10.4.1   *Goal Achievement Matrix and the rational planning process*

Within Chapter 1, the importance and centrality of rational decision making to the process of project appraisal was emphasised. Within that five-stage rational process, the setting of objectives was identified as the first step, within which the overall purpose of the proposed project was defined and agreed. Hill, when putting forward his method, argued that neither CBA nor the Planned Balance Sheet satisfied the requirements of 'rational' planning. In the context of decisions relating to the allocation of resources to proposed projects, he described the rational process as the determination of appropriate future action by 'utilising scarce resources in such a way as to maximise the attainment of ends held by the system' (Hill, 1968). In other words, it was his belief that the evaluation of alternative plans should involve a comparative assessment of them based on the extent to which each achieves the specified 'ends' or objectives of the system under consideration. While Hill believed that the costs and benefits within the Planned Balance Sheet did relate to a broad range of community objectives, and thus was a great improvement on CBA, he criticised Lichfield's method on the basis of its failure to give an *explicit* statement of these objectives, which he believed to be of central importance to the appraisal process. This criticism was based on the belief that the balance sheet method ignored the dependence of each of the costs and benefits for their basis and validity on the achievement of certain objectives, thereby implying that they in some way existed independently and had their own implicit value. Hill believed that Lichfield's overall criterion for choosing a particular option based on the maximisation of net benefit to the community was meaningless if specified in the abstract. The idea that objectives should be stated explicitly within the evaluation, thereby insuring its basic rationality, was accepted by Lichfield as a valid criticism.

10.4.2   The basic steps used to form the matrix

The Goal Achievement Matrix has a format similar in nature to the Planned Balance Sheet. The main difference lies in its use of two weighting systems. One set of weights is applied to the different groups within the community while the other set is used to weight the objectives. (This is in marked contrast to the Planned Balance Sheet where no weights are used.) It is a rational approach, concentrating on the step-by-step assessment of optional means of meeting goals/objectives. An outline of the steps of Goal Achievement Matrix analysis is as follows:

(1)   Formulate planning objectives and define them.
(2)   Specify alternative courses of action.
(3)   Prepare a cost–benefit account for each objective which considers monetary effects, quantified but non-monetary effects, and intangible/non-quantifiable effects. Costs and benefits are compared and aggregated, where possible, for each objective.
(4)   Develop a weighting scheme which reflects the relative importance to the community of each objective, as well as reflecting the incidence of costs and benefits associated with each alternative. (This is achieved using survey and consultation techniques.)
(5)   If achievement of objectives cannot be measured in single units (monetary units where possible), develop transformation functions, where possible, which demonstrate the trade-offs between outcomes measured in different units and facilitate the aggregation of such measures of achievement on a single scale.
(6)   Sum up the weighted indices of goal achievement. The preferred plan from among the competing schemes is the one with the largest overall index.

10.4.3   Formulating the objectives

The three most important aspects that must be addressed when forming the list of objectives for a given problem are:

(1)   Whose preferences or interests must primarily be taken into consideration within the study? From whose perspective or viewpoint is the evaluation being carried out?
(2)   What categories of consequences do the objectives relate to? Are economic consequences the overriding priority or are environmental, social and technical ones being considered on an equal footing?
(3)   What person or persons should formulate the objectives, and what procedures should be employed to devise them?

Within any given appraisal process, it will be necessary to ascertain whose interests are being taken into consideration. Once this has been achieved, it is necessary to ascertain exactly what those interests and concerns are and to what extent they are likely to be affected by the proposed project.

Most proposals will involve objectives other than the purely economic ones being considered. Hill believed that objectives should be derived explicitly from a consideration of ideals, with the scope of the objectives determined by the nature of the ideals deemed relevant to the project under evaluation. Hill, together with Schechter (1971), put together a set of ideals that could be used in the appraisal of an engineering development project:

- Physical and mental health
- Enjoyment
- Equity
- Economic welfare
- Social stability
- Ecological balance.

Within this listing, economic welfare is just one of six ideals used to form the relevant set of objectives which will thus be broadly based in nature. This is in marked contrast to the cost–benefit analysis framework where 'lip service' is paid to the intangible elements of the analysis that are kept outside the main monetary-based framework, and thus relegated to a minor role in the overall analysis

To derive a set of objectives from a set of specified ideals, direct approaches consisting of consultations with the developers, elected officials, members of community interest groups, together with the sampling of public opinion in the community through surveys and public meetings, should be employed. Indirect approaches involving analysis of both the behaviour patterns of community groups and the previous allocations of public investments can also be employed to determine the goal priorities for the community as a whole.

The estimation of the relative weightings of the objectives should also be undertaken within this process of consultation, public opinion sampling and/or behaviour observation. Hill believed it to be desirable to approach the determination and relative valuation of community objectives from a number of different perspectives simultaneously in order to make the result as robust as possible.

10.4.4 Evaluating costs and benefits within the Goal Achievement Matrix

When the point in the analysis is arrived at where the objectives for evaluation have been specified, it becomes essential to measure the degree to which the options put forward will attain these yardsticks of achievement. These measurements will provide evidence of the advantages and disadvantages (or costs and benefits) of each individual proposal. Within the analysis, these costs and benefits are thus always defined in terms of goal achievement. Benefits can be seen in terms of movement towards desired objectives, while costs can be seen as movement in the opposite direction. Where the goal is defined in quantitative units, the costs and benefits will be stated in the same quantitative terms. Where quantitative measurement is not feasible and only a qualitative description of the goal is possible, benefits will indicate movement towards these qualitative objectives, while costs indicate movement away.

For a given goal/objective, costs and benefits are always defined in terms of the same units if the objectives are quantifiable.

10.4.5   *The structure of the matrix*

The following is the framework on which the matrix is based:

(1)   The set of goals is known and the relative importance of each goal is established.
(2)   The objectives are defined operationally, rather than in abstract terms (note that an objective denotes an attainable goal).
(3)   The consequences of each alternative course of action are determined for each objective.
(4)   The incidence of the benefits and costs of each course of action, measured in terms of the achievement of the goal, is determined for each goal.
(5)   The relative weight to be attached to each group is also established.

The final Matrix format is shown in Table 10.10. Within this table, the following should be noted:

- G1, G2, G3 denote the goal descriptions.
- Each goal has a weighting 1, 2, 3 and so on. These are previously determined.
- Various groups a, b, c, d, e are identified as affected by the given courses of action. These groups can be combined to indicate cost and benefit incidences.
- A relative weight is determined for each group, either for each goal individually or for all goals together.
- The letters A, B and so on are the costs and benefits, which may be defined in monetary, non-monetary/quantitative or in purely qualitative terms. Costs and benefits are recorded for each objective according to the parties affected. A dash '—' in a cell indicates that no cost or benefit that is related to that objective would be incident on that group if the plan were implemented.
- For certain goals, 'sum' indicates that the summation of costs and benefits is both meaningful and useful. The total costs and benefits with respect to that goal can then be compared. This is the case where all costs and benefits are evaluated in quantitative units, but is not so in the case of the qualitative assessment of intangible / non-quantifiable costs and benefits. In such cases, as with goal G2 in Table 10.10, the costs and benefits are so diverse in nature that their summation is not possible.

Hill recommends that, where possible, the overall weightings of the objectives should be set together with the valuations associated with the full achievement of

Table 10.10 Goal Achievement Matrix – general format.

Goal Description		G1			G2			G3		
Relative Description		2			3			5		
Incidence	*Rel. Wt.*	*Cost*	*Ben.*	*Rel.Wt.*	*Cost*	*Ben.*	*Rel. Wt.*	*Cost*	*Ben.*	
Group a	1	A	D	5	E	—	1	Q	R	
Group b	3	H		4	—	R	2	S	T	
Group c	1	L	J	3	—	S	1	V	W	
Group d	2	—		2	T	—	2	—	—	
Group e	1	—	K	1		U	1	—	—	
		Sum	Sum		(No sum)			Sum	Sum	

these objectives and should be set in advance of the analysis of each of the individual options. Furthermore, the weightings of the various incident groups should be established before their actual performance on the various costs and benefits associated with each objective have been identified.

10.4.6   Uncertainty within Goal Achievement Matrix

In general, allowance for uncertainty should be made indirectly through the use of conservative estimates, the requirement of safety margins, continual feedback and adjustment, and through the existence of a risk component in any discount rate used. (Estimates made at low discount rates are very sensitive to variations in the estimates of subsequent events, while higher discount rates result in less sensitivity to variations in the estimation of subsequent events.)

Uncertainty concerning anticipated consequences is, however, best treated by probability formulation, where a range of possible outcomes, rather than one unique outcome, is predicted.

10.4.7   The simplified Goal Achievement method

The most simplistic strategy is to present the decision maker with the entire matrix of valuations and request that they use it to assess the relative performance of the different options, rather than attempt to evaluate the overall level of goal achievement of each one in absolute terms. This approach is relatively undemanding for the matrix compiler, but requires a substantial input from the decision maker, who, armed with the level of achievement of objectives for each relevant grouping, along with their relative weightings, must still trade off the extent of achievement of the set of objectives against their relative importance weightings. That is, the decision maker must compare the weights ascribed to the objectives with the differential achievement of these objectives.

Thus, the combined relative weights of objectives and their relative incidence on the defined groupings are assigned to the measurement of achievement of those objectives. The weighted indices of goal achievement are then summed and the preferred plan is the one with the largest index.

The essential problem in this instance is to take various impacts that have been measured in various different ways and convert these into common units of value to which the importance weightings can be applied. The simplest approach to evaluating indices of goal achievement is to measure all objectives on the least demanding scale – an ordinal one. (An ordinal scale is one which ranks elements on it as 1, 2, 3, etc.)

The relevant project would be evaluated with respect to each objective to determine whether it increases, decreases or leaves goal achievement at its pre-existing level both for the community as a whole and the relevant incident groupings.

The following arbitrary values could be assigned:

'**+1**'   for Goal Achievement being enhanced
'**-1**'   for Goal Achievement being lessened
'**0**'   for zero net effect.

The two sets of weights could then be introduced and the index of goal achievement determined.

Example 10.3   The Simplified Method

Let us make the following assumptions regarding the comparison of two road schemes, P1 and P2:

- Comparison is on the basis of the ability of the two options to decrease overall noise levels while minimising community severance.
- The relative weights of these objectives and the relative importance levels of incidence on the two community groups of relevance, Group *a* and Group *b*, are measurable.

Project P1 decreases noise levels for Group *a*, thus giving a score of +1, and increases them for Group *b*, thus scoring −1. P2, on the other hand, increases noise levels to Group *a* (−1), while decreasing them to Group *b* (+1). P1 leads to an increase in community severance for Group *a* (−1), but has no effect on Group *b* (0). P2, in contrast, does not affect severance for Group *a* (0), but does lead to increased severance for Group *b* (−1). The system of weightings is shown in Table 10.11, with the final outcome of the evaluation shown Table 10.12.

Weighted Indices of Goal Achievement:

Project P1 = +1
Project P2 = −6

Thus, P1 is preferable to P2. (Note: because the evaluation uses a relatively low-level ordinal scale, the analysis may not use all the information available on the extent of achievement of objectives.)

	Noise	Severance
Importance weightings (overall)	2	1
Incidence weightings		
group *a*	3	3
group *b*	1	2

Table 10.11 Incidence and importance weightings.

	Weight	Weighted noise impact		Weight	Weighted severance impact	
		P1	P2		P1	P2
Community (overall)	2	P1	P2	1	P1	P2
Group *a*	3	+6	−6	3	−3	0
Group *b*	1	−2	+2	2	0	−2
		+4	−4		−3	−2

Table 10.12 Goal Achievement by measurement on an ordinal scale.

10.4.8   An advanced version of the Goal Achievement Matrix

Within this advanced version, once the objectives have been set, the appropriate evaluation criteria (*j* in number) for each objective listed, the performances of the various options (*i* in number) on each of the criteria measured and explicit points of 'success' and 'failure' for each criterion identified, two important tasks must be undertaken:

(1)   The importance weighting, (w_j), of each criterion must be assessed by the decision maker on a scale ranging from 0 to 3.
(2)   Each criterion score is translated onto an artificial scale ranging from −5 (worst) to +5 (best), used as a common means of measuring each option's performance. The highest score is assigned to the 'success' benchmark of each criterion and the minimum score is assigned to the 'failure' benchmark. Once these boundary points have been labelled, a linear relationship defines the path between these two extremes. The performance of a given project option will fall somewhere between these two boundary points. A positive score indicates a performance consistent with achieving the objective defined by the criterion in question, with a negative score indicating the opposite. By this means, a performance score for each option (f_{ij}) is obtained.

If *n* is the number of selected decision criteria and $x(=5)$ is the maximum score on the artificial scale used, a normalised un-weighted score ($f_{(unwt)i}$) is calculated using the formula:

$$f_{(unwt)i} = \sum_j f_{ij} / (n \times x) \tag{10.4}$$

The weighted score for each option on each of the decision criteria is estimated using the formula:

$$f'_{ij} = f_{ij} \times w_j \times \left(n \div \sum w_j\right)$$ (10.5)

In conclusion, the overall weighted score for each option $f_{(wt)i}$ is calculated using the formula:

$$f_{(wt)i} = \sum_j f'_{ij} \div (n \times x)$$ (10.6)

The score obtained will reflect, in each case, the option's compliance or non-compliance with the objectives set by the decision maker and forms the basis for the selection process.

Example 10.4   The Advanced Method

Three options for a toll motorway linking two European countries are being considered. The evaluation is concerned with one incidence group only – the central government decision makers – and is based on the extent to which they meet the following set of six appropriate goals:

(1)   Minimisation if environmental impact.
(2)   Maximisation of accessibility/regional development.
(3)   Increase volume in cross-border traffic.
(4)   Reduce 'price' of transport.
(5)   Optimise attractiveness of proposal to private investors.
(6)   Maximise public acceptability of scheme.

These goals are translated into the following six criteria of evaluation:

(1)   Reduction in carbon monoxide levels.
(2)   Travel-time savings.
(3)   Cross-border traffic growth relative to internal growth in both countries.
(4)   User benefits as a proportion of total investment costs.
(5)   Internal rate of return from project relative to minimum required for private sector involvement.
(6)   Level of public acceptability of scheme.

Table 10.13 indicates how each of these criteria is measured, what level is equivalent to the median score of zero on the scale of – 5 to + 5, and what constitutes the maximum score of +5.

Having defined the success and failure benchmarks in Table 10.13, Table 10.14 indicates the actual score for each option on each of the six criteria in the scale appropriate to each criterion. Using the information in Table 10.13 an equivalent unweighted score on the agreed common scale from – 5 to + 5 is shown on Table 10.15.

Contd

Example 10.4 Contd

The scores from Table 10.15 are then normalised using Equation 10.4 as follows:

$$f_{(unwt)1} = -0.9/(6 \times 5) = -0.03$$
$$f_{(unwt)2} = +12.2/(6 \times 5) = +0.41$$
$$f_{(unwt)3} = +12.0/(6 \times 5) = +0.40$$

If the weightings for the six criteria are set as follows:

CO_2 reduction	= 1
Travel-time savings	= 2
Cross-border traffic growth	= 2
User benefits	= 1
Financial acceptability	= 1
Public acceptability	= 1

The weighted score for each of the options is calculated using Equation 10.5. These are given in Table 10.16. The scores from Table 10.16 are then normalised using Equation 10.6 as follows:

$$f_{(wt)1} = +2.4/(6 \times 5) = +0.08$$
$$f_{(wt)2} = +12.2/(6 \times 5) = +0.41$$
$$f_{(wt)3} = +12.2/(6 \times 5) = +0.41$$

The analysis shows that Options 2 and 3 are inseparable regardless of whether the criteria are weighted or not, and both are substantially better than Option 1. More detailed analysis would be required to separate the two best performing options.

10.4.9   Concluding notes on the Goal Achievement Matrix

The key to plan evaluation by the Goal Achievement Matrix is the identification and weighting of the decision criteria derived from the objectives identified as relevant to the problem under examination. In practice, Hill feels that goals and their relative weights may have to be determined iteratively, as a result of a complex process of interaction among all stakeholders reacting to explicitly stated objectives and the alternative courses of action proposed to meet these objectives.

Also, the effect of changes in weights on the relative desirability of alternative project options (i.e. a sensitivity analysis) should be analysed. The effect of different sets of weights on the choice of the preferred project can thus be determined.

The Goal Achievement Matrix, like the Planned Balance Sheet, cannot determine whether a project should proceed or not. All are designed for the comparison and ranking of alternatives and should be used to justify the need for a project only with great care.

Table 10.13 Performance measures for each criterion.

Criteria	Method of measurement	Median performance	Optimum performance
Reduction in levels of CO_2	Annual reduction in levels of CO_2	0% change in CO_2 emission levels	EU target of 0.1% emission reduction per annum reached
Travel-time savings	Percentage reduction in time savings relative to comparable projects	12.5% reduction in travel time for motorists	25% reduction in travel time for motorists
Level of cross-border traffic growth	Cross-border growth as % of internal traffic growth	Cross-border growth as % of internal growth is 103.5% of EU average	Cross-border growth as % of internal growth is 139.6% of EU average
Level of user benefits	User benefits as % of investment cost	User benefits are 2.05% of total costs	User benefits are 6.15% of total costs
Financial acceptability	Subjective valuation based on various financial indicators (debt/service and gearing ratios)	Not applicable	Not applicable
Public acceptability	Subjective valuation based on results of opinion surveys carried out on members of local community	Not applicable	Not applicable

Table 10.14 Option performances.

Criterion	Option performances		
	Option 1	Option 2	Option 3
CO_2 reduction	0.066% increase in CO_2	0.038% increase in CO_2	0.044% increase in CO_2
Travel-time savings	10.25% reduction	10.25% reduction	10.5% reduction
Cross-border traffic growth	140% EU average	140% EU average	140% EU average
User benefits	1.76% of total costs	6.2% of total costs	6.2% of total costs
Financial acceptability	Extremely unacceptable	Neutral	Neutral
Public acceptability	Very acceptable	Extremely acceptable	Extremely acceptable

Table 10.15 Un-weighted option scores on common scale.

Criterion	Option scores (on −5 to +5 scale)		
	Option 1	Option 2	Option 3
CO_2 reduction	−3.3	−1.9	−2.2
Travel-time savings	−0.9	−0.9	−0.8
Cross-border traffic growth	5.0	5.0	5.0
User benefits	−0.7	5.0	5.0
Financial acceptability	−5.0	0.0	0.0
Public acceptability	4.0	5.0	5.0
Sum of un-weighted scores	−0.9	12.2	12.0

Criterion	Option scores (on −5 to +5 scale × Weight)			Table 10.16 Weighted option scores.
	Option 1	Option 2	Option 3	
CO_2 reduction	−2.475	−1.425	−1.650	
Travel-time savings	−1.350	−1.350	−1.200	
Cross-border traffic growth	7.500	7.500	7.500	
User benefits	−0.525	3.750	3.750	
Financial acceptability	−3.750	0.000	0.000	
Public acceptability	3.000	3.750	3.750	
Sum of weighted scores	+2.400	+12.225	+12.150	

The Goal Achievement Matrix, along with CBA, differs from the Planned Balance Sheet in that they are devised for the evaluation of a single functional sector, although, as pointed out above, in cases where the optimum solution is not sought, Goal Achievement may be used to assess a set of multisectoral proposals. (The Planned Balance Sheet was devised to evaluate multiple sector development projects.)

The Goal Achievement Matrix can, in some cases, be exceedingly complex. The information contained within it can be so diverse that it becomes difficult for the decision maker to digest. However, Hill believes it to be preferable to both the other methods. Cost–benefit analysis bases itself on the sole criterion of economic efficiency, while the Planned Balance Sheet omits to explicitly identify overall community goals and the community's valuations of them. Hill feels, in overall terms, that the Goal Achievement Matrix adequately expresses the complexity of the consequences of major development projects, particularly in urban areas. Its ability to make explicit the effects of alternative courses of action with regard to all valued objectives, together with the level of incidence of these effects on various relevant community subgroups, is designed to help decision makers to make more rational decisions.

The methodology of the Goal Achievement Matrix, in its most complex form, involves proceeding from the identification of goals, with the focus throughout being on levels of goal achievement for the community as a whole and for groups within it. Objectives are derived from a set of pre-determined 'higher-level' ideals whose formulation is primarily the responsibility of the strategic decision makers. In contrast, the Planned Balance Sheet commences with the identification of all relevant effects and the preferences of those groups that are affected. The objectives are then formulated on the basis of *their* preferences. The Goal Achievement Matrix could, therefore, be interpreted as being directed at examining whether the proposals put forward have achieved certain aims that the strategic planners and decision makers have consciously set out to achieve. The Planned Balance Sheet, on the other hand, has a different emphasis, concentrating on the impacts of the proposals in question on the welfare of all groups and individuals affected.

Figure 10.1
Hong Kong Tunnel
(Source: W.S. Atkins).

10.5 Summary

The cost effectiveness technique is an important variant of cost–benefit analysis, particularly useful in situations where the relevant benefits of the competing scheme options are quantifiable but not in monetary terms. Choice of the correct measure or measures of effectiveness is crucial to the achievement of meaningful results.

The Planned Balance Sheet allows qualitative criterion scores to be evaluated alongside monetary valuations and allows the distribution of all costs and benefits across the relevant sectors of the community to be made known. It is, however, a rather unwieldy technique. Interpretation of the data contained within it requires a high level of technical expertise.

The Goal Achievement Matrix forms an important link between the first and second parts of this textbook. Although it is seen as a variant of the optimising cost–benefit analysis technique, its inclusion of non-monetary valuations results in a methodology that is best suited to assessing the relative strengths and weaknesses of competing options, rather than to assessing the absolute viability of each alternative. Thus, although it has its origins in the optimisation principle on which cost–benefit analysis is based, its structure and format are very much consistent with the compromise principle inherent in multicriteria models. Indeed, the Goal Achievement Matrix shares common features with one particular form of a multicriteria model detailed in Part II of the book, the Simple Additive Weighting (SAW) Model. This technique, along with a number of other multicriteria models used in the appraisal of engineering projects, is dealt with within the succeeding chapters.

References

Carley, M. (1979) *Rational Techniques in Policy Analysis*. Policy Studies Institute, London.

Hill, M. (1966) *A Method for Evaluating Alternative Plans: The Goal Achievement Matrix Applied to Transportation Plans*. PhD dissertation, University of Pennsylvania.

Hill, M. (1968) A Goal Achievement Matrix for Evaluating Alternative Plans. *Journal of the American Institution of Planners*, **34**, 19–29.

Hill, M. (1973) *Planning for Multiple Objectives: An Approach to the Evaluation of Transportation Plans*. Technion, Philadelphia, PA.

Hill, M and Schechter, M. (1971) Optimal Goal Achievement in the development of outdoor recreational facilities, in *Urban and Regional Planning* (ed. A.G. Wilson), Pion, London, pp. 110–120.

Lichfield, N. (1969) Cost Benefit Analysis in Urban Expansion – a case study: Peterborough. *Regional Studies*, **3**, 123–155.

Lichfield, N. (1971) Cost Benefit Analysis in Planning: A Critique of the Roskill Commission. *Regional Studies*, **5** (3), 157–183.

Lichfield, N., Kettle, P. *and Whitbread, M.* (1975) *Evaluation in the Planning Process*, Pergamon, Oxford, UK.

Massam, B. (1980) *Spatial Search*. Pergamon International Library of Science and Technology, Engineering and Social Studies, Pergamon Press, London.

PART 2

NON-ECONOMIC-BASED PROJECT APPRAISAL TECHNIQUES

Chapter 11

Multicriteria Analysis

11.1   Introduction

The book has thus far concentrated on appraisal procedures where the evaluation is predominantly an economic one. A number of techniques have been examined that seek to augment the somewhat narrow focus of this type of analysis, so that attributes that are not readily valued in money terms can be included within the analysis. This was done either by subjecting the attribute to a technique that allowed it to be expressed in monetary terms or by altering the technique to allow the inclusion of the attribute in its non-monetary form. In both situations, the analysis remains essentially an economic one, with certain modifications or adjustments carried out in order to permit a more inclusive evaluation. Many of the simpler decision problems in the engineering field involve quantitative attributes that are relatively easy and straight-forward to both define and measure in money terms.

We now move firmly away from the concept of the appraisal process being a purely economic one, with the possibility of making some adjustments in order to widen the focus of the study to the point where the analysis is seen as multicriteria-based, where all attributes, be they monetary or non-monetary, economic or environmental, are assessed on an equal basis. Many of the more complex decision problems involve attributes that are difficult to both define and measure in monetary terms. While some attributes may be quantifiable, it may prove impossible to translate these into monetary values. Others may be intangibles/qualitative attributes, with no attainable quantitative measure of their effect. The appraisal of many public sector engineering development projects involves consideration of such decision criteria. The multicriteria framework allows such factors to be presented in a comprehensive and consistent format. The goal achievement technique explained at the end of Chapter 10 forms an important bridgehead between the two parts of this book. While it is a methodology derived from the cost–benefit analysis framework, it has evolved into a decision system that allows all types of attributes to be assessed on an equal footing, possessing many of the characteristics of an effective multicriteria model.

Engineering Project Appraisal: The Evaluation of Alternative Development Schemes, Second Edition.
Martin Rogers and Aidan Duffy.
© 2012 John Wiley & Sons, Ltd. Published 2012 by John Wiley & Sons, Ltd.

The overall strategy within multicriteria decision models involves *decomposition* followed by *aggregation*. The decomposition process divides the problem into a number of smaller problems involving each of the individual criteria. This breaking down of a problem into a number of smaller problems makes it easier for the decision maker to analyse the information coming from diverse origins. The process of aggregation allows all the individual pieces of information to be drawn together to allow a final decision to be made. Within multicriteria models, the process of aggregation involves either the use of information or the making of certain assumptions concerning the relative importance weightings of the different criteria.

This type of decision model is particularly relevant to the appraisal process for a public sector based engineering proposal where environmental and social criteria must be assessed on a somewhat equal footing to the economic considerations. Use of a process that is based on an ability to include all possible factors, rather than one that leads to the exclusion or marginalisation of certain classes of attribute, is much more likely to lead to public acceptance of whatever the appraisal reveals.

11.2   Multicriteria evaluation models

The models examined within the book can be broken down into four groups:

(1)   Simple 'non-compensatory' methods
(2)   Simple Additive Weighting Method
(3)   Analytic Hierarchy Process
(4)   Concordance Analysis techniques

Methods such as dominance, satisficing, sequential elimination and attitude-oriented techniques are termed 'simple' because selection of the preferred option does not involve trading-off the disadvantages on one criterion against the advantages on another. There is no process of decomposition followed by aggregation. Superiority in one criterion cannot be offset by an inferior performance on some other one. It is, therefore, a simpler process but one which leads to less rational outcomes than some of the more complex 'compensating' multicriteria techniques referred to subsequently.

With dominance techniques, options are denoted as being members of an exclusive 'non-dominated' set if no other option exists that performs better than it on any of the criteria. Otherwise the option is deemed to be 'dominated' by another. Satisficing techniques do not actively choose the best options but merely eliminate those that do not meet certain minimum performances associated with each of the criteria. In the case of sequential elimination techniques, the decision is made by reference to only one criterion, with option choice based on the relative performance of this criterion alone. After the initial evaluation on the first chosen criterion, if more than one option remains, a second criterion can be used to separate them. Attitude-oriented methods take into account the decision maker's attitude towards the environment within which the choice is being made, this attitude being either pessimistic, in which case selection of an option is based on how badly it performs on its worst scoring criterion, or optimistic, in which case selection is based on how well it performs on its best scoring criterion.

After these simple decision models, all the approaches explained in detail in this book involve the processes of decomposition and aggregation, and all entail some level of compensation of performances between the different criteria.

The *Simple Additive Weighting Method* is a classical 'compensatory' model. To allow compensation to take place, the performances of a given option on the different criteria must be put onto one common scale of measurement. The individual criterion scores for each option can then be manipulated mathematically to compute an index of overall performance for each one. These indices enable the options to be ranked relative to each other. The Additive Weighting Method is the simplest form of a more general decision model based in multi-attribute utility theory (MAUT). MAUT, or utility theory, is widely used to solve project appraisal problems, not only in the engineering sector but also in the financial and actuarial spheres. It involves devising a function, U, that will express the utility of a project option in terms of a number of agreed relevant decision criteria (Keeney and Raiffa, 1976).

Checklists are one of the oldest forms of the Simple Additive Weighting Method, and are most often applied to the appraisal of projects where environmental concerns are of overriding importance. Their form varies from the straightforward use of a set of weights to combine criterion performances measured on identical linear scales to the employment of more complex mathematically based predictive techniques (Bissett, 1978).

The *Analytic Hierarchy Process* (*AHP*) has its basis in MAUT. It works by establishing hierarchies within the problem under examination. It breaks down the decision into a number of discrete elements within each level of the hierarchy and then uses a pairwise comparison methodology to establish priorites both within and between the hierarchies put in place by the decision maker. On each criterion in question, the relative merit of each project option is determined from a pairwise analysis of the scores for all combinations of options. A similar pairwise analysis is also used to determine their relative importance. A combination of these two processes yields a relative ranking of all options (Saaty, 1977, 1980).

Concordance Analysis is a 'partially-compensating' decision modelling technique. It is described as such because there is no question of the 'trading-off' of one criterion directly against another for each project option so that each can be given a cumulative score indicating its 'attractiveness' (Rogers and Bruen, 1995). It is a pairwise method where, on each criterion, the level of dominance of one given option over another determines its concordance score. For every pair of options, the scores on the different criteria are combined, using the importance weightings chosen for the different criteria, to give a concordance index for that pair. As this can be seen as an indirect form of 'trading-off' scores from one criterion against another, the process is called partial compensation. The end result is a ranking or grouping of the options concerned rather than the determination of a score for each. Calculation of the criterion weightings is again a very important step within this type of model.

The Simple Additive Weighting Method, the Analytic Hierarchy Process and Concordance Analysis techniques are dealt with in detail in separate succeeding chapters. The remainder of this chapter outlines the 'simple' multicriteria methods referred to above.

11.3   Simple non-compensatory methods

11.3.1   Dominance

An option is dominated if another option exists that performs better than it on one or more of the decision criteria and equals it on the remainder. It is, in effect, a screening or 'sieving' process, where the options being evaluated are reduced to a shortlist by eliminating all those that are dominated.

The sieving process proceeds as follows:

(1)   Compare the first two options; if one is dominated by the other, discard the dominated one.
(2)   Next, compare the retained option with the third option and again discard the dominated one.
(3)   Then introduce the fourth option and so on as before.
(4)   If the process involves n project options the process requires $n - 1$ steps.

Example 11.1   Selecting a site for a paper mill
Developers wishing to build a new paper mill are considering seven potential sites. An appropriate site, in their view, should possess the following attributes (Yoon and Hwang, 1995):

- A good supply of water and manpower.
- A positive community attitude to water pollution.
- A small likelihood of union formation.

These three attributes are translated into three criteria, C_1, C_2 and C_3 respectively. C_1 and C_2 are measured on a four-point scale: *poor – fair – good – excellent*. C_3 is measured as a probability from zero to one, with preference being in favour of the lower score.

Table 11.1 gives the score for the seven sites, S_1 to S_7, on each of the three criteria.

Solution

The first option S_1 is compared with S_2. S_1 is better than S_2 on criteria C_2 and C_3 but worse than S_2 on criterion C_1. Therefore, neither dominates the other. Now compare

Contd

Example 11.1 Contd

S_3 with S_1. S_3 is worse than S_1 on criteria C_2 and C_3 and equal to S_1 on C_1. S_3 is thus dominated by S_1. S_4 is now compared with S_1 and S_2. It is not dominated by either of them. The process is continued until S_7 is reached. The dominance relationships are shown in Figure 11.1. Three of the options, S_2, S_4 and S_5, are non-dominated and constitute the shortlist of preferred options. All others are dominated: S_1 and S_6 by S_5, S_7 by S_2 and S_5, and S_3 by S_4.

Table 11.1 Decision matrix for the seven sites.

Site	Criteria		
	C_1	C_2	C_3
S_1	Poor	Good	0.5
S_2	Excellent	Fair	1.0
S_3	Poor	Poor	1.0
S_4	Fair	Fair	0.1
S_5	Good	Excellent	0.2
S_6	Fair	Good	0.9
S_7	Good	Fair	1.0

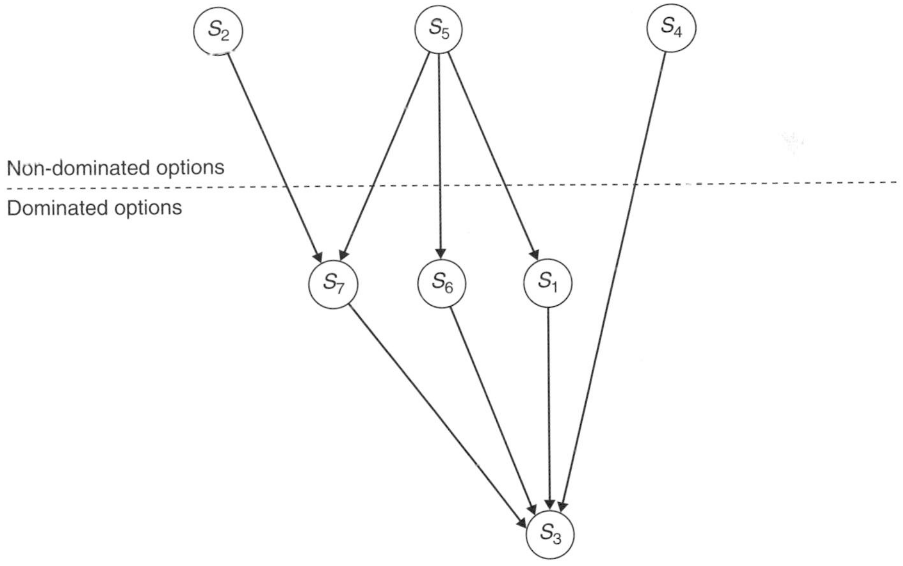

Figure 11.1 Dominance graph for the seven options.

While the shortlisting of non-dominated options may appear to be a rational attempt to isolate the better performing proposals under examination, a study by McAnarney (1987) found that some dominated options might, in overall terms, be better than some of the non-dominated options. For example, consider three options, A_1, A_2 and A_3, evaluated on the basis of four criteria, C_1, C_2, C_3 and C_4 as shown in Table 11.2.

Option	Criteria			
	C_1	C_2	C_3	C_4
A_1	8	9	6	9
A_2	8	8	6	9
A_3	3	3	7	4

Table 11.2 Performance matrix for four options.

Option A_2 is dominated by A_1 although the difference between the two is marginal. A_3, however, is non-dominated, even though it is considerably worse than A_2. A_2 would therefore not be on the shortlist, even though it performs considerably better than one that is (A_3). The chances of such an error being made can be reduced by using one of the other simple methods referred to below (such as the Conjunctive Method) immediately prior to use of the dominance method.

11.3.2   Satisficing methods

These methods are not used to identify a chosen best option. They work on the basis of dividing proposals into two categories, acceptable and unacceptable. An option is acceptable if it meets certain minimum standards set by the decision maker. There are two distinct methods within this system – conjunctive and disjunctive.

Conjunctive method

Using this approach, any option that has a criterion valuation less than the standard level will be rejected. For an option to be chosen, it must exceed a minimum value on all criteria. Options are thus easily rejected using this technique; they need to fail on only one criterion. The approach requires the decision maker to generate the minimum cut-off criterion valuations. The setting of these cut-off values is critical to the proper working of the technique. If they are set too high, all options will be deemed unacceptable, whereas if they are set too low too many will remain. Sometimes the standard levels are increased gradually and the group of acceptable options is narrowed down to one choice.

Example 11.2
Decision makers for the military wish to purchase a new fleet of off-road vehicles. The choice is between four models, A_1, A_2, A_3 and A_4. Six criteria (C_1 to C_6) are used to assess the options; maximum speed (km/h), operating range (km), maximum payload (kg), purchase cost (£000), reliability (four-point scale), and technical specification (four-point scale). Table 11.3 gives the performance of each option on the six criteria. The decision maker specified the following minimum cut-offs for each of the criteria:

Contd

Example 11.2 Contd

$C_1 = \geq 100$

$C_2 = \geq 1500$

$C_3 = \geq 2000$

$C_4 = \leq 60$

$C_5 = \geq fair$

$C_6 = \geq fair$

Solution

Given these minimum acceptable values, A_1 and A_4 are deemed acceptable. They satisfy all the requirements. A_2 is unacceptable because its maximum payload is too low and its cost is too high. The maximum speed of A_3 is too low.

Table 11.3 Decision matrix for off-road vehicles.

Option	Criteria					
	C_1	C_2	C_3	C_4	C_5	C_6
A_1	100	1500	2000	55	Fair	Excellent
A_2	125	2700	1800	65	Poor	Fair
A_3	90	2000	2100	45	Good	Good
A_4	110	1800	2000	50	Fair	Fair

Disjunctive method

An option is chosen in this approach if it exceeds some specified cut-off value on at least one of the criteria. In effect, an option is deemed acceptable if it performs to a high level on one of the criteria. Those that do not perform exceptionally well on any of the decision criteria are not selected under this process.

Example 11.3

Taking the same problem as in Example 11.2, the decision maker specifies the following desirable scores for each of the six criteria:

$C_1 = \geq 130$

$C_2 = \geq 2800$

$C_3 = \geq 2200$

$C_4 = \leq 40$

$C_5 = \geq excellent$

$C_6 = \geq excellent$

Contd

Example 11.3 Contd

Solution

Given these desirable levels, only A_1 is deemed acceptable because of its excellent technical specification. None of the other options attains any of the minimum levels assigned to the decision criteria.

11.3.3   *Sequential Elimination Methods*

There are two approaches that eliminate options in a sequential way; the Lexicographic and the Elimination by Aspects techniques. Both use one criterion at a time to evaluate the options, excluding all except the best performing one. If a tie exists between options, it is resolved by examining their relative performance on another criterion.

Lexicographic method

Within this method, the options are examined initially on the criterion deemed to be the most important. It is used mostly in cases where one criterion predominates. If one option has a better score on the chosen criterion than all other options, it is chosen and the process ends. If, however, a number of options are tied on the most important criterion, this subset of tied options is then compared on the next most important criterion. The process continues in a sequential manner until a single option is identified or all criteria have been gone through and complete separation proves impossible.

For example, take the situation where a number of project options are being evaluated and cost is seen as the most important criterion. Only where the cost for two options is equal does some other criterion come into play to separate them. This method can be altered slightly by stating that differences in scores between options on a given criterion must be greater than a certain threshold before they are deemed significant. This variation is called a 'lexicographic semi-order'.

Example 11.4   (Yoon and Hwang, 1995)

A building contactor wishes to buy a number of new dumper trucks for its site operation. Eight different makes of truck are considered, M_1 to M_8, on four criteria, list price in £ (C_1), resale value in £ (C_2), handling on a 10-point scale from 1 (worst) to 10 (best) (C_3) and acceleration (seconds for 0 to 60 mph) (C_4).

The performance matrix for the eight options is given in Table 11.4.

Contd

Example 11.4 Contd

The decision maker has placed the criteria in the following order of importance:

(1)   Handling (C_3)
(2)   List price (C_1)
(3)   Acceleration (C_4)
(4)   Resale value (C_2)

Identify the option to be selected using:

(i)   The basic lexicographic method
(ii) The lexicographic semi-order, with the following thresholds of significance specified:
  - Handling   – more than two grades
  - List price   – more than £3000
  - Acceleration – more than three seconds
  - Resale value – more than £1500

Solution

Part (i)
On the most important criterion, C_3, five options (M_1, M_2 M_3, M_4 and M_7,) are tied for first position. On the second most important criterion, C_1, M_1 is selected as it has the lowest price, £1300 less than the next cheapest option.

Part (ii)
Let us now approach the problem using the lexicographic semi-order method, where the same basic logic is used but the difference in performance between any pair of options on a given criterion must exceed a certain threshold before they can be separated. Using the thresholds of significance referred to above and the same order of importance for the criteria, the same five options, M_1, M_2 M_3, M_4 and M_7, are tied on the most important criterion, handling. No other option is within the threshold set, a two-point difference. Examining these options on the second ranked criterion, list price, M_1 is the lowest, with three of the remaining options, M_2 M_3, and M_7, within £3000 of its list price. Thus, four options still remain. On the third most important criterion, acceleration, M_3 is the best rated at 10.8 seconds. None of the three remaining options fall within the threshold value of significance of two seconds, with M_1, M_2 and M_7 being 3.9, 8.3 and 4.5 seconds slower, respectively. Since only one model remains, M_3 is chosen dumper truck.

Option	Criteria				
	C_1	C_2	C_3	C_4
M_1	8 300	3 000	7	14.7
M_2	9 600	3 600	7	19.1
M_3	10 580	3 600	7	10.8
M_4	13 700	6 000	7	13.0
M_5	29 850	12 000	4	13.7
M_6	11 050	3 600	4	16.2
M_7	9 800	3 600	7	15.3
M_8	27 650	12 000	1	13.5

Table 11.4 Performance matrix for dumper truck options.

Elimination by aspects

This technique is very similar to the lexicographic method. Put forward by Tversky (1972). It also examines performances on one criterion at a time when making comparisons between proposals. In this case, however, options that do not meet some specified standard are gradually eliminated until all except one have been removed. Also, criteria are ranked in terms of their probabilistic discrimination power rather than in terms of their importance. This means that those criteria that will be more successful in eliminating options from the process will be used first.

Within the technique devised by Tversky, each option is viewed as a set of aspects represented by criteria that may be quantitative or qualitative in nature. In an effort to describe the model, Tversky described a television commercial advertising a computer course:

'There are more than 24 companies in the San Francisco area offering training in computer programming'.

The announcer puts two-dozen eggs and one walnut on the table to represent the options and continues:

'Let us examine the facts. How many of these schools have on-line computer facilities for training?'

The announcer removes several eggs.

'How many of these schools have placement services that would help you find a job?'

The announcer removes some more eggs.

'How many of these schools are approved for veterans' benefits?'

This continues until the walnut alone remains.

The announcer cracks the nutshell revealing the name of the company and concludes:

'This is all you need to know in a nutshell'.

Thus, the following aspects are used to eliminate options:

- On-line computer facilities for training.
- Placement services.
- Approval for veterans' benefits.

11.3.4 Attitude-Oriented Methods

The following two non-compensatory models take into account the decision maker's attitude towards the environment within which the choice is being made. In the case of the Maximin technique, the decision maker has a pessimistic attitude, therefore the worst performing criterion for each option is identified and the option that scores best on its 'worst criterion' is chosen. With Maximax, an optimistic attitude prevails. Here the option scoring best on its 'best criterion' is chosen.

Maximin

Any team is only a strong as its weakest link. In the context of an engineering development project, there may be circumstances where the strength of any given proposal will be gauged by how well it performs on its weakest decision criterion.

Where a decision maker does not have any prior information regarding which criterion will have the greatest influence on overall performances, a pessimistic attitude should be adopted, where the option whose worst score is better than the worst performances of the others is chosen. In other words, the 'best of the worst' is being selected.

An option is thus represented by its single worst criterion performance. All other criterion scores are disregarded. Since we may be comparing scores from different criteria, the technique only works where all criteria are measured on a common scale that can be quantitative or qualitative.

The selection process involves two steps:

(1)   Determine the worst criterion score for each option.
(2)   Choose the option with the best score on its worst criterion.

Example 11.5

A local authority wishes to select a site for a major regional landfill facility. Six sites, A_1 to A_6, are put forward for consideration on five criteria, C_1 to C_5:

(1)   Road access (C_1)
(2)   Effect on the landscape (C_2)
(3)   Proximity to centres of population (C_3)
(4)   Ecology (C_4)
(5)   Archaeological significance (C_5)

All criteria are scored on a 10-point scale from 1 (worst) to 10 (best). Table 11.5 gives the decision matrix for the six sites.

Contd

> **Example 11.5 Contd**
> *Solution*
>
> The manager of the local authority cannot decide which criterion will prove most important and is afraid that any option, no matter how good its overall performance, may attract widespread public opposition if it performs badly on one criterion. He therefore opts for choosing a site that will be a consistent performer over all the necessary attributes rather than one that performs brilliantly on some but relatively poorly on others.
>
> Examining Table 11.5, the seventh column lists the worst scores for each option. A_4 is then selected as having the highest 'worst' score. As its performance does not dip below five on any criterion, it is seen to fulfil the objectives set down by the manager of the local authority. While option A_6 scores extremely well on four out of five criteria, its negative effect on the existing landscape leads to it performing very poorly on C_2. Use of this model implies that the decision maker would see this poor score as a fundamental barrier to acceptance of option A_6.

Table 11.5 Performance matrix for the six sites.

Site	C_1	C_2	C_3	C_4	C_5	Min	
A_1	4	6	3	2	3	2	
A_2	7	2	8	2	4	2	
A_3	8	5	4	6	3	3	
A_4	6	7	5	5	6	5	← Max
A_5	3	5	6	8	7	3	
A_6	8	9	2	8	8	2	

Maximax

In contrast to Minimax, this technique chooses an option based on its best criterion score rather than its worst. The method is useful in situations where options can be chosen for specific functions based on its performance any one of the particular criteria in question.

Again, within this method, only one criterion score represents each option under consideration. All other criterion performances are disregarded. Also, it too requires that all criteria be scored on a common scale.

Maximax has two operating procedures also:

(1) Determine the best criterion score for each option.
(2) Choose the option with the best score on its best criterion.

Example 11.6

An engineering company wish to select a new chief executive. The board of directors feels that the person selected should have 'star quality' in at least one of a set of attributes on which all candidates are evaluated. Seven candidates are shortlisted, A_1 to A_7, each evaluated on the following five criteria, C_1 to C_5:

(1)   Decision making skills (C_1)
(2)   Profile within the profession (C_2)
(3)   Academic and professional record (C_3)
(4)   People skills (C_4)
(5)   International reputation (C_5)

All criteria are scored on a 10-point scale from 1 (worst) to 10 (best). Table 11.6 gives the decision matrix for the seven candidates.

Solution

The board of directors wishes to select a candidate who is the 'best of the best' in one of the attributes relevant to the job. Any weakness in another criterion can be dealt with by a judicious choice of deputy chief executive. The Maximax technique is thus appropriate in this case.

Within Table 11.6, the seventh column lists the best scores for each candidate. A_2 is then selected on the basis of having a superb profile international reputation. Selecting a 'second-in-command' with good academic credentials can compensate for the chosen candidate's poor academic record. Candidate A_5 would be such a person.

Table 11.6 Performance matrix for the seven candidates.

Candidate	Criteria					Max	
	C_1	C_2	C_3	C_4	C_5		
A_1	3	4	6	3	2	6	
A_2	8	8	2	8	9	9	← Max
A_3	7	4	6	5	5	7	
A_4	7	2	2	8	4	7	
A_5	4	6	8	2	7	8	
A_6	5	7	4	5	6	7	
A_7	8	7	2	8	5	8	

11.4   Summary

The models described in the chapter provide many of the basic elements required for a sound decision process. The data requirements in most cases are relatively modest; some only requiring detailed data on the most important criteria. Some, such as the lexicographic semi-order allow uncertainty regarding criterion valuations to be brought into the model by the use of thresholds. In the context of the appraisal of complex engineering projects, where all decision criteria need to be examined in some detail, use of the attitude-oriented or lexicographic methods, which concentrate on each option's performance on one sole criterion, may be inappropriate. Furthermore, use of a dominance method to analyse a complex performance matrix containing options whose performances on the various criteria are conflicting, would not yield worthwhile results, as it is unlikely that very many options would be dominated. The method would thus not partition the proposals effectively into a relatively small kernel of shortlisted proposals. Satisficing techniques tend to identify options that meet some minimum acceptable standard set by the decision maker rather than one or more proposals that perform best within the overall decision matrix.

While more complex decision methods may be required to analyse a complex engineering project appraisal problem effectively, these simple techniques nonetheless provide an excellent basis for understanding many of the fundamental theoretical aspects of the more intricate models discussed later in the book.

References

Bissett, R. (1978) Quantification, decision-making and environmental impact assessment in the United Kingdom. *Journal of Environmental Management*, **7**, 43–58.

Keeney, R.L. and Raiffa, H. (1976) *Decisions with Multiple Objectives*. John Wiley & Sons, Inc., New York.

McAnarney, D.K. (1987) *Multiple Attribute Decision Making Methods: A comparative Study*. Masters Degree Thesis (unpublished), Kansas State University.

Rogers, M.G. and Bruen, M.P. (1995) Non-monetary based decision-aid techniques in EIA – an overview. *Proceedings of the Institution of Civil Engineers, Municipal Engineer*, **109** (June), 98–103.

Saaty, T.L. (1977) A scaling for priorities in hierarchical structures. *Journal of Mathematical Psychology*, **15**, 234–281.

Saaty, T.L. (1980) *The Analytic Hierarchy Process*. McGraw Hill, New York.

Tversky, A. (1972) Elimination by Aspects: A theory of choice. *Psychological Review*, **79**, 281–299.

Yoon, K.P. and Hwang, C-L. (1995) *Multiple Attribute Decision Making: An Introduction*. Sage Publications, Thousand Oaks, CA.

Chapter 12

The Simple Additive Model

12.1 Background

The Simple Additive Weighting Model has its basis in Multi-Attribute Utility Theory (MAUT). MAUT was devised by Keeney and Raiffa (1976) and is a methodology that is readily applicable to the appraisal of complex engineering projects. It permits possible consequences to be traded off against each other. It involves devising a function, U, that expresses the 'utility' of a project option in terms of a number of relevant decision criteria. Utility is a concept expressing a decision maker's level of satisfaction with a given outcome. It is used to determine the existence or absence of preference between the outcomes of a set of options under examination. Each outcome, or criterion as we have so far called it, has its own utility function.

A utility function is expressed on an ordered metric scale. The numbers of this scale have no absolute physical meaning and the scale is constructed by assigning numbers to any two points at its extremes. Usually these points correspond to the best possible outcome and the worst possible outcome for the attribute in question. Very often the best possible outcome is assigned a utility of one and the worst a utility of zero.

In decision making terms, a decision maker, when examining the performance of any two options on a given utility function will either prefer one to the other, in which case one will outscore the other, or be indifferent between them, in which case both will be assigned the same score. These are the only possible results. The utility function will monotonically increase from zero to one. This increase can be linear or non-linear.

In the vast majority of appraisals for complex engineering projects, the decision maker must take into account a large number of different attributes or types of consequences. These consequences relate to the economic, environmental, social and technical performance of the various options under examination. In principle, the same Utility Theory developed for the single decision attribute can be directly

Engineering Project Appraisal: The Evaluation of Alternative Development Schemes, Second Edition.
Martin Rogers and Aidan Duffy.
© 2012 John Wiley & Sons, Ltd. Published 2012 by John Wiley & Sons, Ltd.

extended to cover such cases. All the separate utility functions for the individual criteria are combined within one mathematical expression, called a multi-attribute utility function.

Any decision maker implicitly attempts to maximise some function, U, that aggregates all the different points of view to be taken into account. Therefore, if a decision maker is requested for information regarding a range of options, his reply will be both coherent and consistent with a certain unknown function, U. This function is now expressed in terms of a number of relevant criteria. Estimating the form of this function is the basis for the problem solving process within MAUT.

In the simplest approach, if the utility of each criterion is independent of that of the others (a property known as 'utility independence'), then the multi-attribute utility function can be constructed as a weighted average of the utility functions for each individual attribute (consequence), that is:

$$U(X) = \sum_{all\,i} w_i U(x_i) \tag{12.1}$$

where X is a vector of containing the n criteria:

$$X = (x_1, x_2, x_3, \ldots, x_n) \tag{12.2}$$

and w_i is the weight for criterion, i, which specifies the relative contribution of each criterion in the final decision.

In certain situations, the utility of certain criteria can be influenced by other criteria and utility independence cannot be assumed. Say, for example, two highway proposals are being compared on a pair of criteria, visual effects and political acceptability. If a change in the relative performance of the two options in terms of visual effects has an effect on the relative political acceptability of the two, utility independence is absent, and a more complex multi-attribute model must be used. One such form is the multiplicative model, which allows a cross influence between criteria.

The multiplicative form can be demonstrated for three criteria as follows:

$$U(X) = \begin{array}{l} k_1 u_1(x_1) + k_2 u_2(x_2) + k_3 u_3(x_3) \\ + K[k_1 k_2 u_1(x_1)u_2(x_2) + k_1 k_3 u_1(x_1)u_3(x_3) + k_2 k_3 u_2(x_2)u_3(x_3)] \\ + K^2 k_1 k_2 k_3 u_1(x_1)u_2(x_2)u_3(x_3) \end{array} \tag{12.3}$$

The first three terms correspond to the simple additive model, as all the other terms arise out of possible interactions between the different criteria arising out of an absence of utility independence.

The simple additive model is a perfectly valid model if one selects the criteria carefully in order to minimise the possibility of such intercriterion interactions. It is used in many project appraisals (Vincke, 1992) and is the form of the MAUT model dealt with in detail within this text. More detailed information on MAUT can be found in Keeney and Raiffa (1976).

12.2   Introduction to the Simple Additive Weighting (SAW) Method

The simple additive method is probably one of the prominent and most widely used multicriteria methods. In effect, a score is obtained by adding the contributions from each of the chosen criteria. Since two criteria with different measurement scales cannot be added, a common numerical scaling system, such as the normalised one referred to above, is required to allow the addition of the different criterion performances for each option. The total score for each option is estimated by multiplying the comparable normalised rating for each criterion by its importance weighting and then summing these products over all the criteria in question.

The overall weighted score, V_i, for a project option, i, using this method can be written as:

$$V_i = \sum_{j=1}^{j=n} w_j r_{ij} \tag{12.4}$$

where w_j is the weight for criterion j, r_{ij} is the rating for option i on criterion j. The option with the largest value of V_i is selected.

The additive weighting method thus converts the multicriterion problem to a single-dimension problem. Again, in the vast majority of complex engineering project appraisals, the different criterion types being assessed within the process, because of their diverse origins (economic, environmental, technical), are expressed in various units of measurement. These must then be converted to a common scale before the additive model can be used. A common procedure employed is to convert all criterion scores to a normalised linear scale going from zero (worst) to one (best), though this is often magnified to a 10-point or a 100-point scale.

It is important to emphasise the fundamental difference between the economic analysis of a proposal and a multicriteria evaluation. Within an economic study, all attributes are economics based and are measured in their original monetary units. Within the multicriteria analysis for an engineering proposal, while cost may well be included as a criterion because of the overall importance of its economic consequences, it is treated the same as any of the other criteria, whatever their origin or scale of measurement. Therefore, the original monetary value of the cost cannot be used. A feasible range of possible costs must be assigned to this criterion denoting the minimum (0) and maximum (1, 10 or 100) ratings for it on the linear scale. Each option examined can then be given a score on the standard scale based on its raw monetary score. A weighting for the cost criterion is then derived which reflects its importance relative to the others deemed relevant to the study. It is then in a form that can be included within the additive weighting model, with a single overall score being obtainable for each option examined.

Example 12.1   The Basic Simple Additive Weighting Model

Consider a comparative evaluation of three proposed designs for a new car. They are compared on the basis of the following five decision criteria:

(1)   Purchase cost (£)
(2)   Level of safety (10-point scale)
(3)   Aesthetic appearance (10-point scale)
(4)   Mass (kg)
(5)   Reliability (%)

Table 12.1 indicates the raw scores for each proposal on the five criteria.

   Table 12.2 shows the relative normalised weights for each criterion. All criterion scores are converted to dimensionless numbers by calculating the ratio of each criterion score relative to the best overall score for it over all the available options under examination. The best raw score in each case is assigned a rating of 10. In the case of cost and mass, lowest is best; with all others, highest is best.

   Take the cost criterion. The best score is for Option 2 (£11 000). This is assigned a rating of 10, with the other two being assigned the following ratings:

$$\text{Rating}_{(\text{Option 1})} = (11\,000 \div 18\,000) \times 10 = 6.1$$

$$\text{Rating}_{(\text{Option 3})} = (11\,000 \div 15\,000) \times 10 = 7.3$$

The same process is used to derive the ratings for mass.

   For the reliability criterion, the best score is for Option 1 (95%). This is assigned the 10-rating, with the other ratings as follows:

$$\text{Rating}_{(\text{Option 2})} = (80 \div 95) \times 10 = 8.4$$

$$\text{Rating}_{(\text{Option 3})} = (88 \div 95) \times 10 = 9.3$$

Safety and appearance need no conversion as they are already scored on a 10-point scale. The full set of ratings for each option is given in Table 12.3.

   The final step involves the multiplication of the ratings by the relevant criterion weights. This calculation is shown in Table 12.3. It can be seen that the preferred proposal is Option 1.

Table 12.1 Raw criterion scores.

Option	\multicolumn{5}{c}{Criteria of evaluation}

Option	Cost (£)	Safety (0–10)	Appearance (0–10)	Mass (kg)	Reliability (%)
Car_1	18 000	9	8	970	95
Car_2	11 000	8	4	720	80
Car_3	15 000	7	6	600	88

Criterion	Weight
Cost	0.25
Safety	0.20
Appearance	0.25
Mass	0.10
Reliability	0.20

Table 12.2 Criterion weightings.

Criterion	Weighting	Option 1	Option 2	Option 3
Cost	0.25	6.10	10.00	7.30
Safety	0.20	9.00	8.00	7.00
Appearance	0.25	8.00	4.00	6.00
Mass	0.10	6.20	8.30	10.00
Reliability	0.20	10.00	8.40	9.30
Score	—	*7.95*	*7.61*	*7.59*

Table 12.3 Overall scores for the three options.

12.3   Sensitivity testing

The data that is input into a decision model is seldom known with complete certainty. A sensitivity analysis allows the decision maker to gauge the effect that incremental changes in criterion weights and valuations will have on the final result. The importance weightings are of particular interest within a sensitivity analysis, as their valuation is the result of subjective judgements by experts who, because of their diverse backgrounds, may disagree about their correct value.

With criterion ratings, sensitivity testing arises from errors that may derive from the actual estimation of the valuations themselves. They may be based on incomplete data or, in the case of qualitative assessments, may be derived from subjective judgments by specialists from the relevant field. Their exact value may therefore not be known with complete confidence.

The process can be arduous and time consuming, involving a large number of iterations, as different criterion and weighting scores are varied incrementally and the effect of these changes on the original ranking of options is gauged. Incremental changes in criterion valuations possessing a high weighting score are more likely to alter the baseline ranking than changes in the scores of less significant criteria. Particular attention must also be paid to criterion scores that involve a high degree of uncertainty and subjectivity. A competent decision maker should identify such criteria and analyse the effect of their variation on the overall ranking.

Example 12.2   Using a sensitivity analysis within the SAW Method
A timber manufacturing firm wishes to open a new plant and six alternative locations have been identified. The sites, all in different countries, are assessed on the basis of the following five criteria:

Contd

Example 12.2 Contd

(1) C_1 – Cost of land purchase (£m)
(2) C_2 – Building costs (£k)
(3) C_3 – Labour costs (£)
(4) C_4 – Ease of transportation (10-point scale)
(5) C_5 – Suitability of climate (10-point scale)

The performance matrix for the six sites is indicated in Table 12.4.

(i)   Rank the six options using the simple additive weighting model.
(ii) Gauge the effect a $\pm 50\%$ variation in weightings for the first three criteria (C_1, C_2 and C_3) will have on the baseline rankings.

Solution

Part (i)
Because three of the criteria have zero valuations on one of the options, the transformation process employed in the previous example cannot be used for these three criteria to translate the raw scores onto a common 10-point scale. Instead, the best score is assigned the 10-rating and the worst score assigned the 0-rating. All other valuations are assigned scores in between these two boundary points by means of linear interpolation, that is:

- For criterion 1 – land costs
- Site No. 1 (cost £0; 10-rating)
- Site No. 5 (cost £17.9m; 0-rating)

Therefore:
Sites No. 2, 3 and 4 (cost £2 million)

$$\text{Rating} = 10 \times ((17.9 - 2.0) \div 17.9)$$
$$= 8.9$$

Site No. 6 (cost £0.6 million)

$$\text{Rating} = 10 \times ((17.9 - 0.6) \div 17.9)$$
$$= 9.7$$

The same calculations are completed for the second and third criteria. In the case of 'building costs', Site No. 5 is assigned the 10-rating and Site No. 2 is assigned the 0-rating. For 'labour costs' Site No. 6 is assigned the 10-rating with the 0-rating again going to Site No. 2. The baseline scores for the six options are given in Table 12.5. Site 1 is the preferred option in the baseline case.

Contd

Example 12.2 Contd

Part (ii)

The analysis can be broken down into the following six sensitivity tests (ST_1 to ST_6):

$ST_1 - C_1 + 50\%$, all other weights the same
$ST_2 - C_1 - 50\%$, all other weights the same
$ST_3 - C_2 + 50\%$, all other weights the same
$ST_4 - C_2 - 50\%$, all other weights the same
$ST_5 - C_3 + 50\%$, all other weights the same
$ST_6 - C_3 - 50\%$, all other weights the same

The weighting systems derived from the above adjustments are shown in Table 12.6. Table 12.7 shows the overall scores and rankings for each of the options on each of the six sensitivity tests. Graphical illustrations of the results of the baseline and sensitivity tests are shown in Figures 12.1a and 12.1b.

Site 1 and Site 4 are virtually inseparable. Although Site 1 is very marginally better within the baseline test, the sensitivity tests show the two running very close. These two sites are well ahead of the other four sites. Site 3 is ranked third, marginally ahead of Site 6, which is itself marginally better than Site 5. Site 2 is definitely sixth and last.

In overall terms, the data indicate that Sites 1 and 4 are the best options, well ahead of Sites 3, 5 and 6. Site 2 is definitely the worst. These relative positions are reinforced by use of Borda's Sum of Ranks Method (Vansnick, 1986). It sums the individual ranking positions of each option for each of the sensitivity tests and calculates the average, placing the lowest score in first position and so on. Sites 1 and 4 have equal scores (1.83) with Sites 3, 5 and 6 all clustered together between 3.5 and 4. Site 2 is last with a score of 6.0. The average rank for each option is given in Table 12.7.

Final Result
Sites 1, 4 → Sites 3, 6, 5 → Site 2

Table 12.4 Raw criterion scores.

	Criteria of evaluation				
Option	Land (£m)	Building (£m)	Labour (£k)	Transport (0–10)	Climate (0–10)
Site$_1$	0.00	1.20	12.50	2.50	2.50
Site$_2$	2.00	12.00	72.00	5.00	5.00
Site$_3$	2.00	5.50	5.00	5.00	5.00
Site$_4$	2.00	3.00	5.00	5.00	5.00
Site$_5$	17.90	0.00	3.80	10.00	7.50
Site$_6$	0.60	10.90	0.00	7.50	10.00
Weight	*0.28*	*0.28*	*0.25*	*0.15*	*0.04*

Table 12.5 Baseline rankings of the six options.

Criterion	Weighting	Site 1	Site 2	Site 3	Site 4	Site 5	Site 6
Land	0.28	10.00	8.900	8.900	8.900	0.0	9.700
Building	0.28	9.00	0.000	5.500	7.500	10.0	0.900
Labour	0.25	8.30	0.000	9.300	9.300	9.5	10.000
Transport	0.12	2.50	5.000	5.000	5.000	10.0	7.500
Climate	0.07	2.50	5.000	5.000	5.000	7.5	10.000
Score		7.87	3.442	7.307	7.867	6.9	7.068
Rank		*1st*	*6th*	*3rd*	*2nd*	*5th*	*4th*

Table 12.6 Criterion weightings within the six sensitivity tests.

Criterion	Normalised criterion weightings					
	ST_1	ST_2	ST_3	ST_4	ST_5	ST_6
Land	0.369	0.163	0.246	0.326	0.249	0.320
Building	0.246	0.326	0.369	0.163	0.249	0.320
Labour	0.219	0.291	0.219	0.291	0.333	0.143
Transport	0.105	0.139	0.105	0.139	0.107	0.137
Climate	0.061	0.081	0.061	0.081	0.062	0.080

Table 12.7 Results of sensitivity tests.

Option	Sensitivity Tests						Avg. Rank
	ST_1	ST_2	ST_3	ST_4	ST_5	ST_6	
$Site_1$	8.13 (1st)	7.52 (3rd)	8.00 (1st)	7.68 (3rd)	7.91 (2nd)	7.80 (1st)	1.83
$Site_2$	4.11 (6th)	2.55 (6th)	3.01 (6th)	4.00 (6th)	3.05 (6th)	3.93 (6th)	6.00
$Site_3$	7.50 (3rd)	7.04 (4th)	7.08 (4th)	7.60 (4th)	7.52(3rd)	7.02(3rd)	3.50
$Site_4$	7.99 (2nd)	7.69 (2nd)	7.82 (2nd)	7.92 (2nd)	8.02 (1st)	7.66 (2nd)	1.83
$Site_5$	6.05 (5th)	8.02 (1st)	7.28 (3rd)	6.39 (5th)	7.18 (5th)	6.52 (5th)	4.00
$Site_6$	7.39 (4th)	6.63 (5th)	6.31 (5th)	8.07 (1st)	7.39 (4th)	6.64 (4th)	3.83

12.4   Probabilistic Additive Weighting

Variations in results from a decision model are usually a consequence of the variability associated with estimates of the various criteria under consideration. As shown in Section 12.3, a sensitivity analysis is one way to evaluate the effects of such changes in criteria valuations on the final result. Another method for evaluating the effect of alterations in criterion scores on the final result is to directly incorporate the uncertainty associated with the criterion estimates into the additive weighting model.

The two basic probability concepts required within this analysis are *expected value* and *variance*. Let us examine these two briefly.

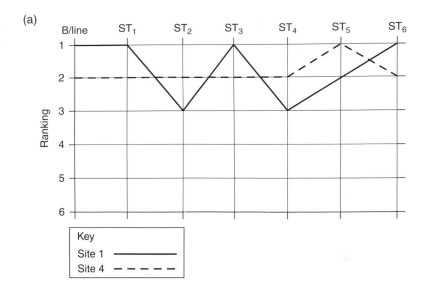

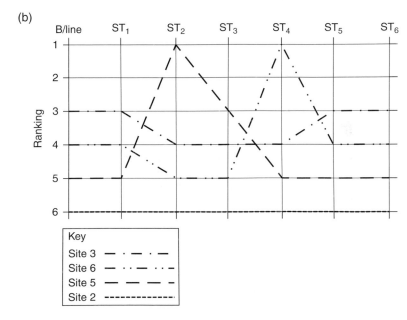

Figure 12.1 Illustration of baseline and sensitivity test results: (a) two best-performing options; and (b) other four options.

12.4.1 *Expected value*

This is a standard measure for comparisons involving uncertainty and incorporates the effect of uncertainty on criterion scores by use of a weighted average formula. Results are weighted based on their probability of occurrence. The sum of the products of all criterion scores for a given alternative by their respective probabilities yields its expected value. The expected value of criterion i for option j can be denoted by the formula:

$$EV(v_{ij}) = \sum_{i=1}^{z} p_{ij} x_{ij} \qquad (12.5)$$

where z is the number of possible results, x_{ij} is the value of the result for option j on criterion i and p_{ij} is the probability of the result for option j on criterion i.

For a given option, once the expected value has been estimated for each criterion, the aggregate score is computed using the equation:

$$EV(U_j) = \sum_{i=1}^{n} w_i EV(x_{ij}) \qquad (12.6)$$

where U_j is the value of the result for option j on criterion i, n is the number of criteria and w_i is the importance weight for each criterion i.

12.4.2   Variance

The variance is a measure of the dispersion or spread around the expected value as computed above. The standard deviation is the square root of the variance. It is estimated for option j on criterion i using the equation:

$$VAR(v_{ij}) = \sum_{i=1}^{z} p_{ij} x_{ij}^2 - [EV(v_{ij})]^2 \qquad (12.7)$$

Again, once this value has been estimated on each of the criteria in question, the aggregate variance for option j over n criteria is estimated:

$$VAR(U_j) = \sum_{i=1}^{n} w_i^2 \, VAR(x_{ij}) \qquad (12.8)$$

Thus, both the expected values and variances for each of the options under consideration can be estimated and used to identify the best proposal.

Example 12.3

A car manufacturer is assessing three vehicle assembly systems, S_1, S_2 and S_3. Three criteria form the basis for the evaluation, C_1 (reliability), C_2 (ease of maintenance) and C_3 (cost). The assessors assigned the following normalised importance weightings to the decision criteria:

- $C_1 = 0.16$
- $C_2 = 0.36$
- $C_3 = 0.48$

Contd

Example 12.3 Contd

The basic scores of each option on the three criteria are given in Table 12.8, along with their aggregate scores.

The decision maker believes that there is a substantial level of uncertainty associated with the reliability and maintenance criteria. To reflect this uncertainty, a pessimistic, optimistic and most likely estimate for each of the two criteria are generated, together with their associated probabilities. These are given in Table 12.9. These enable the expected value and variance to be calculated for the two criteria as follows.

Expected Value for Option S_1

$$
\begin{aligned}
EV_{C1} &= (0.2)(30) + (0.6)(43) + (0.2)(58) &&= 43.4 \\
EV_{C2} &= (0.1)(32) + (0.75)(44) + (0.15)(60) &&= 45.2 \\
EV_{C3} &= (0)(0) + (1)(30) + (0)(0) &&= 30 \\
EV_{AGG} &= (0.16)(43.4) + (0.36)(45.2) + (0.48)(30) &&= \mathbf{37.616}
\end{aligned}
$$

Expected Value for Option S_2

$$
\begin{aligned}
EV_{C1} &= (0.25)(23) + (0.5)(31) + (0.25)(38) &&= 30.75 \\
EV_{C2} &= (0.2)(26) + (0.7)(34) + (0.1)(43) &&= 33.33 \\
EV_{C3} &= (0)(0) + (1)(25) + (0)(0) &&= 25 \\
EV_{AGG} &= (0.16)(30.75) + (0.36)(33.33) + (0.48)(25) &&= \mathbf{28.919}
\end{aligned}
$$

Expected Value for Option S_3

$$
\begin{aligned}
EV_{C1} &= (0.1)(32) + (0.8)(37) + (0.1)(44) &&= 37.2 \\
EV_{C2} &= (0.08)(19) + (0.84)(22) + (0.08)(24) &&= 21.92 \\
EV_{C3} &= (0)(0) + (1)(44) + (0)(0) &&= 44 \\
EV_{AGG} &= (0.16)(37.2) + (0.36)(21.93) + (0.48)(44) &&= \mathbf{34.963}
\end{aligned}
$$

Variance of Option S_1

$$
\begin{aligned}
Var_{C1} &= (0.2)(30^2) + (0.6)(43^2) + (0.2)(58^2) - (43.4)^2 &&= 78.64 \\
Var_{C2} &= (0.1)(32^2) + (0.75)(44^2) + (0.15)(60^2) - (45.2)^2 &&= 51.36 \\
Var_{C3} &= (1)(30^2) - (30)^2 &&= 0 \\
Var_{AGG} &= (0.16^2)(78.64) + (0.36^2)(51.36) + (0.48^2)(0) &&= \mathbf{8.669}
\end{aligned}
$$

Contd

Example 12.3 Contd

Variance of Option S_2

$$
\begin{aligned}
\text{EV}_{C1} &= (0.25)(23^2) + (0.5)(31^2) + (0.25)(38^2) - (30.75)^2 = 28.19 \\
\text{EV}_{C2} &= (0.2)(26^2) + (0.7)(34^2) + (0.1)(43^2) - (33.33)^2 \quad = 20.41 \\
\text{EV}_{C3} &= (1)(25) - (25)^2 \qquad\qquad\qquad\qquad\qquad\quad = 0 \\
\text{Var}_{AGG} &= (0.16^2)(28.19) + (0.36^2)(20.41) + (0.48^2)(0) \quad = \mathbf{3.367}
\end{aligned}
$$

Variance of Option S_3

$$
\begin{aligned}
\text{Var}_{C1} &= 0.1(32^2) + 0.8(37^2) + 0.1(44^2) - (37.2)^2 \qquad = 7.36 \\
\text{Var}_{C2} &= 0.8(19) + 0.84(22) + 0.08(24) - (21.92) \qquad = 1.033 \\
\text{Var}_{C3} &= (1)(44^2) - (44)^2 \qquad\qquad\qquad\qquad\qquad = 0 \\
\text{Var}_{AGG} &= (0.16)^2(7.36) + (0.36^2)(1.0344) + (0.48^2)(0) = \mathbf{0.322}
\end{aligned}
$$

The expected values for the three options give a ranking consistent with the basic raw scores listed in Table 12.8. The expected values represent the scores one can assign to each option with a confidence level of 50%, in other words an expected mean/average score. If one wishes to increase the confidence level associated with these scores, the variances must be used.

Firstly, the square root of the variance yields the standard deviation of each mean score. For confidence levels greater than 50%, the option score is obtained by decreasing the expected value by z times the standard deviation, where z is a value obtained from a standard normal distribution curve. Values of z for confidence levels between 50% and 99% are given in Table 12.10. The z values associated with 50%, 85% and 99% are given in Table 12.11.

50% confidence

$$
\begin{aligned}
z &= 0 \\
S_1 &= 37.62 \\
S_2 &= 28.91 \\
S_3 &= 34.96
\end{aligned}
$$

85% confidence

$$
\begin{aligned}
z &= 1.037 \\
S_1 &= 37.62 - (1.037 \times 2.944) = 34.37 \\
S_2 &= 28.91 - (1.037 \times 1.835) = 27.01 \\
S_3 &= 34.96 - (1.037 \times 0.56) = 34.38
\end{aligned}
$$

Contd

Example 12.3 Contd

99% confidence

$z = 2.327$

$S_1 = 37.62 - (2.327 \times 2.944) = 30.77$

$S_2 = 28.91 - (2.327 \times 1.835) = 24.64$

$S_3 = 34.97 - (2.327 \times 0.56) = 33.64$

What can be seen is that while the expect values indicate that S_1 is approximately 10% better than S_3 (37.62 > 34.96), their score are practically the same at 85% confidence level (34.56 $\cong$ 34.38). At 99%, S_3 outperforms S_1 by 10% (33.64 > 30.77).

Thus, while S_1 has a larger expected value than S_3, its standard deviation is greater, therefore the uncertainty associated with its aggregate score is also higher, therefore the larger the confidence level sought the less attractive S_1 becomes relative to S_3.

Attribute	Weighting	Option		
		S1	S2	S3
Reliability	0.16	43	31	37
Maintenance	0.36	44	34	22
Cost	0.48	30	25	44
Total		*37.12*	*29.2*	*34.96*

Table 12.8 Option scores.

Table 12.9 Scores and associated probabilities of the options.

	Criterion C_1						Criterion C_2					
	Score			Probability			Score			Probability		
	S_1	S_2	S_3	S_1	S_2	S_3	S_1	S_2	S_3	S_1	S_2	S_3
Pessimistic	30	23	32	0.2	0.25	0.1	32	26	19	0.1	0.2	0.08
Most likely	43	31	37	0.6	0.5	0.8	44	34	22	0.75	0.7	0.84
Optimistic	58	38	44	0.2	0.25	0.1	60	43	24	0.15	0.1	0.08

Table 12.10 Standard normal distribution data.

z	Probability	z	Probability	z	Probability	z	Probability
0	0.50	−0.45	0.6736	−0.90	0.8159	−1.60	0.9452
−0.05	0.5199	−0.50	0.6915	−1.00	0.8413	−1.70	0.9554
−0.10	0.5398	−0.55	0.7088	−1.05	0.8531	−1.80	0.9641
−0.15	0.5596	−0.60	0.7257	−1.10	0.8643	−1.90	0.9713
−0.20	0.5793	−0.65	0.7422	−1.15	0.8749	−2.00	0.9772
−0.25	0.5987	−0.70	0.7580	−1.20	0.8849	−2.10	0.9821
−0.30	0.6179	−0.75	0.7734	−1.30	0.9032	−2.20	0.9861
−0.35	0.6368	−0.80	0.7881	−1.40	0.9192	−2.30	0.9893
−0.40	0.6554	−0.85	0.8023	−1.50	0.9332	−2.40	0.9918

	Statistical data		Confidence level		
	Variance	Standard deviation	50% $z = 0$	85% $z = 1.04$	99% $z = 2.33$
S_1	8.669	2.994	37.62	34.37	30.77
S_2	3.367	1.835	28.91	27.01	24.64
S_3	0.314	0.560	34.96	34.38	33.64

Table 12.11 Option scores for different confidence levels.

12.5   Assigning weights to the decision criteria

The assignment of weights to the constituent criteria is central to the working of the additive model. Their purpose is to express the importance of each chosen decision criterion relative to all others. A number of basic techniques exist for calculating criterion weightings within the additive model. All require the input of judgements from the decision makers.

The following methods are outlined in the next sections:

- Presumption of equal weights
- Ranking system for obtaining weights
- Ratio system for obtaining weights
- Basic pairwise analysis
- Resistance-to-change grid

12.5.1   *Presumption of equal weights*

If the decision makers are not in a position to assign weights to the criteria, or are unwilling to do so, the process can proceed initially on the basis that all criteria are treated as being of equal importance, with each attribute given an equal weight.

In these circumstances, it is particularly important to carry out an extensive sensitivity analysis to gauge the effect of varying the weightings away from their equal valuations on the baseline performances of the project options.

12.5.2   *Ranking system for obtaining weights*

This procedure involves the decision maker initially ranking the criteria in order of importance. Each criterion is then assigned a score based on its rank, with the one ranked first assigned the score '1', the one ranked second assigned the score '2', and so on. In the case of criteria tying for the same rank, an average score is assigned to them. For example, take the case where one criterion is ranked first and assigned the score '1'. Three criteria are ranked equal second. Their score is obtained by averaging the three scores '2', '3' and '4'. Each criterion will thus be given the score '3'. The criterion ranked below these three in fifth place will be assigned the next highest score of '5'. The least important criterion will thus end up with the score n, where n is the number of criteria.

The normalised importance weight for each criterion can be calculated using the formula:

$$w_i = \frac{n - r_i - 1}{\displaystyle\sum_{i=1}^{n}(n - r_i + 1)} \qquad (12.9)$$

where w_i is the normalised weighting for the ith criterion, r_i is the ranking score for the ith criterion and n is the number of decision criteria.

Example 12.4

Five options for the siting of a wastewater treatment plant are considered on the basis of the following six criteria:

(1)   Construction and maintenance Cost (C_1)
(2)   Construction disturbance (C_2)
(3)   Impact on birdlife (C_3)
(4)   Visual impact (C_4)
(5)   Water quality impact (C_5)
(6)   Political acceptability (C_6)

The decision makers agreed a ranking of these criteria from most important to least important. This information is given in Table 12.12.

Use the rank-sum weighting method to derive importance scores for these six criteria.

Solution

The total number of criteria under consideration is six, therefore:

$$n = 6$$

As can be seen from Table 12.12, the cost criterion (C_1) is deemed to be the most important by the decision makers and is assigned the rank score of '1'. Visual impact (C_4) and water quality impact (C_5) are deemed equal second and are assigned the average of the rank scores '2' and '3'. Impact on birdlife (C_3) is ranked fourth and assigned the score '4' and political acceptability (C_6) is ranked fifth and given the score '5'. Construction disturbance (C_2) is adjudged the least important criterion and is assigned a score of '6', equivalent to the total number of criteria n.

Using Equation 10.8, each rank score is divided by the sum of all rank scores to give normalised weights, which can be given as:

Construction and maintenance Cost (C_1)	= 0.29
Construction disturbance (C_2)	= 0.05
Impact on birdlife (C_3)	= 0.14
Visual impact (C_4)	= 0.21
Water quality impact (C_5)	= 0.21
Political acceptability (C_6)	= 0.10

Criteria	Rank position	Rank score (r_1)	$n - r_1 + 1$	w_1
C_1	1st	1	6	0.286
C_4, C_5	2nd	$(3 + 2) \div 2$	4.5($\times$2)	0.214
C_3	3rd	4	3	0.143
C_6	4th	5	2	0.095
C_2	5th	6	1	0.048
			$\Sigma = 21$	

Table 12.12
Computation of normalised weights from ranking method.

12.5.3   Ratio system for obtaining weights

This technique is similar in nature to the ranking system outlined in Section 12.5.2. The computation starts by assigning a score of '1' to the criterion or criteria adjudged by the decision makers to be least important. They are then requested to give the other criteria scores greater than '1', with the number of times the assigned score being greater than unity reflecting the importance of that criterion relative to the least important one or ones. Sometimes the decision makers will place a ceiling on the multiple separating the least and most important criteria, for example deciding that the most important criterion will be four times the weighting of the least important one. These weights can then be normalised so that they sum to unity by using the equation:

$$w_i = \frac{z_i}{\sum_{i=1}^{n} z_i} \qquad (12.10)$$

where w_i is the normalised weighting for the ith criterion, z_i is the weight score assigned to ith criterion and n is the number of decision criteria.

Example 12.5
Analysing the same problem as outlined in Example 12.4, the decision makers evaluate the same six criteria but this time the ratio method is used. They decide that the most important criterion should be separated from the least important one by a factor of four. Thus, construction impact (C_1) is assigned the score '1' and cost (C_2) is assigned the score '4'. The water quality and visual impacts are both assigned a score of '3.5', with impact on birdlife (C_3) given the score '2' and political acceptability (C_6) assigned '1.5'. Use the ratio weighting method to derive importance scores for these six criteria.

Solution
Table 12.13 outlines the computation of the normalised weights for each criterion from these raw scores. It can be seen that the resulting weights are very similar to those obtained by the rank sum method:

Contd

Example 12.5 Contd

Construction and maintenance Cost (C_1) = 0.26
Construction disturbance(C_2) = 0.07
Impact on birdlife (C_3) = 0.13
Visual impact (C_4) = 0.22
Water quality impact (C_5) = 0.22
Political acceptability (C_6) = 0.10

Criteria	Rank Position	Ratio Score (z_1)	W_1
C_1	1st	4	0.258
C_4, C_5	2nd	3.5 (×2)	0.226
C_3	3rd	2	0.129
C_6	4th	1.5	0.097
C_2	5th	1	0.065
		$\Sigma = 15.5$	

Table 12.13 Computation of normalised weights by ratio method.

12.5.4 Pairwise comparison weighting system

Within this technique, each criterion is pairwise compared with each of the others. In each case, the decision maker judges whether the criterion in question is less, equally or more important than the other criterion with which it is being directly compared. The scale is used for each pairwise analysis is shown in Table 12.14.

The relative weight for each criterion is obtained by adding the scores it obtains against the other $n - 1$ criteria (n = number of criteria). These are then normalised to give a final weighting.

Result of comparison A vs. B	Score for Option A
A less important than B	0
A equally important as B	1
A more important than B	2

Table 12.14 Performance scale for Criterion A within pairwise analysis.

Example 12.6
Again analysing the same problem as outlined in Example 12.4, the decision makers analyse the same six criteria, this time using the pairwise comparison method. Table 12.15 details the results of the pairwise analysis of the six criteria. Use this information to estimate their importance weights.

Contd

Example 12.6 Contd

Solution

Table 12.15 outlines the computation of the raw and normalised weights from the full set of pairwise comparisons. It can be seen that the resulting weights are again quite similar to those obtained by the previous two methods.

Construction and maintenance Cost (C_1) $= 0.27$
Construction disturbance (C_2) $= 0.03$
Impact on birdlife (C_3) $= 0.17$
Visual impact (C_4) $= 0.23$
Water quality impact (C_5) $= 0.23$
Political acceptability (C_6) $= 0.07$

Note: A situation may arise where the least important criterion may have a row sum of zero. Since a zero weight is not allowable, in this situation a score of '1' should be added to the row sum of each criterion. The least important criterion will then have a raw score of one, while maintaining the differential between it and the other criterion row scores.

Table 12.15 Pairwise comparison matrix for the six criteria.

	C_1	C_2	C_3	C_4	C_5	C_6	Row sum	Norm Wt
C_1		2	2	1	1	2	8	0.267
C_2	0		0	0	0	1	1	0.033
C_3	0	2		1	1	1	5	0.167
C_4	1	2	1		1	2	7	0.233
C_5	1	2	1	1		2	7	0.233
C_6	0	1	1	0	0		2	0.067
							Total = 30	

12.5.5 The resistance-to-change grid

This technique, devised by Rogers and Bruen (1998), is almost identical to the above pairwise method in the way the weightings are computed. However, the previous three techniques do not explicitly define what 'importance' means. The 'resistance-to-change' technique has a firm methodological basis which ensures that the weights obtained reflect the decision maker's actual preferences in terms of what he/she deems to be important. The first three techniques, where decision makers spontaneously award weight scores to each of the criteria, do not necessarily relate the actual scores given for each criterion directly back to the basic concept of its importance weighting. The resistance-to-change methodology results in decision makers automatically placing the criteria into a hierarchy of relative importance. This hierarchy can then be used to compute directly the weighting for each.

The key to using this method lies in deriving two terms for each criterion, one expressing its most desirable aspect and the other expressing its least desirable aspect. This is termed expressing the criterion in 'bipolar form'. The six criteria from the last three examples can be expressed in bipolar form as follows:

- C_1 – economical waste treatment/costly waste treatment
- C_2 – low level of construction disturbance/high level of construction disturbance
- C_3 – negligible impact on birdlife/significant impact on birdlife
- C_4 – negligible visual impact/significant visual impact
- C_5 – high quality water/low quality water
- C_6 – highly acceptable politically/politically unacceptable

The left-hand side of each of the above bipolar expressions represents the criterion at its most desirable, with the right-hand side representing it at its least desirable. For an economic criterion, the preferred side is assumed to be the one minimising cost. For an environmental/social criterion, the preferred side is assumed to be the one minimising environmental/social impact.

To derive a pairwise comparison matrix of the type illustrated in Table 12.15 for each pair of criteria, the decision maker is required to examine the bipolar expressions for both. He/she is then asked the question that, if he/she had to change one of these criteria from its desirable side to its undesirable side, which one would he/she be least willing to change. For each comparison, the criterion resisting change is given the score '1', while the criterion which the decision maker is willing to change to its undesirable side is given the score '0'. If the decision maker has an equal resistance to changing either criterion, each is given the score '0.5'. When a given criterion has been compared for resistance to change with all others, its scores are totalled to give a final weighting. All weightings obtained are then summed and each is divided by this figure to attain normalisation.

The theoretical basis for the grid originated from Hinkle (1965), who found that the more important a criterion is, the more likely it will be that the decision maker resists changing it to its more undesirable state.

An example of the 'resistance-to-change' grid that yields weightings identical to the basic pairwise technique in Example 12.6 is illustrated in Table 12.16.

Looking in detail at criterion C_3 (impact on birdlife), the overall score of 2.5 out of a possible five indicates the following:

- The decision maker preferred the idea of an economical waste plant having a significant impact on birdlife than a costly one having minimal impact on birdlife. Hence, in this case, C_3 was assigned the score '0'.
- The decision maker preferred the idea of a waste plant having a minimal impact on birdlife but creating significant construction impacts rather than one having significant impact on birdlife but creating minimal construction disturbance. Hence, in this case, C_3 was assigned the score '1'.

Table 12.16 Sample of 'resistance-to-change' grid.

	C_1	C_2	C_3	C_4	C_5	C_6	Row sum	Norm Wt
C_1		1	1	0.5	0.5	1	4	0.267
C_2	0		0	0	0	0.5	0.5	0.033
C_3	0	1		0.5	0.5	0.5	2.5	0.167
C_4	0.5	1	0.5		0.5	1	3.5	0.233
C_5	0.5	1	0.5	0.5		1	3.5	0.233
C_6	0	0.5	0.5	0	0		1	0.067
							Total = 15	

- The decision maker found the idea of a waste plant having a minimal impact on birdlife but generating significant visual impacts equally desirable to one having significant impact on birdlife but generating minimal visual impacts. Hence, in this case, C_3 was assigned the score '0.5'.
- The decision maker found the idea of a waste plant generating a minimal impact on birdlife but having a major impact on water quality equally desirable to one generating significant impact on birdlife but having a minimal impact on water quality. Hence, in this case, C_3 was again assigned the score '0.5'.
- The decision maker found the idea of a waste plant having a minimal impact on birdlife but lacking in political acceptability equally desirable to one having significant impact on birdlife while being politically acceptable. Hence, in this case, C_3 was assigned the score '0.5'.

This method, in terms of the way in which its final results are presented, is virtually identical to the basic pairwise comparison technique as illustrated in Example 12.6. Its major advantage, however, lies in giving the stakeholder making the preference decisions a precise definition of what the term 'criterion importance'.

12.5.6 Hierarchy of weights

Thus far, all the multicriteria decision problems have involved a set of criteria, all of which are at the same level. If, however we wish to create a hierarchy of criteria, starting with main criteria, which defines the criterion types under assessment, each of which has its own set of subcriteria that are used to actually assess the relative performance of the options under consideration, the relative weightings must be derived at each level in the hierarchy using one of the techniques detailed above.

The process of weighting different levels of criteria and showing the effect that varying the weights at a given level in the hierarchy has on criteria at the lower levels is shown in Example 12.7.

Example 12.7
A number of design options for the construction of a bridge in a major conurbation are being considered on the basis of three different main criteria:

(1) Economic impact (Cr_1)
(2) Environmental impact (Cr_2)
(3) Political impact (Cr_3)

Each of these main criteria is expressed in terms of a number of subcriteria. These will be used to measure the actual performance of the options under consideration. They can be listed as:

(1) Economic impacts
  - Construction cost ($Cr_{1.1}$)
  - Efficiency of construction ($Cr_{1.2}$)
(2) Environmental impacts
  - Noise pollution impacts ($Cr_{2.1}$)
  - Air pollution impacts ($Cr_{2.2}$)
  - Landscape impacts ($Cr_{2.3}$)
  - Impact on local amenities ($Cr_{2.3}$)
(3) Political impacts
  - Political acceptability at local level ($Cr_{3.1}$)
  - Political acceptability at national level ($Cr_{3.2}$)

The relative weightings of the main criteria as given by the relevant decision maker are shown in Table 12.17, and the relative weightings for the three sets of sub-criteria are shown in Table 12.18.

(i) Calculate the normalised weightings for each of the subcriteria.
(ii) What would these values be if the main criterion C_1 is doubled in importance to 60%, with C_2 and C_3 maintaining their relative weights?

Solution (i)

(1) Economic criteria
The economic criteria have an overall normalised weight of 0.3. This is divided 70% in favour of cost and 30% to efficiency of construction. The final weighting for the two subcriteria are therefore:

Construction cost $=0.3\times0.7=0.21$
Efficiency in construction $=0.3\times0.3=0.09$
Total $=0.30$

(2) Environmental criteria
The environmental criteria have an overall normalised weight of 0.4. This is divided 30% in favour of noise impact, 10% to air impact, 40% to landscape

Contd

Example 12.7 Contd

impact and 20% to amenity impact. The final weighting for the four subcriteria are therefore:

```
Noise pollution impact =0.4×0.3 =0.12
Air pollution impact   =0.4×0.1 =0.04
Landscape impact       =0.4×0.4 =0.16
Amenity impact         =0.4×0.2 =0.08
Total                           =0.40
```

(3)   Political criteria
The poltical criteria have an overall normalised weight of 0.3. This is divided 25% in favour of political acceptance locally and 75% to political acceptance nationally. The final weighting for the two subcriteria are therefore:

```
Level of political acceptance locally    =0.3×0.25 =0.075
Level of political acceptance nationally =0.3×0.75 =0.225
Total                                             =0.30
```

Thus all options are assessed on the basis of the above eight criteria of evaluation, whose normalised weights are:

Construction cost ($Cr_{1.1}$) = 0.21
Efficiency of construction ($Cr_{1.2}$) = 0.09
Noise pollution impacts ($Cr_{2.1}$) = 0.12
Air pollution impacts ($Cr_{2.2}$) = 0.04
Landscape impacts ($Cr_{2.3}$) = 0.16
Impact on local amenities ($Cr_{2.3}$) = 0.08
Political acceptability at local level ($Cr_{3.1}$) = 0.07
Political acceptability at national level ($Cr_{3.2}$) = 0.23
Total = ***1.00***

Solution (ii)
In this case, the economic criteria have a total weighting of 60%, with the other criteria sharing the remaining 40% in a ratio of 4:3. Thus environmental criteria have at total weight of 40×(4/7) and political criteria have a total weight of 30×(3/7). The relative weights of the main criteria are thus:

C_1 = 0.60
C_2 = 0.23
C_3 = 0.17
Total = 1.00

Contd

Example 12.7 Contd

The relative weights of the sub-criteria remain the same. The weightings for the sub-criteria are thus as follows:

$$(Cr_{1.1}) = 0.6 \times 0.7 \quad = 0.42$$
$$(Cr_{1.2}) = 0.6 \times 0.3 \quad = 0.18$$
$$(Cr_{2.1}) = 0.229 \times 0.3 = 0.07$$
$$(Cr_{2.2}) = 0.229 \times 0.1 = 0.02$$
$$(Cr_{2.3}) = 0.229 \times 0.4 = 0.09$$
$$(Cr_{2.3}) = 0.229 \times 0.2 = 0.05$$
$$(Cr_{3.1}) = 0.171 \times 0.25 = 0.04$$
$$(Cr_{3.2}) = 0.171 \times 0.75 = 0.13$$
Total $\qquad = 1.00$

Table 12.17 Weightings for main criteria.

Main criterion	Weighting (%)
Economic (C_1)	30
Environmental (C_2)	40
Political (C_3)	30

Table 12.18 Weightings for subcriteria.

Subcriterion	Weighting (%)
Construction cost ($C_{1.1}$)	70
Efficiency of construction ($C_{1.2}$)	30
Noise pollution ($C_{2.1}$)	30
Air pollution ($C_{2.2}$)	10
Landscape ($C_{2.3}$)	40
Amenities impact ($C_{2.4}$)	20
Acceptability locally ($C_{3.1}$)	25
Acceptability nationally ($C_{3.2}$)	75

Using the above example as a case in point, while each option will be assessed on the basis of its measured quantitatively or qualitatively based performance on the eight subcriteria, the relative weighting of each main criterion is of great importance, as each one sets the basis for the total weighting that will be shared out among its constituent subcriteria. The nature of this share-out is determined by the relative weights of each set of subcriteria. As seen above, altering the relative importance of the main criteria directly affects the weights assigned to the eight subcriteria.

The hierarchy of criteria are shown schematically in Figure 12.2.

12.5.7 Multiple weighting systems

In many circumstances within the assessment of an engineering development project, there may be more than one set of decision makers. Each of these groups should be allowed to generate its own set of criterion weightings. The model is then run using each set, noting, in each case, the final position of the options under examination.

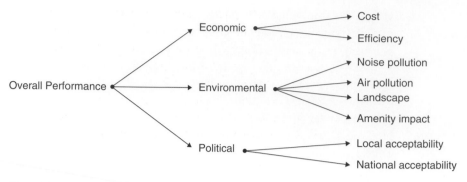

Figure 12.2 Hierarchy of criteria in Example 12.7.

The level of agreement between the groupings regarding the best performing options can then be gauged.

These different decision groupings can arise out of different circumstances. The following is a list of three possible situations where multiple decision groupings, each generating their own set of importance weights, can arise:

(1)   geographical location,
(2)   area of professional expertise, or
(3)   position in the decision making hierarchy.

Geographical location

Often a development project spans a very wide geographical area. Different localities may be affected to varying degrees by the project and, as a result, will differ in the criteria they deem most important. In this situation, the geographical area directly affected by the proposal should be divided up logically into zones, each one with a set of actors inputting their views directly regarding the relative weightings of the criteria. For example, in the case of a land reclamation project affecting a number of small urban centres in close proximity to the flood plain of the river in question, each town will be affected to different degrees depending on factors such as their population, the amount of agricultural and industrial activity in the locality and their proximity to the river. The weightings from each individual urban centre will reflect these differing concerns.

Area of professional expertise

When the overall viability of a development project is being assessed on the basis of a range of criteria – economic, environmental, social and political – it is often desirable to obtain importance weightings from decision makers with different perspectives on the relative importance of the attribute groupings. A grouping which comprises project promoters/financiers will tend to emphasise the economic aspects of the proposal, and their weighting system would reflect this. Importance weightings from project designers, on the other hand, would highlight the environ-mental/aesthetic considerations of the project, while a set from a group with a political background will stress the importance of the social/political attributes. Again, the model would be run for the three different weighting systems and the level of agreement regarding the best performing options assessed.

The case study in the following chapter details the application of the SAW model in deciding the best transport strategy for a major city where the decision makers involved in the formulation of the competing strategies are from three distinct types of background:

(1)   administrators/promoters
(2)   designers
(3)   politicians

Three weighting systems are used within the subsequent analysis, reflecting the emphasis on different attributes of each of these three groupings

Position in the decision making hierarchy

In the case of a development that is highly sensitive politically, or where different levels of the political hierarchy are responsible for different aspects of the project, it may be necessary to obtain attribute weightings from each relevant layer of administration.

In the case of the planning of a proposed incineration plant to cater for the waste disposal needs of a major urban centre, three levels of government may be involved. The national administration may be funding the project, with the regional govern-ment responsible for its design and the operation of the plant handled at local level. A multicriteria analysis of this proposal may require weights from decision makers from these three different levels of administration, reflecting their different political and operational perspectives on the project.

12.5.8   Scoring systems for the criteria

The assessment of the performance of criteria depends to a very large extent on the nature of the criterion being considered. Where the criterion is assessed in quantita-tive terms in its own units, the process is relatively straightforward. As can be seen from the worked examples at the start of this chapter, the transformation of these scores onto a 10-point scale involves assigning '10' to the option performing best on the criterion in question, with the other options being given a score somewhere along the 10-point scale between zero and ten, depending on each one's performance relative to the best one.

The process of assessment of qualitatively-based criteria is less straightforward. On the basis that the same 10-point scale as is used to score the quantitatively-based criteria must be used to assess intangible criteria, some efforts have been made to compile a standardised

Table 12.19 Assignment of scores to the qualitatively-based criterion 'landscape impacts'.

Score	Assessment	Description
8 to 9	Negligible	Almost imperceptible changes to locally important views. All can be ameliorated.
6 to 7	Minor	Some minor changes to locally important views, but these can, to a large extent, be ameliorated.
4 to 5	Moderate	Substantial alteration to views of local importance. Any alteration to views of district-wide importance can be ameliorated.
2 to 3	Major	Major alteration of existing views or landscapes of importance to the district, and a degrading of views of citywide importance.
1	Severe	Visual intrusion to the extent that views of importance to the city are badly affected, with most of the value of the view being lost

set of descriptions for the level of desirability associated with each point on this scale. These attempt to define in straightforward language, for each option under consideration, the outcome that is equivalent to each numbered score on the scale. The actual descriptions will vary depending on the nature of the qualitative criterion being assessed.

Take for example a qualitatively assessed environmental criterion such as the landscape impacts of a structure such as a high-level bridge in the centre of a large conurbation. While it is obvious that the score '10', the most desirable outcome, is equivalent to 'no impact on the landscape', and the score '0', the least desirable outcome, is equivalent to 'extreme intrusive impacts on the landscape' with all value of the view being lost, Table 12.19 defines how the scores from '1' to '9' might be defined:

One can see that there is a high level of description associated with these qualitatively assessed criteria, allowing quite a precise valuation to be assigned to each of the competing options. To compile rating information for all the qualitative criteria associated with a given multicriteria evaluation, experts in the relevant field must be employed to compile the descriptions associated with each scale point. Once this tabular information is put together for each attribute in question, the decision maker can use it to assign criterion valuations to the various options.

12.6 Checklists

Checklists are applications of the Simple Additive Weighting Model to option choice problems in engineering where the decision criteria are predominantly environmental in nature. They are thus systematic methods that can be used for comparing and evaluating different development proposals on the basis of their environmental impact. Checklists can exist in simple form, where the decision maker is concerned purely with the identification of those environmental criteria of relevance to the decision in question. The checklists of interest to the text are 'scale-weighted', where numerical scores are assigned to each project option on each identified environmental factor. These individual scores are then weighted based on the relative importance of the criterion in question, allowing a single score to be generated for each proposal.

The scale-weighted checklist is thus, in effect, a simple additive weighting model. For each option under consideration, a composite index or score can be derived using the following formula:

$$\text{Index}_j = \sum_{i=1}^{n} W_i R_i \qquad\qquad (12.11)$$

where Index_j is the composited index for jth option, n is the number of decision criteria, W_i is the importance weighting for ith criterion and R_i is the rating or score for jth option on ith criterion.

In the context of the overall project planning process, the scale-weighted checklist is seen as an invaluable aid to the selection of the 'best practicable environmental option'. The simple additive framework allows a trading-off of each option's scores on the different criteria using the relative weights to deliver an overall score that permits a direct comparison between them to be made.

The checklist methodology is generally used within the framework of an Environmental Impact Statement, where two or more alternative schemes are required to be assessed on relative terms. This is particularly the case in the United States where the Federal regulatory body, the Council for Environmental Quality (CEQ), has stressed the importance of the use of such decision models to assess different alternatives, stating that such processes lie at the heart of the environmental impact statement (CEQ, 1978). In the United States, environmental impact statements can involve the assessment of up to 50 project alternatives and decision methods are seen as central to the selection process. Numerous checklist methodologies have been developed within Environmental Impact Assessment (EIA). EIA is the formal process within which the potential effect of different development options on the environment is gauged.

Let us look in some detail at the following types of scale-weighting checklist used to assess development proposals on the basis of their environmental impact:

- The Environmental Evaluation System (EES) (Dee, 1972)
- Sondheim's Methodology (Sondheim, 1978)
- Mongkol's Methodology (Mongkol, 1982)

12.6.1   *Environmental Evaluation System*

EES was developed for assessing water resource projects and is based on a checklist of 78 environmental and socio-economic parameters (e.g. dissolved oxygen, employment opportunities) chosen for their relevance to water developments.

One thousand importance-weighting units are distributed amongst these 78 parameters (e.g. dissolved oxygen = 31, employment opportunities = 13). Quantitative estimates of all parameters are normalised on a common scale of environmental quality (zero to one) using value functions. The final score for each parameter is determined by multiplying its normalised score by its importance rating. The combined score for each alternative is determined by summing final scores for each individual parameter.

Detailed methodology

(a) Checklist format

The checklist of environmental parameters is conceived in a hierarchical format. The first level of the hierarchy consists of four general environmental categories:

(1) Ecology
(2) Environmental pollution (physical/chemical)
(3) Aesthetics
(4) Human interest/social

Each of these is subdivided into a number of intermediate components, with five each under both ecology and human interest, four under environmental pollution and six under aesthetics, giving a total of twenty.

For example, the environmental pollution category consists of the following four components:

(1) Water quality
(2) Air quality
(3) Land pollution
(4) Noise pollution

These intermediate components are further subdivided into specific environmental parameters, 78 in total. For example, air quality is divided into the following seven individual parameters:

(1) Carbon monoxide
(2) Hydrocarbons
(3) Nitrogen oxides
(4) Particulate matter
(5) Photochemical oxidants
(6) Sulfur oxides
(7) Others

(Note: noise pollution, on the other hand, consists of just one specific environmental parameter – noise pollution.)

These 78 parameters constitute the crucial level of impact assessment. Many of them can be expressed in quantitative terms. They were selected by a screening process on the basis of their significance for water resource development. Similar lists could be devised for other types of development.

The final level comprises impact measurements of the environmental parameters resulting from each option. For example, a mathematical representation of noise levels in dB(A) from each source point to the reception point or points of relevance to the study can be calculated to establish the 'noise parameter', while chemical testing will enable nitrogen oxide levels to be measured in the vicinity of the proposed project.

(b) Impact weightings

To aid the problems of comparison and summation of impacts, parameters are weighted so that they relate to one another in terms of relative importance. Once a weighting system is in place, it is possible to sum impacts for each alternative and compare totals (Dee used a panel of experienced water project evaluators to derive appropriate weights).

One thousand Parameter Importance Units (PIU) were distributed among the parameters. For example, the PIUs for the seven air quality parameters are:

(1)   Carbon monoxide – 5
(2)   Hydrocarbons – 5
(3)   Nitrogen oxides – 10
(4)   Particulate matter – 12
(5)   Photochemical oxidants – 5
(6)   Sulfur oxides – 10
(7)   Others – 5

These weights, once established, represent, at a given place and time, the relative importance of the listed parameters for water resource developments.

(c) Measurement of environmental quality

The next step in the process is the 'transformation' of pre-development parameter estimates into measures of environmental quality, which are then used for comparison purposes with the post-development situation. Within EES, Environmental Quality (EQ) is measured on a scale from zero (very bad) to one (very good). For quantitatively measured parameters, such as BOD (BOD is one of the fourteen parameters listed under water quality and is allotted 25 PIU), 'very bad' implies levels at or near those minimum levels permissible by law, while 'very good' implies levels similar to those in water not receiving effluent. Thus, given a pre-development BOD score of 0.5 on the environmental quality scale, a project option that resulted in lowering the BOD levels, thus giving a post-development score nearer to 1.0, would lead to an improvement in environmental quality. In this way, both beneficial and adverse impacts can be quantified on the same scale.

The 'transformation' of a parameter estimate into an environmental quality score is achieved by using a 'value function'. Value functions are obtained for all parameters and relate parameter estimates to environmental quality. Value functions are devised for all 78 parameters by a panel of experts. The procedure is summarised as follows:

● Information is obtained on the general relationship between the parameter and the quality of the environment; for example, BOD is widely used as an index of water quality and can thus be easily related to environmental quality.
● The quality scale of the value functions is divided into equal intervals between zero and one, and parameter estimates are determined for each of these intervals.
● The process is continued until a curve is drawn.

For example, in the case of the value function for dissolved oxygen (a parameter under water quality with a PIU of 31), the relationship can be summarised as shown in Table 12.20.

Dissolved oxygen (mg/l)	0.0	4.0	6.0	10.0
Environmental quality	0.0	0.35	0.70	1.0

Table 12.20 Value function for dissolved oxygen.

Attribute	Max. score	Options		
		A_1	A_2	A_3
Ecology	240	215	200	220
Environmental pollution	402	340	350	310
Aesthetics	153	100	120	135
Human interest	205	200	175	175
Total	*1000*	*855*	*845*	*840*

Table 12.21 Analysis of three competing options on EES.

(d) Final estimation of environmental impact

The value functions allow environmental impacts to be quantified as follows:

- Existing data or data from special studies are used to relate the existing measurements of each parameter to the scales of environmental quality. This measures each parameter at its pre-development level and can be used as a basis for estimates of likely effects from various project actions.
- Impacts for each relevant parameter are predicted, each in terms of its own measurement units. In the case of dissolved oxygen, these units are mg/l, while noise measurement would be in dB(A). Aesthetic factors, however, such as visual and topographical effects, would be expressed qualitatively (high:medium:low:)
- These projected parameter valuations are used to determine the environmental quality score. If, using the value function for dissolved oxygen, levels fall from 6 to 4 mg/l, then the value function enables the differential to be transformed into a difference in environmental quality (from 0.7 to 0.35). Similar calculations are carried out for all relevant parameters.
- Each parameter quality score is multiplied by the number of Parameter Importance Units (PIUs) allotted to that parameter, giving a final score for each parameter in Environmental Impact Units (EIU). (Note: EIU=EQ×PIU)
- Since all parameters can be converted into these units, the individual values can be summed to provide a total score for all impacts.

There is no question that a project will be adjudged non-viable purely on the basis of its EES score. Professional judgement must also be a factor, with the relevant experts examining not just the overall score of the options but also their performance on each of the four main categories of impact.

Table 12.21 indicates an analysis of three development options, A_1, A_2 and A_3, using the Environmental Evaluation System (Dee, 1972). The overall score for each is given together with their scores within the four general environmental categories. The example shows option A_1 marginally outperforming the other two. However, it is, in fact, ranked first on only one of the four environmental categories. All three options are ranked first on at least one of the general attributes. Given that the trade-off analysis yields three scores in such close proximity, and given their differing relative positions on the four main criteria, the expert judgement would be that the difference in performance between them is negligible and that the analysis cannot separate them.

Comment

EES uses predictions regarding the probable environmental conditions that would result if the proposed project were to proceed. It compares the projected non-development and post-development scenarios. A difference in Environmental Impact Units (EIUs) between the two project options indicates the most preferable course of action.

This expression can be represented in mathematical form as follows:

$$E = \sum_{i=1}^{m}(v_i)_1 w_i - \sum_{i=1}^{m}(v_i)_2 w_i \qquad (12.12)$$

where E is the environmental impact, $(v_i)_1$ is the value in environmental quality units of parameter i with-project $(0 \rightarrow 1)$, $(v_i)_2$ is the value in environmental quality units of parameter i without-project $(0 \rightarrow 1)$, w_i is the relative weight of i and m is the total number of parameters.

Alternative projects can be evaluated in a similar fashion, with impacts summed for each individual design. Projects can then be ranked, with the project having the largest number of EIUs being first choice and that with the least number being ranked last.

The main disadvantages of EES can be listed as:

- As the method is a checklist, EES is open to criticism over its selection of parameters – the list may appear arbitrary. Also, interactions between parameters are not explored.
- In EES, weights are not allowed to vary between similar projects. If it were to be adapted to other project types, a new parameter listing and accompanying list of weightings would have to be devised.
- EES does not consider uncertainty and risk, nor does it identify distributional effects, that is those sections of the public that might be affected by the proposed project. However, it does help in 'scoping', ensuring that all relevant effects are considered within the evaluation.

12.6.2   Sondheim's Environmental Assessment Methodology

This method can simultaneously evaluate a large number of project alternatives, is applicable to a wide variety of situations and can include direct public participation in the assessment process.

Methodology

The method requires three separate panels:

(1)   a coordinating panel
(2)   a rating panel
(3)   a weighting panel.

(1)   The coordinating panel

This grouping is established by the organisation managing the project. It has four main tasks:

(1)   To list all realistic and viable project alternatives, coding them from 1 to '*m*'.
(2)   To define 'environment' as a function of independent or quasi-independent aspects or parameters, coding them from 1 to '*n*'. Each of the n parameters has some bearing on the environment. As with the EES, these parameters can be biological, chemical, aesthetic and so on in nature. No evaluations of their relative importance are needed at this point. (These first two points constitute a scoping process.)
(3)   To choose a rating panel whose professional expertise is in the areas of the parameters listed in 2. It is usual to have *n* members of the panel, with each member responsible for one of the relevant parameters.
(4)   To form a weighting panel, consisting of *y* members, representing Government, industry, public interest groups, community organisations and others potentially affected by the development. It is not expected that any member will be expert in a given area, merely knowledgeable, as all members' discussions would be viewed as being equally valid.

Note: the weighting panel must find the membership of the rating panel acceptable. Furthermore, the listings in (1). and (2). must be agreed before the process can get underway.

(2)   The rating panel

The members of the rating panel judge each project alternative according to specific criteria formulated for each parameter by the coordinating panel. This area of operation is looser than the EES Method. Whether the panel members working on a given aspect decide to construct a model of the relevant task, to conduct experiments to collect relevant data, use the EES approach whereby environmental quality indices (0 to 1) are calculated or make use of simple point assignment procedures is entirely their decision, depending on available data, time and money considerations and individual abilities.

Whatever scheme is used should be made explicit, and explained as fully as possible.

Result = *n* rating schemes for each of the *m* options

All ratings are on ratio or interval scales.

(3)   The weighting panel

Each of the *y* members evaluates all the *n* parameters on an independent non-negative interval or ratio scale of his or her own devising. Weightings, in each member's case, are determined by consideration of the extent to which each of the parameters is significant to his or her own organisation's analysis of the project – the judgements are more personal.

Obtaining a preference listing of project alternatives

The rating for each parameter is standardised by firstly subtracting it from the mean value of all scores on that parameter and then dividing by the standard deviation of all scores for that parameter. This process can be expressed mathematically as:

$$r_{i,j} = \left[\frac{x_{i,j} - \mu_j}{\sigma_j} \right] \tag{12.13}$$

where $r_{i,j}$ is the standardised score of project alternative i for parameter j, $x_{i,j}$ is the unstandardised score of project alternative i for parameter j, μ_j is the mean value of all raw scores for parameter j and σ_j is the standard deviation of unstandardised scores for parameter j.

This equation ensures that the mean of the scores for each parameter will be zero and its standard deviation will be one.

These scores are arranged into an m by n rating matrix R, with standardised ratings arranged by columns for each parameter. The raw weights for each parameter given by each member of the weighting panel are arranged in columns in an n by y matrix W. Matrix post-multiplication of $R \times W$ is then performed, resulting in matrix D – an m by y matrix. This procedure gives an overall weighted score for each parameter, for each project, as follows:

$R \times W = D$, i.e.,

$$\begin{bmatrix} r_{11} & r_{12} & .. & r_{1n} \\ r_{21} & r_{22} & .. & r_{2n} \\ .. & .. & .. & .. \\ r_{m1} & r_{m2} & .. & r_{mn} \end{bmatrix} \times \begin{bmatrix} w_{11} & w_{12} & .. & w_{iy} \\ w_{21} & w_{22} & .. & w_{2y} \\ .. & .. & .. & .. \\ w_{n1} & w_{n2} & .. & w_{ny} \end{bmatrix} = \begin{bmatrix} d_{11} & d_{12} & .. & d_{1y} \\ d_{12} & d_{22} & .. & d_{2y} \\ .. & .. & .. & .. \\ d_{m1} & d_{m2} & .. & d_{my} \end{bmatrix} \tag{12.14}$$

Matrix D is then standardised column by column using the same procedures. The resulting matrix is then reduced to a m by one-column matrix by summing through the rows to give a single index value for each project option.

These final indices form the basis of a decision making framework. By comparing scores from any two project alternatives, an assessment of the degree of preference of one option over the other can be readily obtained.

12.6.3   Mongkol's Methodology

Mongkol's Methodology is another variant of the EES concept. It includes some important characteristics which are not considered comprehensively in the other two above methods, such as impact magnitude, prevalence, duration and frequency, risk, importance/significance and mitigation.

The main points of the system are:

- The method uses a matrix to take account of mitigation. The matrix is three-dimensional, the variables being environmental cause, environmental effect and environmental impact. For example, the 'cause' could be the construction of a new highway, the 'effect' could be the increased traffic levels resulting from the road and the 'impact' could be the high noise levels due to increased traffic. Any cells where the coincidence of the three parameters results in serious environmental impacts are noted. This indicates that mitigation measures are required to reduce the magnitude of the relevant impact.
- The method also uses more flexible value functions to take account of local conditions and uses maximum and minimum allowable levels for selected environmental parameters.
- Mongkol advocates the use of an error term to accommodate the risk of the wrong decisions being taken.
- Instead of using concepts such as 'net environmental cost' or 'net environmental benefit', Mongkol uses the concept of an 'environmental benefit/cost ratio' as follows:

$$\text{Environmental cost} - \text{benefit ratio} = \frac{\sum \text{Beneficial Impacts}}{\sum \text{Adverse Impacts}} \qquad (12.15)$$

Methodology

The method consists of four consecutive steps in identifying, evaluating and comparing environmental impact magnitude, prevalence, duration, risk, importance and mitigation in addition to making decisions about the viability of a project based on environmental and economic considerations.

Step 1: Tracing pathways of cause–effect–impact
A three-dimensional matrix illustrating the quantity and pathway of environmental impacts keeps decision makers and the public informed. It can be expressed as three two-dimensional matrices:

(1)   a cause–effect matrix
(2)   a cause–impact matrix
(3)   an effect–impact matrix.

Figure 12.3 illustrates the layout of a cause–effect matrix. (The layouts for cause–impact and effect–impact are identical.) Any serious impact is expressed numerically within the appropriate cell of the matrix, indicating that mitigation measures are required to reduce its magnitude. The actual impact magnitude is noted in the matrix.

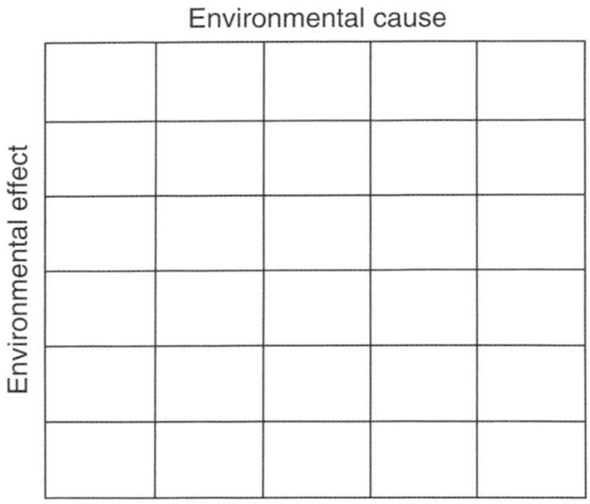

Environmental cause

Figure 12.3 Layout of cause–effect matrix.

Step 2: Transforming impact magnitude into a value of environmental quality
Unlike the EES system, where the actual units of the environmental parameters are used, in this instance the actual impact magnitude is divided by its maximum/minimum allowable or desirable level. The relationship between these values and environmental quality values reflects the views of the public as well as those of the professionals/experts. (The original EES does not combine the public system with the professional system.)

This value function is sufficiently flexible for use by any group of people for any environmental parameter produced by any type of project, unlike the EES which is more limited and rigid.

Step 3: Comparison of aggregated losses and gains in environmental quality units
Mongkol uses a benefit/cost ratio instead of net benefits obtained by subtracting costs from benefits. Impacts that have with-project environmental quality values less than without-project ones are defined as adverse impacts. Conversely, those that have with-project environmental quality values greater than without-project ones are termed beneficial environmental impacts.

The benefit/cost ratio is obtained by summing the positive values and dividing this figure by the sum of the negative values. This ratio will help decision makers judge the environmental viability of a development effectively, even in the case of a one-alternative project.

Step 4: Decisions based on economic and environmental criteria.
Having obtained the environmental benefit/cost ratio (EnvBC) for the development project, the economic benefit/cost ratio (EcBC) can now be used in conjunction with it to assess the overall viability of the project. Four overall cases are summarised by Mongkol:

(1)   if **EnvBC < 1.0** and **EcBC < 1.0**, the project should be rejected.
(2)   if **EnvBC > 1.0** and **EcBC > 1.0**, the project should be accepted and proceed.
(3)   if **EnvBC > 1.0** and **EcBC < 1.0**, the project is economically remedial.
(4)   if **EnvBC < 1.0** and **EcBC > 1.0**, the project is environmentally remedial.

12.7 Case Study: Using the Simple Additive Weighting Model to choose the best transport strategy for a major urban centre

The decision makers charged with deciding transport policy for a major urban centre wish to select a transportation strategy to suit the needs of the city over the next twenty years. The group of decision makers is subdivided into three groups, one representing budget holders/administrators, the second representing transport planners/designers and a third comprising a selection of elected public representatives from the locality.

After a period of deliberation, the group decides that the chosen strategy should concentrate on one form of transport mode, be it road, rail, bus or pedestrian/cyclist. As a result, they come up with six core strategic options; these are shown in Table 12.22.

Added to these are two non-mode specific options, one broadly setting out not to develop any new transport infrastructure and try to make the best use of existing resources; and the other, demand management based, setting out to reduce congestion charges by charging tolls for all motorists entering the inner city area (Table 12.23).

Table 12.22 Six core strategic options.

Option	Title	Brief description
A_1	Extensive bus development	Development of a network of radial bus corridors moving out from the city centre to the suburbs.
A_2	Incremental Highways Development	Development of a limited number of relatively minor road projects.
A_3	Extensive highways development	A widespread package of major urban road schemes including a bypass of the city.
A_4	Extensive cycling and pedestrian development	Completion of a city-wide cycle network and related measures including measures to give priority to pedestrians and cyclists.
A_5	Extensive light rail development	Development of three light rail lines gong to the north, south and western suburbs, plus a line to the airport and main bus depot.
A_6	Extensive heavy rail/metro development	Development of city-wide heavy rail including an underground section in the city centre

Table 12.23 Benchmark and demand management options.

Option	Title	Brief description
A7	Do-minimum	A 'benchmark' option against which all others can be assessed in relative terms.
A8	Demand management	A set of measures to reduce car use, including road tolling and parking restraint within the inner city area.

Table 12.24 Listing of main and subcriteria.

Main criterion	Title of main criterion	Listing of constituent subcriteria linked to each main criterion
Cr_1	Regional development	• Foster economic development ($Cr_{1.1}$) • Strengthen existing industrial base ($Cr_{1.2}$) • Attract new investment ($Cr_{1.3}$) • Improve physical access ($Cr_{1.4}$)
Cr_2	Quality of transport	• Improve comfort/reliability of travel ($Cr_{2.1}$) • Reduce accident levels ($Cr_{2.2}$) • Reduce environmental impact of transport ($Cr_{2.3}$) • Improve security to transport users ($Cr_{2.4}$)
Cr_3	Industrial development	• Improve transport access for manufacturers ($Cr_{3.1}$) • Improve access to outer regions of city ($Cr_{3.2}$) • Increase consistency with national plan ($Cr_{3.3}$) • Promote sustainable development ($Cr_{3.4}$)
Cr_4	City centre trade	• Maintain primacy of city centre ($Cr_{4.1}$) • Promote development of new towns ($Cr_{4.2}$) • Maximise coherence with plans of nearby regions ($Cr_{4.3}$) • Promote heritage conservation ($Cr_{4.4}$)
Cr_5	Plan Implementation	• Optimise use of existing infrastructure ($Cr_{5.1}$) • Minimise timescale for implementation ($Cr_{5.2}$) • Minimise construction disturbance ($Cr_{5.3}$) • Ability to implement strategy within existing legislation ($Cr_{5.4}$)

The decision makers compiled a set of criteria in hierarchical form. The following five main criteria were identified:

(1)   Development of regional economy (Cr_1)
(2)   Improvement in quality of transport (Cr_2)
(3)   Development of industry in the region (Cr_3)
(4)   Development of the city centre trade (Cr_4)
(5)   Efficiency in implementation of strategy (Cr_5)

Each is measured by means of a number of subcriteria of evaluation, which are listed in Table 12.24.

12.7.1   Assessing the importance weightings for the decision criteria

Main criteria

Each group of decision makers compiled its own set of weights for the main criteria. Using the ratio method, the three groupings each derived a set, as illustrated below in Tables 12.25, 12.26 and 12.27.

The weights for the main criteria derived from the three decision groupings are summarised in Table 12.28. These weights reflect that the budget holders view the

Main criterion	Rank position	Ratio score (z_1)	W_1
Cr_1, Cr_5	1st	2 (×2)	0.26
Cr_3	2nd	1.5	0.19
Cr_4	3rd	1.25	0.16
Cr_2	4th	1	0.13

Table 12.25 Weightings of main criteria from budget holders group.

Main criterion	Rank position	Ratio score (z_1)	W_1
Cr_2	1st	2.5	0.32
Cr_4	2nd	1.75	0.23
Cr_3	3rd	1.5	0.19
Cr_1, Cr_5	4th	1 (×2)	0.13

Table 12.26 Weightings of main criteria from designers group.

Main criterion	Rank position	Ratio score (z_1)	W_1
Cr_4	1st	2.25	0.29
Cr_1	2nd	1.75	0.23
Cr_3	3rd	1.5	0.19
Cr_5	4th	1.25	0.16
Cr_2	5th	1	0.13

Table 12.27 Weightings of main criteria from political grouping.

Main criterion	Budget	Design	Political
Cr_1	0.26	0.13	0.23
Cr_2	0.13	0.32	0.13
Cr_3	0.19	0.19	0.19
Cr_4	0.16	0.23	0.29
Cr_5	0.26	0.13	0.16

(Decision groupings)

Table 12.28 Summary of main weights from the three decision groups.

economic criteria (Cr_1,Cr_5) as being most important, the designers/planners see the criterion relating to comfort, security and environmental impact (Cr_2) as being dominant and the politicians see the criterion relating to city centre and sub-urban trade and future plans for these areas (Cr_4) as being of primary importance together with the criterion relating to regional development (Cr_1).

Subcriteria

Within each main group, the relative importance of each subcriterion was decided by consultant transport planners employed by the decision makers. These weights were then agreed and adopted by each of the three groups. The weights of the subcriteria within each main cluster are outlined in Table 12.29.

To obtain the normalised weights for each of the subcriteria from each decision grouping, the information from Tables 12.28 and 12.29 are combined as shown in Table 12.30.

Table 12.29 Relative weights of each set of subcriteria within each main group.

Cr_1	Wt	Cr_2	Wt	Cr_3	Wt	Cr_4	Wt	Cr_5	Wt
$Cr_{1.1}$	0.25	$Cr_{2.1}$	0.20	$Cr_{3.1}$	0.30	$Cr_{4.1}$	0.35	$Cr_{5.1}$	0.30
$Cr_{1.2}$	0.25	$Cr_{2.2}$	0.40	$Cr_{3.2}$	0.20	$Cr_{4.2}$	0.35	$Cr_{5.2}$	0.25
$Cr_{1.3}$	0.25	$Cr_{2.3}$	0.30	$Cr_{3.3}$	0.20	$Cr_{4.3}$	0.15	$Cr_{5.3}$	0.20
$Cr_{1.4}$	0.25	$Cr_{2.4}$	0.10	$Cr_{3.4}$	0.30	$Cr_{4.4}$	0.15	$Cr_{5.4}$	0.25

Table 12.30 Three sets of normalised weightings for each subcriterion.

	Decision groupings		
Subcriterion	Budget	Design	Political
$Cr_{1.1}$	0.065	0.0325	0.0575
$Cr_{1.2}$	0.065	0.0325	0.0575
$Cr_{1.3}$	0.065	0.0325	0.0575
$Cr_{1.4}$	0.065	0.0325	0.0575
$Cr_{2.1}$	0.026	0.0640	0.0260
$Cr_{2.2}$	0.052	0.1280	0.0520
$Cr_{2.3}$	0.039	0.0960	0.0390
$Cr_{2.4}$	0.013	0.0320	0.0130
$Cr_{3.1}$	0.057	0.0570	0.0570
$Cr_{3.2}$	0.038	0.0380	0.0380
$Cr_{3.3}$	0.038	0.0380	0.0380
$Cr_{3.4}$	0.057	0.0570	0.0570
$Cr_{4.1}$	0.056	0.0805	0.0102
$Cr_{4.2}$	0.056	0.0805	0.0102
$Cr_{4.3}$	0.024	0.0345	0.0435
$Cr_{4.4}$	0.024	0.0345	0.0435
$Cr_{5.1}$	0.078	0.0390	0.0480
$Cr_{5.2}$	0.065	0.0325	0.0400
$Cr_{5.3}$	0.052	0.0260	0.0320
$Cr_{5.4}$	0.065	0.0325	0.0400
	Total = 1.000	Total = 1.000	Total = 1.000

12.7.2 Assessment of each option on each of the subcriteria

All eight options (A_1 to A_8) were assessed on each of the 20 subcriteria of evaluation using a 10-point scale. The format of this scale was slightly at variance with the type outlined above in Table 12.19. The scale here ranges from very positive effect (10) to very negative effect (0) with neutral effect on the given criterion assigned the score 5. Details of the scale are given in Table 12.31.

Option valuations on each subcriterion

Using the scale in Table 12.31, the eight options were assessed on the 20 subcriteria. The assessment was undertaken by the transport planning consultants working in conjunction with the project decision making group. The valuations are given in matrix form in Table 12.32.

Score	Assessment
10	The option has a strong positive effect on the subcriterion in question.
9 to 8	The option has a moderately positive effect on the subcriterion in question.
7 to 6	The option has a slightly positive effect on the subcriterion in question.
5	The option is neutral/has no discernable effect on the subcriterion in question
4 to 3	The option has a slightly negative effect on the subcriterion in question.
2 to 1	The option has a moderately negative effect on the subcriterion in question.
0	The option has a strong negative effect on the subcriterion in question.

Table 12.31 Subcriterion measurement scale.

Subcriterion	A_1	A_2	A_3	A_4	A_5	A_6	A_7	A_8
$Cr_{1.1}$	5	9	10	4	6	10	0	5
$Cr_{1.2}$	9	9	10	4	10	10	0	3.5
$Cr_{1.3}$	8	8	9	5	9	9.5	0	5
$Cr_{1.4}$	9	8	8	4	10	10	0	5
$Cr_{2.1}$	8	6	9	5	10	10	0	9
$Cr_{2.2}$	5	7	9	10	6	9	0	6.5
$Cr_{2.3}$	7	6	9	9	8	9	2	9
$Cr_{2.4}$	7	5	6	5	8	9	5	7
$Cr_{3.1}$	6	9	5	9	6	5	5	6
$Cr_{3.2}$	7	5	10	4	7.5	8	0	6
$Cr_{3.3}$	6	6	10	5	6	9	0	7
$Cr_{3.4}$	8	7	6	6	9	9	2	6.5
$Cr_{4.1}$	9	5	5	5	10	10	2	2
$Cr_{4.2}$	8	7	9	5	9	10	3.5	9
$Cr_{4.3}$	8	9	9	6	9	9	0	2
$Cr_{4.4}$	5	5	5	9	5	5	0	6
$Cr_{5.1}$	8	6	7	6	9	6	10	9
$Cr_{5.2}$	2	5	0	9	2	0	2	10
$Cr_{5.3}$	0	2	2	4	0	2	10	10
$Cr_{5.4}$	2	5	5	3	2	4	5	0

Table 12.32 Valuation matrix for the eight options.

12.7.3   Results of the multicriteria assessment

For each system of weights, the valuations for each option given in Table 12.32 are multiplied by the appropriate set of weights as given in Table 12.30. The result, in each of the three weighting cases, is a score for each option that allows them to be ranked in relative terms. Table 12.33 gives the baseline ranking obtained for each weighting system.

It can be seen that option A_6 (heavy rail) is the outstanding option, with options A_3 (extensive roads) and A_5 (light rail) close behind in joint second. A_1 (bus)/A_2

Rank	Budget Option	Budget Score	Design Option	Design Score	Political Option	Political Score
1	A_6	7.5	A_6	8.2	A_6	8.1
2	A_3	7.0	A_3	7.5	A_5	7.5
3	A_5	6.9	A_5	7.4	A_3	7.3
4	A_2	6.6	A_1	6.6	A_1	6.8
5	A_1	6.3	A_2	6.5	A_2	6.7
6	A_8	6.1	A_4	6.4	A_8	5.9
7	A_4	5.7	A_8	6.4	A_4	5.8
8	A_7	2.6	A_7	2.1	A_7	2.2

Table 12.33 Ranking of options for three weighting systems.

Option	Rank number Budget	Rank number Design	Rank number Political	Sum	Position
A_6	1	1	1	3	1st
A_3	2	2	3	7	2nd
A_5	3	3	2	8	3rd
A_1	5	4	4	13	4th
A_2	4	5	5	14	5th
A_8	6	7	6	19	6th
A_4	7	6	7	20	7th
A_7	8	8	8	24	8th

Table 12.34 Summing the ranks for each option from the three decision groups.

(incremental roads) are in joint fourth place overall, just ahead of A_8 (demand management), with A_4 (cycling/walking) just behind. The 'do-nothing' option, A_7, is shown not to be a viable strategy.

In some cases where the results from different weighting systems give slightly confusing results, for each option the ranks of the three decision groups can be summed to give an overall score, with the option with the lowest sum being ranked first, the second lowest ranked second and so on. This technique, where a number of ranks are summed together is a form of weighted average, is known as Borda's Method. Within it, the n rankings (in this situation 'n' is the number of decision groups) are aggregated to give an overall unique ranking. Table 12.34 shows this computation for the case study.

This approach is useful as an aid for the decision makers in reaching a final consensus. The closeness of summed rankings three pairs, A_3/A_5, A_1/A_2 and A_4/A_8, indicates the difficulty in separating them.

12.7.4 Sensitivity analysis

Varying the above three weighting systems by either emphasising W or de-emphasising the importance of their most prominent criteria has very little effect on the final results. Examination shows that the only weight formation that would result in A_6 being placed anywhere but first involves criterion Cr_5 being given an importance three times that of all the others. This gives the set shown in Table 12.35.

Main criterion	Rank position	Ratio score (z_i)	W_i
Cr_5	1st	3	0.428
Cr_1, Cr_2, Cr_3, Cr_4	2nd	1 (×4)	0.143

Table 12.35 Weightings with Cr_5 at an elevated level of importance.

The ranking that results from these weights is:

$$A_8 \rightarrow A_6 \rightarrow A_3 \rightarrow A_5 \rightarrow A_2 \rightarrow A_4 \rightarrow A_1 \rightarrow A_7$$

As Cr_5 increases beyond three times the importance of the other four, option A_8 becomes more dominant over the other seven options. For the vast majority of combinations of weights, however, A_6 remains the dominant option, followed by A_3 and A_5.

12.8   Summary

The Simple Additive Weighting Model requires criteria of evaluation, originally measured on different scales, to be evaluated on a common basis and permits multiple weighting systems reflecting the views of different decision making groups to be input into the process. As with other appraisal techniques detailed previously within this text, sensitivity analysis is seen to play a major role within this technique, allowing the robustness of the baseline results to be gauged. If there is a significant level of uncertainty associated with the criterion valuations, the Probabilistic Additive Weighting Model exists for explicitly accounting for this uncertainty within this technique. The basic SAW Model, in the form of the scale-weighted checklist, is the predominant decision model used to assess project options on the basis of their environmental impact. The case study detailed within the chapter shows in detail how the SAW Method can be applied to an engineering decision problem involving economic, environmental, technical and political criteria.

References

CEQ (Council for Environmental Quality) (1978) National Environmental Policy Act – Regulations. *Federal Register*, Vol. 43, No. 230, November 29, pp. 55978–56007, United States.

Dee, N. (1972) *Environmental Evaluation System for Water Resources Planning*. Final Report. Battelle-Columbus Laboratories, Columbus, OH.

Hinkle, D. (1965) The Change of Personal Constructs from the Viewpoint of a Theory of Construct Implications. PhD Dissertation, Ohio State University.

Keeney, R.L. and Raiffa, H. (1976) *Decisions with Multiple Objectives*. John Wiley & Sons, Inc., New York.

Mongkol, P. (1982) A conceptual development of a quantitative environmental impact assessment methodology for decision makers. *Journal of Environmental Management*, **14**, 301–317.

Rogers M.G. and Bruen, M.P. (1998) A New System for Weighting Criteria within Electre. *European Journal of Operational Research*, **107** (3), 552–563.

Sondheim, M.W. (1978) Comprehensive methodology for assessing environmental impact. *Journal of Environmental Management*, **6**, 27–42.

Vansnick, J.C. (1986) *De Borda et Concordacet a l'agrégation multicritère*. Université Paris-Dauphine, France.

Vincke, P. (1992) *Multicriteria Decision-Aid*. John Wiley and Sons Ltd, Chichester.

Chapter 13

Analytic Hierarchy Process (AHP)

13.1   Introduction

The Analytical Hierarchy Process (Saaty, 1980) is a multicriteria technique that permits qualitatively-based data to be transformed into pairwise comparison data. It is essentially a framework within which the decision maker can express a complex engineering problem using a hierarchical structure.

Saaty describes the theme of the Analytic Hierarchy Process (AHP) as a combination of 'decomposition by hierarchies and synthesis by finding relations through informed judgement'. Firstly, therefore, the structure of the decision problem is determined. This structure, in the case of the AHP, is in the form of a hierarchy. The problem is broken down into its constituent parts using this hierarchical structure. Each level of the hierarchy is the subject of its own pairwise analysis. The individual results from each of these comparisons are then joined together or 'synthesised' based on their position within the hierarchy in order to obtain an overall result.

The purpose of the AHP, as with most other decision-aid systems, is to develop a theory and provide a methodology for modelling unstructured decision choice problems in many areas including engineering. It reduces a decision problem to a series of smaller self-contained ones. The relative merit of each project option is determined from a pairwise analysis of the relative performance ratings. This analysis is performed separately for each decision criterion involved. The relative importance of each criterion is also determined from a similar pairwise analysis of decision makers' preferences. The result of the overall process is a ranking of all alternatives on an interval scale, enabling the optimal one to be selected.

The AHP, like a number of decision models referred to previously, such as the Planned Balance Sheet and the Goal Achievement Matrix, explicitly sets out to reject the simplification of a problem's parameters to suit quantitative/monetary models, such as cost–benefit analysis (CBA), believing the multicriteria model to be the best to deal with complex engineering problems.

Engineering Project Appraisal: The Evaluation of Alternative Development Schemes, Second Edition.
Martin Rogers and Aidan Duffy.
© 2012 John Wiley & Sons, Ltd. Published 2012 by John Wiley & Sons, Ltd.

To understand more of the AHP method, we must define the nature of the hierarchies within it and how the priorities within and between the hierarchies are established.

13.2   Hierarchies

A hierarchy enables the decision problem to be broken down into individual elements whose relationships with each other can then be analysed. Stated more formally by Saaty, it is 'an abstraction of the *structure* of a system to study the *functional* interactions of its components and their impacts on the entire system'. The 'structure' and 'function' of a system cannot be separated. The former is the arrangement of its parts, and the latter is the function or duty that the components are meant to serve.

Thus, a hierarchical system is based on the assumption that the entities identified as relevant to the decision can be grouped into disjoint sets, with each set directly affecting the one above it. This is the 'decomposition' referred to in the opening definition. This hierarchical structure illustrates how results at the higher levels are directly influenced by those at the lower levels.

Hierarchies have many advantages. They can be used to describe how changes in priority at upper levels affect priorities of elements in lower levels. They provide detailed information on both the structure and function of the system. They are stable and flexible, accurately mirroring reality, since natural systems are assembled hierarchically.

13.3   Establishing priorities within hierarchies

A hierarchy, by itself, is not a very powerful tool for decision making unless the extent to which the various elements on one level influence those on the next higher level can be determined. In this way, the relative strengths of the impacts of the elements at the lowest level on the overall objectives can be computed. This is the 'synthesis' referred to in the opening definition.

The strengths of priorities of elements on one level relative to the next are determined as follows. Given the elements of one level and one element, X, on the next higher level, the elements of the lower level are compared pairwise in their strength of influence on X. The numbers reflecting these comparisons are inserted in a matrix and the eigenvector with the largest eigenvalue is found. The eigenvector itself provides the priority ordering and the eigenvalue is a measure of the consistency of the judgement.

For example, let us examine four elements – A, B, C and D – within one hierarchy level. These could denote project options that are being pairwise compared on a given criterion, or they could represent criteria whose relative importance is being

	A	B	C	D
A	1	ab	ac	ad
B	1/ab	1	bc	bd
C	1/ac	1/bc	1	cd
D	1/ad	1/bd	1/cd	1

Table 13.1 Example of pairwise comparison matrix.

Pairwise evaluation	Score
A and B are equally important	ab = 1
A is weakly more important than B	ab = 3
B is weakly more important than A	ab = 1/3
A is strongly more important than B	ab = 5
B is strongly more important than A	ab = 1/5
A is very strongly more important than B	ab = 7
B is very strongly more important than A	ab = 1/7
A is absolutely more important than B	ab = 9
B is absolutely more important than A	ab = 1/9

Table 13.2 Example of pairwise comparison matrix for A and B.

assessed. Each pair – AB, AC, AD, BC, BD, and CD – is directly compared with respect to its influence on X and the results placed within a matrix format, as shown in Table 13.1.

Table 13.2 takes a pairwise comparison of A and B for demonstration purposes. The appropriate score is placed in the position 'ab' where row A meets column B. Since an element is always equally important relative to itself (reflexivity), the values in the main diagonal of the matrix are always one. Also, the appropriate reciprocals 1/3, 1/5, 1/7 or 1/9 should be inserted where column A meets row B, that is in position (B, A). Also, the scores 2, 4, 6 and 8 and their reciprocals are used to facilitate compromise between slightly differing judgements.

The requirement for a nine-point scale is based on Saaty's belief that, within the framework of a simultaneous comparison, one does not need more than nine scale points to distinguish between stimuli (Saaty, 1977). Results from psychological studies (Miller, 1956) have shown that a scale of about seven points was sufficiently discerning. Saaty (1977) noted that the ability to make qualitative decisions was well represented by five attributes:

(1)   Equal
(2)   Weak
(3)   Strong
(4)   Very strong
(5)   Absolute

If we make comparisons between adjacent attributes, when greater precision is needed, the scale requires, in total, nine values – thus a nine-point scale. An example of a numerated matrix could be as shown in Table 13.3.

	A	B	C	D
A	1	5	6	7
B	1/5	1	4	6
C	1/6	1/4	1	4
D	1/7	1/6	1/4	1

Table 13.3 Example of a numerated pairwise comparison matrix.

Location	S_1	S_2	S_3
S_1	1	2	4
S_2	1/2	1	2
S_3	1/4	1/2	1

Table 13.4 Performance matrix for three retail outlets based on location.

13.4   Establishing and calculating priorities

The point has now been reached where the comparison matrix is in place and the information contained within it is required to establish priorities between the different elements being examined. To derive these, the judgements made within the pairwise matrix must be pulled together in order that a single number for each element, indicating its priority relative to the others, can be determined.

Take the following example. A developer wants to decide which of three retail outlets he should purchase. The decision is being made on the basis of one criterion – location. Using Saaty's nine-point scale, the developer pairwise compares the three outlets, S_1, S_2 and S_3. The matrix lists the options within the left column and along the top row (Table 13.4). The score along the diagonal positions is always '1', as an option is at all times as equally desirable as itself. The judgements above the diagonal are then input. Once these are decided, all judgements below the diagonal become the reciprocal of those above. If the decision involved n options, then the number of judgements needed to fill the entries is:

$$\text{Number of entries} = \left(n^2 - n\right) \div 2 \tag{13.1}$$

In this case, the number of elements is 3, therefore the number of entries required is:

$$\left(3^2 - 3\right) \div 2 = 3$$

The three required entries in this case are indicated in bold in Table 13.4. All other entries are either '1' along the diagonal positions or are reciprocals of the original three above the diagonal.

The decision maker assesses that the location of site S_1 is marginally better located than site S_2 and assigns it the score '2' on the Saaty scale of nine in this pairwise comparison (between 'equal' and 'weakly preferable'). He assesses site S_1 to be moderately better located than site S_3 and assigns it the score '4' (between 'weakly preferable' and 'strongly preferable'). Site S_2 is assessed as being marginally better located than site S_3 and is thus also given the score '2'. These are all the judgements we need to make. We place '1' along the diagonal positions. The other values are reciprocals of our original judgements. The performance of S_2 against S_1 is the reciprocal of the score for S_1 versus S_2. S_2 is adjudged slightly less preferable in terms of location than S_1 and is assigned the reciprocal of '2', that is '½'. For the same reason S_3 is assigned the score '¼' against S_1 and '½' against S_2.

13.4.1   Deriving priorities using an approximation method

The next step involves translating the scores that have been assigned to each pair of elements within the matrix into a single score for each element, indicating its relative performance. For the particular case illustrated above in Table 13.4, the aim is to establish the relative priorities of the three sites on the basis of location.

Let us first illustrate a procedure for obtaining an approximate estimate for these priorities. (A more widely applicable and comprehensive technique is outlined further below.) Initially, add the scores within each column and then take each of these column totals and divide it into each entry within its column. This gives a set of normalised scores along each column length, that is their sum equals one (Tables 13.5 and 13.6).

To get the final scores, we obtain the average entry score for each row of the normalised matrix, as shown in Table 13.7.

Location	S_1	S_2	S_3
S_1	1	2	4
S_2	1/2	1	2
S_3	1/4	1/2	1
Column total	7/4	7/2	7

Table 13.5 Judgements with column totals.

Location	S_1	S_2	S_3
S_1	4/7	4/7	4/7
S_2	2/7	2/7	2/7
S_3	1/7	1/7	1/7

Table 13.6 Normalised judgement matrix.

Element	Calculation	Result
Site S_1	(4/7 + 4/7 + 4/7) ÷ 3	0.571
Site S_2	(2/7 + 2/7 + 2/7) ÷ 3	0.286
Site S_3	(1/7 + 1/7 + 1/7) ÷ 3	0.143
	Total = 1.000	

Table 13.7 Result for each site.

This simple procedure is sufficient for calculating the priorities of the various elements for the special case where the matrix is consistent. For the example shown above, a quick calculation indicates the consistency of the comparison matrix:

- S_1 versus S_2 yields the result '2' $(S_1 = 2 \times S_2)$
- S_1 versus S_3 yields the result '4' $(S_1 = 4 \times S_3)$

Therefore, for consistency:

- S_2 versus S_3 should yield the result $2 \times S_2 = 4 \times S_3 \Rightarrow S_2 = (4/2 \times S_3) = 2 \times S_3$

We see from the comparison matrix that this is, in fact, the case, because S_2 versus S_3 gives the answer '2'. The matrix is therefore consistent.

In the case where the matrix is not consistent, however, the exact method of determining priorities must be used. Take for example the matrix shown in Table 13.8.

This matrix is not consistent, because although $S_1 = 7 \times S_2$ and $S_1 = 9 \times S_3$, $S_2 \neq 9/7 \times S_3$. The matrix, in fact, judges S_2 to be $5 \times S_3$. If we use the approximate method in this case, we will derive the priorities as seen in Table 13.9.

One can see immediately that, unlike the last example, the columns are not identical. The matrix is thus confirmed as being inconsistent and if one obtains the average value for each row, as shown in Table 13.10, an accurate valuation of the priorities is not obtained:

These are not accurate estimates of the priorities of the elements in question. To obtain these, the method outlined in Section 13.4.2 must be used.

Location	S_1	S_2	S_3
S_1	1	7	9
S_2	1/7	1	5
S_3	1/9	1/5	1

Table 13.8 Example of an inconsistent judgement matrix.

Location	S_1	S_2	S_3		S_1	S_2	S_3
S_1	1	7	9		0.80	0.85	0.6
S_2	1/7	1	5	=	0.11	0.12	0.33
S_3	1/9	1/5	1		0.09	0.02	0.07
Column total	79/63	41/5	15		$\Sigma = 1$	$\Sigma = 1$	$\Sigma = 1$

Table 13.9 Normalised matrix.

Element	Calculation	Result
Site S_1	$(0.80 + 0.85 + 0.60) \div 3$	0.75
Site S_2	$(0.11 + 0.12 + 0.33) \div 3$	0.19
Site S_3	$(0.09 + 0.02 + 0.07) \div 3$	0.06
		Total = 1.00

Table 13.10 Priorities derived using approximate method.

13.4.2   Deriving exact priorities using the iterative eigenvector method

To obtain the exact priorities of the elements where the judgement matrix is not consistent, the principal eigenvector of the matrix must be obtained. It is not intended to go into the theoretical basis for this technique here. This information can be found in Saaty (1994). What this text intends to do is illustrate, for a non-consistent matrix, how the eigenvector of weights can be derived using an iterative mathematical procedure.

The method entails obtaining an initial set of priorities using the approximate method illustrated above. These weights are then input into an n value column vector, where n equals the number of elements under examination. This column vector is then multiplied by the $n \times n$ matrix of judgements. The resulting column vector is then normalised to sum to unity, yielding an updated set of n priorities. The process is then repeated. The updated normalised vector is multiplied by the $n \times n$ matrix of judgements and the resulting column vector is again normalised and compared with the first set of values. If the difference between the two sets of priorities is within a prescribed decimal accuracy, the process is halted and the last set of computed priorities yields what is termed the principal eigenvector of the judgement matrix. If the difference between the two successively derived sets of priorities is not within the prescribed limits, the process is repeated until the required accuracy is obtained.

If the judgement matrix is consistent, the values from the exact method are identical to those from the approximate method. The greater the level of inconsistency throughout the matrix, the greater the divergence between the results from the two methods will be.

Take the judgement matrix in Table 13.8 that has been shown in Table 13.9 to be inconsistent. The priorities derived by the approximate method, given in Table 13.10, are therefore incorrect. These values can, however, be used within the initial iteration of the exact method. These three priorities (0.75, 0.19, 0.06) are placed within a column vector and multiplied by the judgement matrix, as seen in Table 13.11.

1	7	9		0.75		2.6175
1/7	1	5	×	0.19	=	0.5964
1/9	1/5	1		0.06		0.1812

Table 13.11 Column vector computation – Step 1.

The resulting column vector is now normalised to give a revised set of priorities for the three sites S_1, S_2 and S_3 (Table 13.12).

2.6175		0.7710
0.5964	**Normalised** →	0.1756
0.1812		0.0534
Total = 3.3951		1.0000

Table 13.12 Normalisation – Step 1.

These revised set of priorities are now multiplied by the basic judgement matrix and the procedure is repeated as shown in Table 13.13.

1	7	9		0.7710		2.4809
1/7	1	5	×	0.1756	=	0.5527
1/9	1/5	1		0.0534		0.1742

Table 13.13 Column vector computation – Step 2.

The column vector is again normalised, as shown in Table 13.14.

	2.4809			0.7734
	0.5527	**Normalised →**		0.1723
	0.1742			0.0543
Total = 3.2078				1.0000

Table 13.14 Normalisation – Step 2.

The revised set of priorities is again multiplied by the basic judgement matrix, as given within Table 13.15.

1	7	9		0.7734		2.4681
1/7	1	5	×	0.1723	=	0.5542
1/9	1/5	1		0.0543		0.1747

Table 13.15 Column vector computation – Step 3.

The column vector is again normalised, as shown in Table 13.16.

	2.4681			0.7720
	0.5542	**Normalised →**		0.1734
	0.1747			0.0546
Total = 3.1971				1.0000

Table 13.16 Normalisation – Step 3.

The fourth iteration proceeds as given in Tables 13.17 and 13.18.

1	7	9		0.7720		2.4773
1/7	1	5	×	0.1735	=	0.5569
1/9	1/5	1		0.0545		0.1751

Table 13.17 Column vector computation – Step 4.

	2.4773			0.7719
	0.5569	**Normalised →**		0.1735
	0.1751			0.0546
Total = 3.2093				1.0000

Table 13.18 Normalisation – Step 4.

It can be seen that the decimal difference between the priorities from Step 3 and Step 4 is a maximum of 0.0001. The iterative procedure can therefore be terminated and the values from Step 4 taken as the calculated priorities from the exact method. The derived priorities can thus be taken as 0.77, 0.17 and 0.06 for Sites S_1, S_2 and S_3, respectively. These are seen not to be the same as those derived from the approximation method given in Table 13.10.

Note: If the exact method is used to derive the priorities for a totally consistent matrix of judgements, the values from the approximate method are immediately replicated within one step of the iterative procedure and would continue to do so no matter how many times the procedure is repeated.

Taking the consistent matrix from Table 13.5 and the priorities derived for it using the approximation method (4/7, 2/7, 1/7) gives a vector computation as shown in Table 13.19.

The normalisation process yields a set of priorities identical to those given by the approximation method and used within the column vector on the right-hand side of Table 13.19 (Table 13.20).

1	2	4		0.5714		1.7143
1/2	1	2	×	0.2857	=	0.8571
1/4	1/2	1		0.1429		0.4286

Table 13.19 Column vector computation for a consistent matrix.

1.7143		0.5714
0.8571	Normalised →	0.2857
0.4286		0.1429
Total = 3.0000		1.0000

Table 13.20 Normalisation.

13.4.3   Example of the Exact Method for determining priorities

We now look, by way of illustration, at an application of Saaty's method to a simplified environmental evaluation. Within it, three route options for a major motorway proposal are assessed on the basis of three main criteria and nine subcriteria (the actual assessments take place using these subcriteria) in order to determine which of them has the least environmental impact.

A hierarchy of objectives, criteria and alternatives is constructed in order to gauge the extent to which each option contributes to the fulfilment of this overall objective.

The problem involves using AHP at three levels:

(1)   Estimation of the relative importance weights of the main criteria.
(2)   Estimation of the relative importance weights of the subcriteria within each main criterion.
(3)   Determining the relative performance of each option on each of the subcriteria of evaluation.

The hierarchies at the basis of this decision problem can be outlined as:

Level 1

(i) Overall Objective (O)
• Minimisation of environmental impact

Level 2

(i) Main criteria (Cr)

- Environmental impact on occupiers of property (Cr_1)
- Impact on physical environment (Cr_2)
- Environmental impact on road users (Cr_3)

Level 3

(i) Subcriteria to Cr_1, Cr_2 and Cr_3

- *Cr1*
  — Noise impacts $(Cr_{1.1})$
  — Land severance impacts $(Cr_{1.2})$
  — Visual impacts $(Cr_{1.3})$
- *Cr2*
  — Air quality impacts (carbon monoxide level)$(Cr_{2.1})$
  — Air quality impacts (lead level) $(Cr_{2.2})$
  — Nature conservation $(Cr_{2.3})$
- *Cr3*
  — Time savings impacts $(Cr_{3.1})$
  — Vehicle operating cost impacts $(Cr_{3.2})$
  — Driver comfort impacts $(Cr_{3.3})$

Level 4

(1) Project alternatives (A)

- Alignment to the north (A_1)
- Alignment to the south (A_2)
- Central alignment (A_3)

The hierarchical representation of this environmental evaluation problem is given in Figure 13.1.

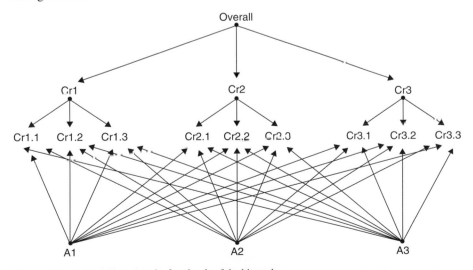

Figure 13.1 Linkages between the four levels of the hierarchy.

(2)   Judgement matrices

(i) Estimation by decision makers of the relative weights for Cr_1, Cr_2 and Cr_3.
Table 13.21 gives the pairwise matrix of judgements from the decision makers regarding the relative importance weightings of the three main criteria.

Criterion	Cr_1	Cr_2	Cr_3
Cr_1	1	5	7
Cr_2	1/5	1	2
Cr_3	1/7	1/2	1

Table 13.21 Judgement matrix for main criteria.

One can see from the normalised matrix in Table 13.22 that not all the columns are identical; therefore, the judgements are not fully consistent.

Location	S_1	S_2	S_3		S_1	S_2	S_3
S_1	1	5	7		0.745	0.769	0.7
S_2	1/5	1	2	=	0.149	0.154	0.2
S_3	1/7	1/2	1		0.106	0.077	0.1
Column total	47/35	13/2	10		$\Sigma=1$	$\Sigma=1$	$\Sigma=1$

Table 13.22 Normalised matrix.

The approximation method yields the initial set of weights shown in Table 13.23.

Element	Calculation	Result
Cr_1	$(0.745+0.769+0.7)\div 3$	0.738
Cr_2	$(0.149+0.154+0.2)\div 3$	0.168
Cr_3	$(0.106+0.077+0.1)\div 3$	0.094
		Total = 1.000

Table 13.23 Derivation of initial estimates of main criteria weights using approximation method.

Two iterations of the exact method yield a stable score for the weightings of each of the main criteria, as shown in Tables 13.24 and 13.25. The final normalised weights are thus as given in Table 13.26.

$$\begin{bmatrix} 1 & 5 & 7 \\ 1/5 & 1 & 2 \\ 1/7 & 1/2 & 1 \end{bmatrix} \times \begin{bmatrix} 0.738 \\ 0.168 \\ 0.094 \end{bmatrix} = \begin{bmatrix} 2.2369 \\ 0.5041 \\ 0.2837 \end{bmatrix} \text{ Normalised} \rightarrow \begin{bmatrix} 0.7396 \\ 0.1666 \\ 0.0938 \end{bmatrix}$$

Table 13.24 Derivation of main criteria weights using exact method – Step 1.

$$\begin{bmatrix} 1 & 5 & 7 \\ 1/5 & 1 & 2 \\ 1/7 & 1/2 & 1 \end{bmatrix} \times \begin{bmatrix} 0.7396 \\ 0.1666 \\ 0.0938 \end{bmatrix} = \begin{bmatrix} 2.2293 \\ 0.5021 \\ 0.2828 \end{bmatrix} \text{ Normalised} \rightarrow \begin{bmatrix} 0.7396 \\ 0.1666 \\ 0.0938 \end{bmatrix}$$

Table 13.25 Derivation of main criteria weights using exact method – Step 2.

$$Cr_1 = 0.74$$
$$Cr_2 = 0.17$$
$$Cr_3 = 0.09$$

Table 13.26 Final weights.

Note: Impacts on occupiers of property is by far the most important of the three in this case.

(ii) Estimations by decision makers of the relative weights within the three main criteria for each of their respective subcriteria

These are given in Table 13.27.

Cr_1	$Cr_{1.1}$	$Cr_{1.2}$	$Cr_{1.3}$
$Cr_{1.1}$	1	7	9
$Cr_{1.2}$	1/7	1	5
$Cr_{1.3}$	1/9	1/5	1

Table 13.27 Judgement matrix – subcriteria for Cr_1.

Again, this can be shown to be an inconsistent judgement matrix. The exact method yields a stable set of weights as shown in Tables 13.28 and 13.29.

$$Cr_{1.1} = 0.77$$
$$Cr_{1.2} = 0.17$$
$$Cr_{1.3} = 0.06$$

Table 13.28 Subcriteria weightings – Cr_1.

Cr_2	$Cr_{2.1}$	$Cr_{2.2}$	$Cr_{2.3}$
$Cr_{2.1}$	1	1/3	1/5
$Cr_{2.2}$	3	1	1/5
$Cr_{2.3}$	5	3	1

Table 13.29 Judgement matrix – subcriteria for Cr_2.

This is an inconsistent judgement matrix. The exact method yields a stable set of weights as shown in Tables 13.30 and 13.31.

$$Cr_{2.1} = 0.10$$
$$Cr_{2.2} = 0.20$$
$$Cr_{2.3} = 0.70$$

Table 13.30 Subcriteria weightings – Cr_2.

Cr_3	$Cr_{3.1}$	$Cr_{3.2}$	$Cr_{3.3}$
$Cr_{3.1}$	1	3	4
$Cr_{3.2}$	1/3	1	1/3
$Cr_{3.3}$	1/4	3	1

Table 13.31 Judgement matrix – subcriteria for Cr_3.

The exact method yields a stable set of weights as shown in Tables 13.32.

$$Cr_{3.1} = 0.62$$
$$Cr_{3.2} = 0.13$$
$$Cr_{3.3} = 0.25$$

Table 13.32 Subcriteria weightings – Cr_3.

(iii) Estimation by decision makers of the relative performance of each option on each of the nine subcriteria

This is as given in Table 13.33.

$Cr_{1.1}$	A_1	A_2	A_3
A_1	1	1/3	8
A_2	3	1	9
A_3	1/8	1/9	1

Table 13.33 Judgement matrix – Performance of each option on $Cr_{1.1}$.

The exact method yields a stable set of performance scores for each option as shown in Tables 13.34 and 13.35.

$$A_1 = 0.30$$
$$A_2 = 0.65$$
$$A_3 = 0.05$$

Table 13.34 Option performances on $Cr_{1.1}$.

$Cr_{1.2}$	A_1	A_1	A_1
A_1	1	1/3	8
A_2	3	1	9
A_3	1/8	1/9	1

Table 13.35 Judgement matrix – Performance of each option on $Cr_{1.2}$.

The exact method yields a stable set of performance scores for each option as given in Tables 13.36 and 13.37.

$$A_1 = 0.30$$
$$A_2 = 0.65$$
$$A_3 = 0.05$$

Table 13.36 Option performances on $Cr_{1.2}$.

$Cr_{1.3}$	A_1	A_1	A_1
A_1	1	1	9
A_2	1	1	9
A_3	1/9	1/9	1

Table 13.37 Judgement matrix – Performance of each option on $Cr_{1.3}$.

The exact method yields a stable set of performance scores for each option as shown in Tables 13.38 and 13.39.

$$A_1 = 0.47$$
$$A_2 = 0.47$$
$$A_3 = 0.06$$

Table 13.38 Option performances on $Cr_{1.3}$.

$Cr_{2.1}$	A_1	A_1	A_1
A_1	1	4	9
A_2	1/4	1	8
A_3	1/9	1/8	1

Table 13.39 Judgement matrix – Performance of each option on $Cr_{2.1}$

The exact method yields a stable set of performance scores for each option as shown in Tables 13.40 and 13.41.

$$\begin{aligned} A_1 &= 0.69 \\ A_2 &= 0.26 \\ A_3 &= 0.05 \end{aligned}$$

Table 13.40 Option performances on $Cr_{2.1}$

$Cr_{2.2}$	A_1	A_1	A_1
A_1	1	1	9
A_2	1	1	9
A_3	1/9	1/9	1

Table 13.41 Judgement matrix – Performance of each option on $Cr_{2.2}$

The exact method yields a stable set of performance scores for each option as given in Tables 13.42 and 13.43.

$$\begin{aligned} A_1 &= 0.47 \\ A_2 &= 0.47 \\ A_3 &= 0.06 \end{aligned}$$

Table 13.42 Option performances on $Cr_{2.2}$

$Cr_{2.3}$	A_1	A_1	A_1
A_1	1	1	9
A_2	1	1	9
A_3	1/9	1/9	1

Table 13.43 Judgement matrix – Performance of each option on $Cr_{2.3}$

The exact method yields a stable set of performance scores for each option as shown in Tables 13.44 and 13.45.

$$\begin{aligned} A_1 &= 0.47 \\ A_2 &= 0.47 \\ A_3 &= 0.06 \end{aligned}$$

Table 13.44 Option performances on $Cr_{2.3}$

$Cr_{3.1}$	A_1	A_1	A_1
A_1	1	3	8
A_2	1/3	1	6
A_3	1/8	1/6	1

Table 13.45 Judgement matrix – Performance of each option on $Cr_{3.1}$

The exact method yields a stable set of performance scores for each option as given in Tables 13.46 and 13.47.

$$A_1 = 0.65$$
$$A_2 = 0.29$$
$$A_3 = 0.06$$

Table 13.46 Option performances on $Cr_{3.1}$

$Cr_{3.2}$	A_1	A_1	A_1
A_1	1	3	7
A_2	1/3	1	5
A_3	1/7	1/5	1

Table 13.47 Judgement matrix – Performance of each option on $Cr_{3.2}$

The exact method yields a stable set of performance scores for each option as given in Tables 13.48 and 13.49.

$$A_1 = 0.65$$
$$A_2 = 0.28$$
$$A_3 = 0.07$$

Table 13.48 Option performances on $Cr_{3.2}$

$Cr_{3.3}$	A_1	A_1	A_1
A_1	1	1/6	7
A_2	6	1	8
A_3	1/7	1/8	1

Table 13.49 Judgement matrix – Performance of each option on $Cr_{3.3}$

The exact method yields a stable set of performance scores for each option as given in Table 13.50.

(iv) Determination of the overall performance for each option

$$A_1 = 0.21$$
$$A_2 = 0.74$$
$$A_3 = 0.05$$

Table 13.50 Option performances on $Cr_{3.3}$

Each of the above nine column vectors can be combined in the 3×9 matrix 'A' given in Table 13.51.

$$A = \begin{matrix} 0.30 & 0.30 & 0.47 & 0.69 & 0.47 & 0.47 & 0.65 & 0.65 & 0.21 \\ 0.65 & 0.65 & 0.47 & 0.26 & 0.47 & 0.47 & 0.29 & 0.28 & 0.74 \\ 0.05 & 0.05 & 0.06 & 0.05 & 0.06 & 0.06 & 0.06 & 0.07 & 0.05 \end{matrix}$$

Table 13.51 Combined performance matrix for each option on all nine criteria.

To obtain a final score for each option indicating its overall performance, the product of the above matrix and a column vector containing the normalised weights for each of the subcriteria must be estimated.

To obtain these values, the normalised weights from each set of subcriteria should be multiplied by the normalised weight of its main criterion, as shown in Table 13.52.

0.74	×	0.77	=	0.57	
		0.17		0.13	
		0.06		0.04	
0.17	×	0.10	=	0.02	
		0.20		0.03	
		0.70		0.12	
0.09	×	0.62	=	0.06	
		0.13		0.01	
		0.25		0.02	

Table 13.52 Overall normalised weights for $Cr_{1.1}$ to $Cr_{3.3}$.

The nine-point column vector, B, containing the normalised weights is thus represented as given in Table 13.53.

		0.57
		0.13
		0.04
B	=	0.02
		0.03
		0.12
		0.06
		0.01
		0.02

Table 13.53 Normalised column vector of weights for the nine subcriteria.

The column vector B is now multiplied by matrix A containing the 3×9 matrix of pairwise comparisons, to get the final overall performance scores for the three options (Table 13.54).

Table 13.54 Estimation of overall performance of the three options.

0.30	0.30	0.47	0.69	0.47	0.47	0.65	0.65	0.21		0.57		0.36
0.65	0.65	0.47	0.26	0.47	0.47	0.29	0.28	0.74	×	0.13	=	0.59
0.05	0.05	0.06	0.05	0.06	0.06	0.06	0.07	0.06		0.04		0.05
										0.02		
										0.03		
										0.12		
										0.06		
										0.01		
										0.02		

Thus, the final scores for each option are as shown in Table 13.55.

A × B =	0.36
	0.59
	0.05

Table 13.55 Final scores for each option.

$A1 = 0.36$, $A2 = 0.59$, $A3 = 0.05$

Overall result

A2 is preferred to A1, with A3 ranking a distant third.

13.5   Relationship between AHP and the Simple Additive Weighting (SAW) model

Although it is usually assumed that the AHP approach is completely distinct from the SAW method detailed in Chapter 12, this is, in fact, not quite the case. Belton (1986) presented the Analytic Hierarchy Process as a variation of the Simple Additive Weighted (SAW) Model, illustrating that Saaty's AHP model can be expressed in the following mathematical form:

$$V_i = \sum_j W_j X_{ij} \qquad\qquad (13.2)$$

where V_i is the overall valuation for alternative i, W_j is the weight assigned to criterion j to reflect its importance relative to the other criteria and X_{ij} is the score of alternative i on criterion j.

But how can the above equation be reconciled to the Saaty model based on pairwise comparison of criteria and options?

Within AHP, the weights W_j and scores X_{ij} are not explicitly distinguished, whereas, within the SAW Model, they are. The weights derived within AHP are the result of the reconciliation of the judgement matrix or matrices of pairwise comparisons of the importance criteria. The final scores obtained stem from pairwise comparisons of the alternatives with respect to each subcriterion of the family of criteria being used for decision making purposes.

This analogy with the SAW Model can be illustrated using the environmental evaluation example shown for three highway options illustrated in the earlier section. In it, firstly, the weighting scores are evaluated for each of the nine environmental criteria using a set of pairwise comparison matrices yielding a nine-dimensional vector. Then, the three options are scored against one another for each of the nine environmental subcriteria, yielding a 3×9 matrix. The final section of the solution, where the 3×9 matrix is multiplied by a transposed nine-point column vector to give the final ratings for the three options, is directly analogous to the summation formula from the SAW Model shown in Chapter 12, where the additive formula combines weights and scores to give an overall valuation for each alternative.

13.6   Summary

Saaty's eigenvector approach to pairwise comparisons provides a framework for calibrating a numerical scale and is useful in engineering problems where quantitative valuations and comparisons do not exist. The participation of a group of decision makers makes it possible to assess trade-offs between diverse criteria or options at a given level of analysis. It can be used as an aid to the decision making process and enables compromise to be attained from the various judgements reached.

It is thus very useful where semantic evaluations for weights and scores are the only type of decision information available. In such situations, AHP is a readily usable format. AHP-based software has been developed by Expert Choice Ltd and is widely used, particularly in the business world.

References

Belton, V. (1986) A comparison of the Analytic Hierarchy Process and a Simple Multi-Attribute Value Function. *European Journal of Operational Research*, **26**, 7–21.

Miller, G.A. (1956) The magical number seven, plus or minus two: some limits on our capacity for processing information. *The Psychological Review*, **63**, 81–97.

Saaty, T.L. (1977) A scaling for priorities in hierarchical structures. *Journal of Mathematical Psychology*, **15**, 207–218.

Saaty, T.L. (1980) *The Analytic Hierarchy Process*. McGraw-Hill, New York.

Saaty, T.L. (1994) *The Fundamentals of Decision Making and Priority Theory with the Analytic Hierarchy Process*, Vol. **VI**, AHP Series, RWS Publishers, Pittsburgh.

Chapter 14

Concordance Techniques

14.1 Introduction

The Simple Additive Weighted Model results in the formation of a function allowing the ranking from best to worst of all project options. The information obtained is thus quite comprehensive, with a cardinal score derived for each element in question. However, this richness of information is a direct consequence of the high quality of data required to be input into it, data that may not be readily to hand if the engineering evaluation of the various options is preliminary in nature. Furthermore, the decision maker may not actually require the high level of information provided by the additive model. What may be required is a decision method that generates a simple ranking of the options involved rather than an actual score for each. In such a situation, Concordance Analysis may be the most appropriate decision model. Furthermore, it is a suitable methodology in situations where it may not be necessary to know the relative positions in the final hierarchy of all options. For example, if it is known that option a is better than options b and c, it may be irrelevant to the decision maker what the relative position of b and c is. It should be possible for the two to remain incomparable without endangering the decision process. In some engineering situations is might actually be quite useful to highlight the incomparability of a number of options because of the absence of sufficient information to allow any meaningful comparison. In such circumstances, the result will reflect a more realistic solution given the quality of data available. If the information is not of a sufficiently high quality to produce a ranking directly connecting all options, then what is termed a 'partial ranking' will be derived where some options are not directly compared. In practice, concordance techniques are generally used to produce a shortlist of preferred options from a relatively large number of project options rather than one single 'best' project option.

Engineering Project Appraisal: The Evaluation of Alternative Development Schemes, Second Edition.
Martin Rogers and Aidan Duffy.
© 2012 John Wiley & Sons, Ltd. Published 2012 by John Wiley & Sons, Ltd.

14.2   Concordance Analysis

Concordance Analysis is termed a non-compensatory multicriteria decision making model. It utilises a number of mathematical functions to indicate the degree of dominance one option or group of options has over others under consideration, while not requiring all options to be directly linked within the ranking tree (Massam, 1988). The conclusion of incomparability between some actions may be quite helpful, since it highlights some aspects of the problem that would perhaps deserve a more thorough study. Within Concordance Analysis, there is no question of the 'trading-off' of one criterion directly against another for each individual option. In contrast, the SAW or AHP Models seek to reduce option evaluation and selection to one in which each of the alternative options is classified using a single score which represents the attractiveness or utility of an option. The selection of a preferred plan is based upon these scores. They rely on the principle that trade-offs between criteria are legitimate so that a utility score for each option can be determined (Rogers and Bruen, 1995).

With Concordance Analysis, the comparison of project options takes place on a pairwise basis with respect to each criterion and establishes the degree of dominance that one option has over another. The main measure of this dominance of one option over another is the *concordance score*. For a given pair of options (a,b), comparison of their relative performance takes place on each individual criterion. As a result, a picture of the level of *dominance* of a over b is built up. Thus, if option a is at least as good as option b on criterion j, then:

$$C_j(a,b) = 1.0$$

If a is *not* at least as good as b on criterion j, then:

$$C_j(a,b) = 0$$

This exercise is done on (a,b) for each criterion j, $j=1$, n

Let us assume there are five criteria, $j=1,5$, and the score for each option on these is as shown in Table 14.1.

Table 14.1 Option scores on each criterion (Concordance calculation).

	Criterion 1	Criterion 2	Criterion 3	Criterion 4	Criterion 5
Option a	4	3	4	2	4
Option b	4	4	2	2	3

On criterion 1, Option a is at least as good as option b, therefore:

$$C_1(a,b) = 1.0$$

On criterion 2, Option a is not at least as good as option b, therefore:

$$C_2(a,b) = 0$$

On criterion 3, Option a is at least as good as option b, therefore:

$$C_3(a,b) = 1.0$$

On criterion 4, Option a is at least as good as option b, therefore:

$C_4(a,b) = 1.0$

On criterion 5, Option a is at least as good as option b, therefore:

$C_5(a,b) = 1.0$

It should be noted that the minimum data required to compile a set of concordance scores on a given criterion are a ranking of the options on that criterion. This can form the basis for all decisions regarding dominance.

Now that we have compiled the concordance scores for all the criteria in question, how do we combine these scores to give an overall indication of the dominance of b by a? We achieve this by use of the relative importance weightings for the criteria.

They are obtained using one of the weighting procedures outlined in a Chapter 12. Use of such methods might result in the scores shown in Table 14.2.

Table 14.2 Criterion weights.

Criterion	Weighting score	Normalised wt
1	1	0.125
2	2	0.25
3	2	0.25
4	1	0.125
5	2	0.125

The overall *concordance index* for (a,b) is obtained by multiplying each criterion concordance score by its normalised weight. In the case of the example given immediately above, the value obtained is:

$$C(a,b) = 0.125(1) + 0.25(0) + 0.25(1) + 0.125(1) + 0.25(1)$$
$$= 0.75$$

The nearer this value is to one, the more certain we are that a outranks or dominates b.

This index is calculated for each pair of options to form the *Concordance Matrix*.

The concordance index can be expressed in mathematical form as follows. Assuming each criterion is assigned a weight w_j ($j=1,n$; n=number of criteria), increasing with the importance of the criterion, the concordance index for each ordered pair (a,b) can be rewritten as follows:

$$C(a,b) = \frac{1}{W} \sum_{j:g_j(a) \geq g_j(b)} w_j \qquad (14.1)$$

where

$$W = \sum_{j=1}^{n} w_j \qquad (14.2)$$

and $g_j(a)$ is the score for criterion j under option a and $g_j(b)$ is the score for criterion j under option b. $C(a,b)$ has a value between zero and one; it measures the strength of the statement 'Project option A outranks Project option B'.

14.3   PROMETHEE I and II

Now that the pairwise dominance relationship between options has been established by means of the indices within the Concordance Matrix, one must now go about establishing a relative ranking of the proposals based on these dominance scores. One of the most straightforward ways of compiling the ranking hierarchy is the system put forward by Brans and Vincke (1985) as part of their PROMETHEE I Decision Method. This model exists at various levels of sophistication (1st Form to 6th Form). Its most basic version, the 1st Form, is described in the next section. A more complex version, the 5th Form, is detailed in Section 14.5.2.

The process put forward by Brans and Vincke is carried out for each pair of n options. Assume firstly that all concordance indices have been calculated and the concordance matrix has been compiled. Let us take a sample concordance matrix with six options pairwise compared as shown in Table 14.3.

	1	2	3	4	5	6
1	—	0.75	0.50	0.63	0.13	0.25
2	0.63	—	0.38	0.63	0.25	0.25
3	0.50	0.63	—	0.75	0.13	0.25
4	0.50	0.50	0.38	—	0.25	0.25
5	1.00	0.75	1.00	0.75	—	0.50
6	1.00	1.00	0.75	0.75	0.63	—

Table 14.3 Sample Concordance Matrix.

For a given option, the sum of the scores along its row indicates the extent to which it is better than all the other options, while the sum of the scores along its column is an indication of the degree to which the other options are better than it. Therefore, in the ideal case, an option being compared against five others would have a row score of five and a column score of zero. The greater the option's row score and the smaller its column score, the better its ranking.

PROMETHEE I uses this property by taking each option's row score and ranking each option, placing the one with the highest score first, the one with the next highest score second, and so on until all are ranked. The result is a *complete pre-order*, as all options are directly connected within the ranking structure, with ties between options allowed (it would be a *complete order* if ties were not permitted). The method then takes the column scores and places the option with the highest score last, the one with the next highest score second last and so on. The result is again a complete pre-order. A final ranking is obtained by finding the intersection between these two separate rankings. This final result can yield a *partial pre-order*, as incomparabilities between options may occur due to possible conflicts between the two rankings.

For example, taking the matrix in Table 14.3, the row and column sums for each option are shown in Table 14.4.

Option	Row sum	Column sum
1	2.25	3.63
2	2.13	3.63
3	2.25	3.00
4	1.88	3.50
5	4.00	1.38
6	4.13	1.50

Table 14.4 Row and column sums for the six options.

Each set of values is used to compile a separate ranking. The first ranks the options from first to last in order of their decreasing row sum score as shown in Figure 14.1.

These scores reflect the degree to which each of the given options dominates the others as indicated by the relevant concordance index.

The second ranks the options from first to last in order of their increasing column sum score as shown in Figure 14.2. For each option, this score reflects the degree to which the other five options dominate it, as indicated by the concordance index of the other five options relative to the option in question.

The overall ranking of the options is obtained by combining the information from the two rankings indicated above as shown in Figure 14.3.

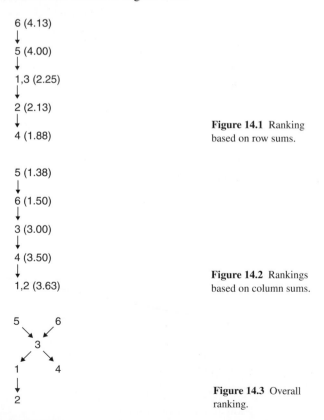

6 (4.13)
↓
5 (4.00)
↓
1,3 (2.25)
↓
2 (2.13)
↓
4 (1.88)

Figure 14.1 Ranking based on row sums.

5 (1.38)
↓
6 (1.50)
↓
3 (3.00)
↓
4 (3.50)
↓
1,2 (3.63)

Figure 14.2 Rankings based on column sums.

Figure 14.3 Overall ranking.

Option	Row sum – Column sum
5	+2.63
6	+2.63
3	–0.75
1	–1.38
2	–1.50
4	–1.63

Table 14.5 The median ranking.

Thus, because of the conflicting information from the two baseline rankings in Figures 14.1 and 14.2, three pairs are incomparable:

(1) 5 and 6
(2) 1 and 4
(3) 2 and 4.

The conflict between each pair of options arises from contradictory information derived from the concordance matrix. Take for example options 5 and 6. Option 6 dominates the other five options more than Option 5. Its row scores are 1, 1, 0.75, 0.75 and 0.63. It is more dominant than Option 5 whose scores are 1, 0.75, 1, 0.75 and 0.5. Hence, on this basis, Option 6 ranks first. Option 5 is dominated by the other options less than Option 6. Its column scores are 0.13, 0.25, 0.13, 0.25 and 0.63. The column scores for Option 6 are 0.25, 0.25, 0.25, 0.25 and 0.5. What can therefore be inferred from the overall ranking?

- Options 5 and 6 are the two best options.
- Option 3 is ranked third.
- Options 1 and 4 are ranked fourth, but are incomparable with each other.
- Option 2 is outranked by all except Option 4, with which it cannot be compared.

If the decision maker wishes to ignore these inconsistencies and obtain an average ranking, the PROMETHEE II Model can be used. Within it, for each option, the column sum is subtracted from the row sum to give an overall score, which is then translated into a straight ranking with no incomparabilities as illustrated in Table 14.5. (A ranking of options is based on the 'average' of the two rankings from Figure 14.1 and 14.2.) This median ranking emphasises the closeness of Options 5 and 6 in terms of their relative performance, with Option 3 firmly in third place. The two options that do not outrank any others – 2 and 4 – occupy the last two places.

Example 14.1 Use of the PROMETHEE I ranking system to evaluate eight potential sites for a regional landfill facility.
Eight sites (1 to 8) were identified by a waste management authority as being potentially suitable locations for the location of a landfill facility. To reduce the number of sites to a shortlist of three, the authority carried out an initial assessment using the following eight decision criteria:

Contd

Example 14.1 Contd

(1) Geology (C_1)
(2) Ecology (C_2)
(3) Landscape (C_3)
(4) Archaeology (C_4)
(5) Land use (C_5)
(6) Road access (C_6)
(7) Proximity to people (C_7)
(8) Haul distance (C_8)

The decision makers involved in the selection process assigned equal weightings to each of the criteria. Assuming a normalised weighting system therefore, each was assigned the value 0.125.

The preliminary nature of the process meant that on each criterion a relative ranking for each rather than a criterion score indicated the relative performance of the different options. Table 14.6 indicates the relative ranking of the eight options on each of the eight decision criteria as judged by the engineering experts employed by the waste management authority.

On the basis of both the derived rankings and the importance weightings assigned to each criterion by the decision makers, a set of concordance indices for each pair of options on each of the eight decision criteria was compiled. These are obtained by multiplying the set of concordance scores on each criterion by its normalized weighting. These are shown in Tables 14.7 to 14.14. For each pair of options, the concordance index on each criterion is summed to give an overall concordance index. These values are shown in Table 14.15. The row sum and column sum scores for each option are given in Table 14.16, and the final partial pre-order obtained from the intersection of the two rankings from Table 14.16 are detailed in Figure 14.4.

The overall ranking shows that the two best options are 6 and 8. The information on these, however, is conflicting. Option 6 dominates the other options more than Option 8, as indicated by their row sum scores, whereas Option 8 is less dominated by the others than Option 6, as indicated by their column scores. They cannot, therefore, be directly compared. They are both ranked above all other six options, with Options 5 and 7 in third and fourth places respectively. Below these are Options 1 and 3, which again are incomparable with each other. Both are ranked above Options 2 and 4, both also incomparable.

The ranking derived from the preliminary information detailed indicates clearly that the shortlist of three proposals should contain Options 6 and 8, with Option 5 completing the selection.

Table 14.6 Ranking of options 1 to 8 on each of the eight criteria.

Geology	Ecology	Landscape	Archaeology	Land use	Road Access	Proximity to people	Haul Distance
2,6,8	8	6,8	8	8	4	6,8	4
↓	↓	↓	↓	↓	↓	↓	↓
7	7	5	5,6	5,6	3,6	3,5,7	5,6,7
↓	↓	↓	↓	↓	↓	↓	↓
1,5	2	1,2,7	1,3	1	1,8	1,2,4	3
↓	↓	↓	↓	↓	↓		↓
3,4	1,6	3	4	3	2,5,7		1,2,8
	↓	↓	↓	↓			
	3	4	7	7			
	↓		↓	↓			
	5		2	2			
	↓			↓			
	4			4			

Table 14.7 Concordance scores for criterion C_1.

	1	2	3	4	5	6	7	8
1	—	0	0.125	0.125	0.125	0	0	0
2	0.125	—	0.125	0.125	0.125	0.125	0.125	0.125
3	0	0	—	0.125	0	0	0	0
4	0	0	0.125	—	0	0	0	0
5	0.125	0	0.125	0.125	—	0	0	0
6	0.125	0.125	0.125	0.125	0.125	—	0.125	0.125
7	0.125	0	0.125	0.125	0.125	0	—	0
8	0.125	0.125	0.125	0.125	0.125	0.125	0.125	—

Table 14.8 Concordance scores for criterion C_2.

	1	2	3	4	5	6	7	8
1	—	0	0.125	0.125	0.125	0.125	0	0
2	0.125	—	0.125	0.125	0.125	0.125	0	0
3	0	0	—	0.125	0.125	0	0	0
4	0	0	0	—	0	0	0	0
5	0	0	0	0.125	—	0	0	0
6	0.125	0	0.125	0.125	0.125	—	0	0
7	0.125	0.125	0.125	0.125	0.125	0.125	—	0
8	0.125	0.125	0.125	0.125	0.125	0.125	0.125	—

	1	2	3	4	5	6	7	8
1	—	0.125	0.125	0.125	0	0	0.125	0
2	0.125	—	0.125	0.125	0	0	0.125	0
3	0	0	—	0.125	0	0	0	0
4	0	0	0	—	0	0	0	0
5	0.125	0.125	0.125	0.125	—	0	0.125	0
6	0.125	0.125	0.125	0.125	0.125	—	0.125	0.125
7	0.125	0.125	0.125	0.125	0	0	—	0
8	0.125	0.125	0.125	0.125	0.125	0.125	0.125	—

Table 14.9 Concordance scores for criterion C_3.

	1	2	3	4	5	6	7	8
1	—	0.125	0.125	0.125	0	0	0.125	0
2	0	—	0	0	0	0	0	0
3	0.125	0.125	—	0.125	0	0	0.125	0
4	0	0.125	0	—	0	0	0.125	0
5	0.125	0.125	0.125	0.125	—	0.125	0.125	0
6	0.125	0.125	0.125	0.125	0.125	—	0.125	0
7	0	0.125	0	0	0	0	—	0
8	0.125	0.125	0.125	0.125	0.125	0.125	0.125	—

Table 14.10 Concordance scores for criterion C_4.

	1	2	3	4	5	6	7	8
1	—	0.125	0.125	0.125	0	0	0.125	0
2	0	—	0	0.125	0	0	0	0
3	0	0.125	—	0.125	0	0	0.125	0
4	0	0	0	—	0	0	0	0
5	0.125	0.125	0.125	0.125	—	0.125	0.125	0
6	0.125	0.125	0.125	0.125	0.125	—	0.125	0
7	0	0.125	0	0.125	0	0	—	0
8	0.125	0.125	0.125	0.125	0.125	0.125	0.125	—

Table 14.11 Concordance scores for criterion C_5.

	1	2	3	4	5	6	7	8
1	—	0.125	0	0	0.125	0	0.125	0.125
2	0	—	0	0	0.125	0	0.125	0
3	0.125	0.125	—	0	0.125	0.125	0.125	0.125
4	0.125	0.125	0.125	—	0.125	0.125	0.125	0.125
5	0	0.125	0	0	—	0	0.125	0
6	0.125	0.125	0.125	0	0.125	—	0.125	0.125
7	0	0.125	0	0	0.125	0	—	0
8	0.125	0.125	0	0	0.125	0	0.125	—

Table 14.12 Concordance scores for criterion C_6.

	1	2	3	4	5	6	7	8
1	—	0.125	0	0.125	0	0	0	0
2	0.125	—	0	0.125	0	0	0	0
3	0.125	0.125	—	0.125	0.125	0	0.125	0
4	0.125	0.125	0	—	0	0	0	0
5	0.125	0.125	0.125	0.125	—	0	0.125	0
6	0.125	0.125	0.125	0.125	0.125	—	0.125	0.125
7	0.125	0.125	0.125	0.125	0.125	0	—	0
8	0.125	0.125	0.125	0.125	0.125	0.125	0.125	—

Table 14.13 Concordance scores for criterion C_7.

	1	2	3	4	5	6	7	8
1	—	0.125	0	0	0	0	0	0.125
2	0.125	—	0	0	0	0	0	0.125
3	0.125	0.125	—	0	0	0	0	0.125
4	0.125	0.125	0.125	—	0.125	0.125	0.125	0.125
5	0.125	0.125	0.125	0	—	0.125	0.125	0.125
6	0.125	0.125	0.125	0	0.125	—	0.125	0.125
7	0.125	0.125	0.125	0	0.125	0.125	—	0.125
8	0.125	0.125	0	0	0	0	0	—

Table 14.14 Concordance scores for criterion C_8.

	1	2	3	4	5	6	7	8
1	—	0.75	0.625	0.750	0.375	0.125	0.500	0.250
2	0.625	—	0.375	0.625	0.375	0.250	0.375	0.250
3	0.500	0.625	—	0.750	0.375	0.125	0.500	0.250
4	0.375	0.500	0.375	—	0.250	0.250	0.375	0.250
5	0.750	0.750	0.750	0.750	—	0.375	0.750	0.125
6	1.000	0.875	1.000	0.750	1.000	—	0.875	0.625
7	0.625	0.875	0.625	0.625	0.625	0.250	—	0.125
8	1.000	1.000	0.750	0.750	0.875	0.750	0.875	—

Table 14.15 Overall Concordance Matrix.

	Row sum	Position	Column sum	Position
1	3.38	Fifth	4.88	Sixth
2	2.88	Seventh	5.38	Eighth
3	3.13	Sixth	4.50	Fifth
4	2.38	Eighth	5.00	Seventh
5	4.25	Third	3.88	Third
6	6.13	First	2.13	Second
7	3.75	Fourth	4.25	Fourth
8	6.00	Second	1.88	First

Table 14.16 Rankings based on row sum and column sum scores.

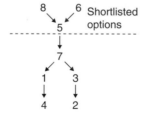

Figure 14.4 Final ranking for the eight options.

14.4   ELECTRE I

The 1st Form of the PROMETHEE I Model, as illustrated in Example 14.1, derives a ranking on the basis of a twin-track approach, with a project option deemed good if there is strong evidence that it dominates the other options (as indicated by a high 'row-sum' score) and, conversely, there is equally strong evidence that the other options do *not* dominate it (as indicated by a low 'column-sum' score).

The ELECTRE I model (Benayoun *et al.*, 1966) brings the twin-track approach a stage further through the introduction of the concept of discordance index as a complimentary indicator to the concordance index. As with the concordance index, it proceeds on a pairwise basis, with the discordance index calculated for each pair of options (a,b). It highlights information that may contradict the statement that 'a dominates b' and measures the degree to which, on any of the criteria, a is worse than b.

$D(a,b)=0$ implies there is no criterion on which b performs better than a. Otherwise $D(a,b)>0$

For a given pair of options (a,b), the difference between the scores for b and a divided by the length of the scale is calculated for each of the relevant criteria. The largest of these valuations determines the discordance index $D(a,b)$ for that particular pair of options. It is thus the maximum discrepancy between b and a on any one criterion. Let us say there are five criteria, $j=1,5$, and the scores for each of six options on these are as shown in Table 14.17.

	Criterion 1	Criterion 2	Criterion 3	Criterion 4	Criterion 5
Option 1	10	20	5	10	16
Option 2	0	5	5	16	10
Option 3	0	10	0	16	7
Option 4	20	5	10	10	13
Option 5	20	10	15	10	13
Option 6	20	10	20	13	13

Table 14.17 Option scores on each criterion (Discordance calculation).

Sample calculations for some of the concordance indices are as follows:

$D(1,2)$ $=6/20$ $=0.3$ (the maximum of 0/20, 0/20, 0/20, 6/20 and 0/20)
$D(1,3)$ $=6/20$ $=0.3$ (the maximum of 0/20, 0/20, 0/20, 6/20 and 0/20)
$D(1,4)$ $=10/20=0.5$ (the maximum of 10/20, 0/20, 5/20, 0/20 and 0/20)
$D(1,5)$ $=10/20=0.5$ (the maximum of 10/20, 0/20, 10/20, 0/20 and 0/20)
$D((1,6)=15/20=0.75$ (the maximum of 10/20, 0/20, 15/20, 3/20 and 0/20)
$D(2,1)$ $=15/20=0.75$ (the maximum of 10/20, 15/20, 0/20, 0/20 and 6/20)
$D(2,3)$ $=5/20$ $=0.25$ (the maximum of 0/20, 5/20, 0/20, 0/20 and 0/20)
$D(2,4)$ $=20/20=1.0$ (the maximum of 20/20, 0/20, 5/20, 0/20 and 3/20)
$D(2,5)$ $=20/20=1.0$ (the maximum of 20/20, 5/20, 10/20, 0/20 and 3/20)
$D(2,6)$ $=20/20=1.0$ (the maximum of 20/20, 5/20, 15/20, 0/20 and 3/20)

The full Discordance Matrix is thus as given in Table 14.18.

	1	2	3	4	5	6
1	—	0.3	0.3	0.5	0.5	0.75
2	0.75	—	0.25	1.0	1.0	1.0
3	0.5	0.25	—	1.0	1.0	1.0
4	0.75	0.3	0.3	—	0.25	0.5
5	0.5	0.3	0.3	0.0	—	0.25
6	0.5	0.15	0.15	0.0	0.0	—

Table 14.18 Full Discordance matrix.

The discordance matrix can be expressed mathematically as:

$$D(a,b) = 0 \text{ if } g_j(a) \geq g_j(b) \forall j \tag{14.3}$$

Otherwise:

$$D(a,b) = \frac{1}{\delta_j} \max\left[g_j(b) - g_j(a) \right] \tag{14.4}$$

where δ_j is the amplitude of the scale associated with the criterion j where maximum discordance occurs, that is:

$$\delta = \max_{c,d,\forall j} \; (g_j(c) - g_j(d)) \tag{14.5}$$

14.4.1   Obtaining a ranking from the concordance and discordance indices

Having obtained the two indices for each pair of options, one must now use these to rank their relative performance. The ELECTRE I Model (Roy, 1968) achieves this by establishing concordance and discordance thresholds.

Within ELECTRE I the relative ranking of the various options is derived by using concordance and discordance thresholds. For each pair of options (a,b), a is deemed to be ranked above b if:

- The concordance index $C(a,b)$ is greater than or equal to a minimum value, termed the concordance threshold, $\hat{c}$.
- The discordance index $D(a,b)$ is less than or equal to a maximum threshold value, termed the discordance threshold, $\hat{d}$.

The strictest thresholds that can be imposed are $\hat{c} = 1$ and $\hat{d} = 0$. These may result in very few ranking relationships being determined. Relaxing the two thresholds from their optimum values may result in more ranking relationships being established but will diminish the robustness of the results obtained. If the concordance and discordance thresholds are set below 0.7 and above 0.3, respectively, the process will lose much of its realism.

Table 14.19 Criterion Weightings.

Criterion	Weighting score	Normalised wt
C_1	3	0.3
C_2	2	0.2
C_3	3	0.3
C_4	1	0.1
C_5	1	0.1

Given the set of option scores in Table 14.17, in order to derive a concordance matrix the set of criterion weightings shown in Table 14.19 is assumed.

Using these in conjunction with the option ratings on each criterion, the concordance matrix given in Table 14.20 can be obtained.

	1	2	3	4	5	6
1	—	0.90	0.90	0.40	0.40	0.30
2	0.40	—	0.80	0.40	0.10	0.10
3	0.10	0.60	—	0.30	0.30	0.30
4	0.70	0.90	0.70	—	0.50	0.40
5	0.70	0.90	0.90	1.0	—	0.60
6	0.80	0.90	0.90	1.0	1.0	—

Table 14.20 Concordance Matrix.

If thresholds of one and zero are chosen, only three ranking relationships can be deduced. Option 6 ranks above both options 4 and 5, with option 5 ranking above option 4. Nothing can be inferred about options 1, 2 and 3. Four options still have the possibility of being the best option – 1, 2, 3 and 6. In outranking terms they form the 'kernel' of the graph (Figure 14.5).

Figure 14.5 Ranking relationships with $\hat{c} = 1$ and $\hat{d} = 0$.

If one reduces the thresholds to 0.7 and 0.3, the resulting graph becomes much richer in terms of the information derived (Figure 14.6)

Figure 14.6 Ranking relationships with $\hat{c} = 0.7$ and $\hat{d} = 0.3$.

14.5   Other Concordance Models

14.5.1   *Indifference and preference thresholds*

The two techniques detailed earlier within this chapter, termed the 1st Form of PROMETHEE I and ELECTRE I, respectively, are two of the simplest forms of the PROMETHEE and ELECTRE models. On a given criterion, *j*, assuming that a higher score is preferable to a lower one at all times (termed a monotonically increasing criterion), if option a is to be assumed as good as option b, that is if a is to outrank b, its performance must be exactly equal to or greater than that of b. Even if the score for option a is very slightly worse than b, no matter how inexact the available data are, the concordance score $C_j(a,b)$ is set equal to zero. More advanced versions of PROMETHEE and ELECTRE take account of potential uncertainties in the valuation of the performances of different options on a given criterion by intro-ducing two thresholds, the indifference threshold, q, and the preference threshold, p. On criterion j, if the score for option a is less that for option b by a value less than or equal to the indifference threshold, then a is assumed to outrank b, that is $C(a,b)=1$. If the score for option a is less that for option b by a value greater than or equal to the preference threshold, then a is assumed not to outrank b, that is $C(a,b)=0$. If the difference between the two values lies between the two thresholds, then the concordance score lies between zero and one, with its exact value estimated on a linear scale dependent upon its proximity to one or other of the defined thresholds. The mathematical expression defining this scoring system for concord-ance scores can be given as:

$$C_j(a,b) = 1 \text{ if } g_j(a) + q_j(g_j(a)) \geq g_j(b) \tag{14.6}$$

or

$$C_j(a,b) = 0 \text{ if } g_j(a) + p_j(g_j(a)) \leq g_j(b) \tag{14.7}$$

Otherwise:

$$C_j(a,b) = \frac{g_j(a) - g_j(b) + p_j(g_j(a))}{p_j(q_j(a)) - q_j(g_j(a))} \tag{14.8}$$

where $C_j(a,b)$ is the Concordance score indicating the dominance of a over b, $g_j(a)$, and $g_j(b)$ are the performance scores for options a and b on criterion j, $q_j(g_j(a))$ is the indifference threshold for criterion j and $p_j(g_j(a))$ is the preference threshold for criterion j.

The value obtained is then multiplied by its normalised weighting. The process is repeated for all criteria $j=1$, n, where n is the number of criteria. The n normalised scores are then added to form the concordance index for (a,b).

The more complex 5th Form of PROMETHEE I uses this double threshold model to form the concordance matrix, with a linear relationship assumed to exist for estimating concordance scores between zero and one. The two thresholds are also used in the most commonly used form of the ELECTRE Model, ELECTRE III, which also explicitly incorporates the concept of discordance within its mechanism.

Example 14.2   Example of the computation of a concordance score using indifference and preference thresholds

Assuming that, for a given pair of options (a,b), their performances on criterion j are:

$$g_j(a) = 100$$

$$g_j(b) = 120$$

Given the following threshold valuations:

$$q_j = 15$$

$$p_j = 40$$

Calculate the Concordance score for (a,b).

Solution

$$g_j(b) - g_j(a) = 20 \Leftrightarrow 0 < C_j(a,b) < 1$$

Therefore:

$$C_j(a,b) = (100 - 120 + 40)/(40 - 15)$$
$$= 20/25 = 0.8$$

14.5.2   *ELECTRE III and PROMETHEE (5th Form)*

This is the most commonly used version of the ELECTRE Model (Roy, 1978). It uses the two thresholds detailed above and defines the discordance index in a different way to ELECTRE I, using it to adjust the concordance index to yield what is termed the credibility index for the pair of options in question. For very low levels of discordance, the credibility index can be taken as equal to the concordance index. As discordance increases above a certain threshold the credibility index decreases below the level of dominance as indicated by the concordance index. When the discordance reaches a certain level, defined as the veto threshold, it neutralises any dominance indicated by the concordance index and the credibility index automatically becomes zero.

The discordance index is defined using a third threshold, the veto threshold, v_j. Any overall outranking of b by a indicated by the concordance index can be overruled if there is any criterion for which option b outperforms option a by at least the veto threshold, even if all the other criteria favour the outranking of b by a. The veto threshold is usually set as a multiple of the preference threshold, p.

In many practical applications, however, estimation of the veto threshold can be difficult. As a result, the veto threshold is often not used and the credibility index for each pair of options is assumed to be equal to the concordance index.

The mathematical expression defining this scoring system for discordance indices, similar in structure to that for the concordance index can be given as:

$$D_j(a,b) = 0 \text{ if } g_j(b) \leq g_j(a) + p_j(g_j(a)) \tag{14.9}$$

or

$$D_j(a,b) = 1 \quad \text{if } g_j(b) \geq g_j(a) + v_j(g_j(a)) \tag{14.10}$$

Otherwise:

$$D_j(a,b) = \frac{g_j(b) - g_j(a) + p_j(g_j(a))}{v_j(q_j(a)) - p_j(g_j(a))} \tag{14.11}$$

The credibility index $S(a,b)$ is defined as:

$$S(a,b) = C(a,b) \quad \text{if } D_j(a,b) \leq C(a,b), \forall j \tag{14.12}$$

Otherwise:

$$S(a,b) = C(a,b) \prod_{j \in J(a,b)} \frac{1 - D_j(a,b)}{1 - C(a,b)} \tag{14.13}$$

where, $J(a,b)$ is the set of criteria for which $Dj(a,b) > cj(a,b)$.

The degree of credibility of outranking is thus equal to the concordance index where no criterion is discordant. Where, however, discordances do exist, the concordance index is lowered in direct relation to the importance of those discordances.

The 5th Form of PROMETHEE I uses the concordance indices derived using the indifference and preference thresholds and then follows the normal PROMETHEE I process detailed in Example 14.1 to derive a relative ranking of all the options. The ELECTRE III model uses a different procedure to derive a final ranking but, as with the PROMETHEE I technique, the ranking procedure is based on a twin-track approach, gauging both the extent to which each option dominates the others and the extent to which the others dominate it. This is achieved using two separate 'distillation' procedures, the first termed the upward distillation, the second the downward distillation. This process is quite complex and has been simplified for the purposes of this text. A fuller explanation of the ELECTRE III model can be found in Rogers *et al.* (1999).

The distillation process is based on first establishing for each option the number of other options it outranks and the number of options that outrank it. This is done using the following rule:

'Option *a* is said to outrank option *b* if $S(a,b)$, the credibility index for the pair of options (*a,b*), is greater than a certain minimum 'threshold' value, λ, usually set at approximately 0.85 and $S(a,b)$ minus $S(b,a)$ is greater than a certain minimum 'discrimination' value, *s*, usually set at approximately 0.15. Each option is given a score of +1 for every option it outranks and a score of −1 for each option that outranks it. The two scores are added together to give that option's 'qualification' score. This value forms the basis for its final ranking that is established using the two distillation procedures.

Downward distillation

The qualification scores for each option are established and the option or options with the highest score are ranked first. Any positive and negative scores assigned to these options are then taken out of the computation and the scores are recalculated. From the remaining options, those having the highest score are ranked second and their positive and negative scores are discarded and the process repeated. The computation proceeds until all options have been ranked from first to last. The downward procedure ranks most strongly those options that most dominate others. During the procedure, where qualification scores are tied, those options can be separated by lowering the λ value in intervals of 0.15.

Upward distillation

The same qualification scores are again employed. The option or options with the lowest score are ranked last. Any positive and negative scores assigned to these options are then taken out of the computation and the scores are recalculated. From the remaining options, those having the lowest score are ranked second last and their positive and negative scores are discarded and the process repeated. The computation proceeds until all options have been ranked from last to first. The upward procedure ranks most strongly those options that are least dominated by the others. Again, during the procedure, where qualification scores are tied, those options can be separated by lowering the λ value in intervals of 0.15.

Final ranking

The final ranking of all options is obtained from the intersection of the rankings from the downward and upward distillation procedures. The complexity of this procedure has led to decision analysts using the PROMETHEE I and II procedures detailed in Example 14.1 to derive a final ranking of options.

Example 14.3 details the distillation procedure used in ELECTRE III and, for a given concordance/credibility matrix, compares the results from it with those obtained using the PROMETHEE I (5th Form) ranking procedure.

Example 14.3
Assuming that a concordance/credibility matrix identical to that given in Table 14.20 was generated using the double threshold concordance model consistent with both ELECTRE III and the 5th Form of PROMETHEE I, use the upward and downward distillation processes to derive a final ranking for the six options and select a shortlist of two. Compare the result with that obtained from the PROMETHEE I ranking system. (*Note*: In this example, the discordance threshold is 'muted', therefore the concordance matrix derived is also the credibility matrix, and ELECTRE III becomes identical to the 5th Form of PROMETHEE I in how this final matrix is computed.)

Contd

Example 14.3 Contd

Solution

Based on a threshold value, λ, of 0.85 and a discrimination value, s, of 0.15, the qualification scores obtained for the six options are given in Table 14.21. The first downward distillation shows Option 6 to have the highest score. It is ranked first, eliminated from the computation, and the qualification scores are recalculated. The second downward distillation (Table 14.22) places Options 1 and 5 best of the remaining proposals. In order to separate them, λ is reduced to 0.7, with s remaining at 0.15. These values place option 5 second and Option 1 third. Both are then eliminated from the computation. The third downward distillation (Table 14.23) places Option 4 fourth. Options 2 and 3 cannot be separated and are placed joint fifth.

The upward distillation also commences using the information in Table 14.21, but this time selecting the lowest scoring option, in this case Option 2, placing it sixth and last, eliminating it from the computation and recalculating the qualification scores for the remaining options. The second upward distillation (Table 14.24) places Option 3 in fifth position. The third upward distillation (Table 14.25) places Option 4 fourth, and the fourth (Table 14.26) places Option 5 in third. Options 1 and 6 cannot be separated and are placed joint first.

The rankings from the upward and downward distillations are given in Figure 14.7, with the overall ranking obtained from the intersection of these two given in Figure 14.8.

Using the PROMETHEE I Model, the row and column sums for each option are shown in Table 14.27, with the final ranking given in Figure 14.9.

The distillation procedure indicates that Option 6 is best, with two options ranked second, incomparable with each other with each ranking above the other four. This is because Option 5 is dominated by Option 6, whereas option 1 is not, hence Option 5 finishes below Option 1 on the upward distillation. The overall stronger performance of Option 5 over Option 1 is emphasised by the PROMETHEE I ranking procedure which places Option 5 alone in second place. Thus, it can be inferred that a shortlist of two would comprise Options 5 and 6.

	1	2	3	4	5	6
	2,3	—	—	2	2,3,4	2,3,4,5
Positive score	2	0	0	1	3	4
Negative score	0	4	3	2	1	0
Qualification	2	−4	−3	−1	2	4

Table 14.21 First downward (and upward) distillation.

	1	2	3	4	5
	2,3	—	—	2	2,3,4
Positive score	2	0	0	1	3
Negative score	0	3	2	1	1
Qualification	2	−3	−2	0	2

Table 14.22 Second downward distillation.

	2	3	4
	—	—	2
Positive score	0	0	1
Negative score	1	0	0
Qualification	−1	0	1

Table 14.23 Third downward distillation.

	1	3	4	5	6
	3	—	—	3,4	3,4,5
Positive score	1	0	0	2	3
Negative score	0	3	2	1	0
Qualification	1	−3	−2	1	3

Table 14.24 Second upward distillation.

	1	4	5	6
	—	—	4	4,5
Positive score	0	0	1	2
Negative score	0	2	1	0
Qualification	0	−2	0	2

Table 14.25 Third upward distillation.

	1	5	6
	—	—	5
Positive score	0	0	1
Negative score	0	1	0
Qualification	0	−1	1

Table 14.26 Fourth upward distillation.

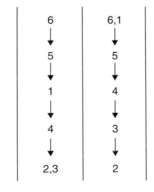

Figure 14.7 Downward and upward distillations.

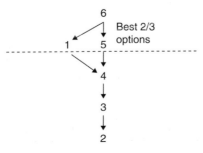

Figure 14.8 Final partial pre-order.

Option	Row sum	Column sum
1	2.9	2.7
2	1.8	4.2
3	1.6	4.2
4	3.2	3.1
5	4.1	2.3
6	4.6	1.7

Table 14.27 Row sum and column sum values for the six options.

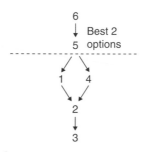

Figure 14.9 Final ranking of options using PROMETHEE I.

Example 14.4 evaluates five options on the basis of four decision criteria using ELECTRE III. This model is most often used to reduce a large number of options to a shortlist for more detailed consideration. This example limits the number of options and criteria purely in the interests of simplicity and clarity. More complex worked examples involving ELECTRE can be found in Rogers and Bruen (2000) and Hokkanen and Salminen (1994).

Example 14.4   Using the ELECTRE III Model to rank four small-scale electrical generation projects

An electricity provider wishes to decide which of five existing generating stations $(St_1, St_2, St_3, St_4, St_5)$ it should upgrade. Each is to be assessed on the basis of the following four criteria:

(1) Financial acceptability (cost and financial return).
(2) Solution delivery (consequences of poor delivery of the technology associated with the proposal).
(3) Strategic contribution (contribution to the overall business plan of the organisation).
(4) Environmental effect (effect of the proposal on the level of access to resources).

The criterion scores for each option are given in Table 14.28. The indifference and preference thresholds for the four criteria are given in Table 14.29 together with their importance weightings. (The organisation's decision makers decided that the first two criteria should be weighted more heavily than the second two.) Veto thresholds were not used. As an example, for the finance criterion, the concordance scores are given in Table 14.30. It can be seen that, unlike ELECTRE I where all the scores are either one or zero, those generated by ELECTRE III can lie between one and zero where the difference in scores between two options is between the two thresholds set for that

Contd

Example 14.4 Contd

criterion. In these situations, the concordance score is estimated using Equation 14.8. These values must then be multiplied by the normalised weighting for that criterion. For each pair of options, the normalised score on each criterion is added together to give a final concordance/credibility index. These are given in Table 14.31.

The qualification scores for each option are then obtained using $\lambda = 0.85$ and $s = 0.15$ (Table 14.32). The downward distillation eliminates St_2 first, followed by St_4, with the other three inseparable in joint third place. The upward distillation eliminates St_3 and St_5 joint first, followed by St_4, then St_1 and finally St_2. The intersection of these two rankings yields St_2 in first place. St_1 and St_4 cannot be compared and are both ranked second, immediately below St_2. St_3 and St_5 cannot be separated and are ranked in joint fourth place. The rankings from the two distillations are given together with the final ranking in Figure 14.10.

This baseline analysis identifies St_2 as the best option, followed by St_1 and St_4 in second place. If, as part of a sensitivity analysis, the weights are changed so that the importance of the finance criterion is gradually made less important, a point is arrived at where Option 1 performs equally well. This occurs at the point where the weighting for F is reduced to just below two, with the others set at their baseline values. At the stage where all weights are equal, St_1 and St_2 are equal first, followed by St_5 and St_3. St_4 at this stage is ranked fifth and last. Any further reduction in the relative weighting of the finance criterion reinforces St_1 in first place with St_2 in second. Lowering the value of both the indifference and preference thresholds for the four criteria has the effect of improving the ranking of St_5 to second place below St_2, with St_4 below both of these (Figure 14.11).

To use the PROMETHEE I Method (5th Form), the rows and columns of the concordance matrix in Table 14.31 must be summed, a ranking obtained from the row sum values (highest first, lowest last) and the column sum values (lowest first, highest last), and a final ranking obtained from their intersection. The row and column sums together with the resulting ranks are given in Table 14.33. It ranks the same options in the first three positions as the baseline ELECTRE ranking (2, 1 and 4).

Overall, St_2 is ranked first, given the particular implied importance of the finance criterion indicated by the decision makers. While the baseline test indicates both St_1 and St_4 in second place, St_1 performs more robustly in the sensitivity tests; hence this should be the reserve choice.

14.6 Summary

Table 14.28 Performance matrix for five options.

	Finance (F) (£000)	Delivery (D) (0 to 100)	Strategic (S) (0 to 100)	Environment (E) (0 to 100)
St_1	−14	70	65	85
St_2	129	100	30	15
St_3	−10	50	40	85
St_4	44	90	35	20
St_5	−14	100	50	40

Table 14.29 Thresholds and weightings for the four criteria.

	F	D	S	E
Indifference threshold (q)	40	20	15	10
Preference threshold (p)	80	40	35	30
Weighting (w)	4	2	1	1
Normalised weight	0.5	0.25	0.125	0.125

Table 14.30 Concordance scores for each criterion (normalised weights in brackets).

	St_1	St_2	St_3	St_4	St_5
St_1	—	0	1	0.55	1
St_2	1	—	1	1	1
St_3	1	0	—	0.65	1
St_4	1	0	1	—	1
St_5	1	0	1	0.55	—

F (0.5)

	St_1	St_2	St_3	St_4	St_5
St_1	—	0.5	1	1	0.5
St_2	1	—	1	1	1
St_3	1	0	—	0	0
St_4	1	1	1	—	1
St_5	1	1	1	1	—

D (0.25)

	St_1	St_2	St_3	St_4	St_5
St_1	—	1	1	1	1
St_2	0	—	1	1	0.75
St_3	0.5	1	—	1	1
St_4	0.25	1	1	—	1
St_5	1	1	1	1	—

S (0.125)

	St_1	St_2	St_3	St_4	St_5
St_1	—	1	1	1	1
St_2	0	—	0	1	0.25
St_3	1	1	—	1	1
St_4	0	1	0	—	0.5
St_5	0	1	0	1	—

E (0.125)

Table 14.31 Concordance matrix for the six options.

	St_1	St_2	St_3	St_4	St_5
St_1	—	0.375	1	0.775	0.875
St_2	0.75	—	0.875	1	0.875
St_3	0.938	0.25	—	0.575	0.75
St_4	0.781	0.5	0.875	—	0.938
St_5	0.875	0.5	0.875	0.775	—

Table 14.32 Qualification scores for the six options.

	St_1	St_2	St_3	St_4	St_5
	—	3,4,5	—	3,5	—
Positive score	0	3	0	2	0
Negative score	0	0	2	1	2
Qualification	0	3	−2	1	−2

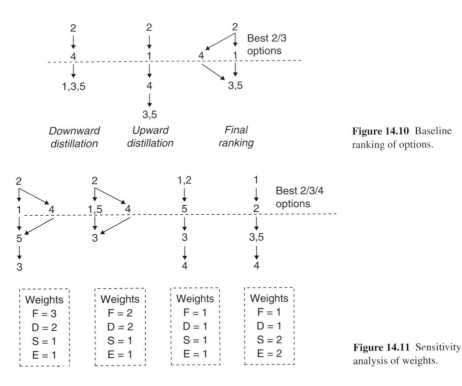

Figure 14.10 Baseline ranking of options.

Figure 14.11 Sensitivity analysis of weights.

Table 14.33
PROMETHEE I ranking.

Option	Row sum	Column sum	Ranking
1	3.025 (3rd)	3.344 (3rd)	3rd
2	3.500 (1st)	1.625 (1st)	1st
3	2.513 (5th)	3.625 (5th)	5th
4	3.094 (2nd)	3.125 (2nd)	2nd
5	3.025 (3rd)	3.438 (4d)	4th

The two concordance models outlined in this chapter, PROMETHEE and ELECTRE, are relatively complex. Both assess the strengths and weaknesses of the options on the basis of their pairwise concordance indices. Versions of both exist that can take account of the error and uncertainty in criterion valuations using indifference and preference thresholds. They differ in the methodology used to derive the ranking set. PROMETHEE uses the summed values for each column and row within the concordance matrix, whereas ELECTRE uses concordance/discordance thresholds or a distillation process to highlight the options that perform best. Concordance models are best used where a relatively large number of options must be reduced to a shortlist.

Despite its complexity, concordance techniques have a number of advantages:

- They are rigorous yet adaptable techniques, allowing criteria measures on different scales to be evaluated within the same framework.
- Unlike the Simple Additive Weighting (SAW) and Analytical Hierarchy Process (AHP) Models, the criterion valuations do not have to be transferred onto a

common scale before the relative performances of the different options can be evaluated.

- The method does not force direct comparison between any two options if insufficient information exists to do so or if the information that does exist gives conflicting indications.

Given their ability to handle data of mixed quality presented in various formats, they have proved to be particularly useful for the preliminary assessment of a wide range of project proposals. Examples of applications of the different ELECTRE Methods are detailed in Hokkanen and Salminen (1994), Maystre *et al.* (1994) and Rogers *et al.* (1999).

References

Benayoun, R., Roy, B. and Sussman, N. (1966) *Manual de Referance du Programme ELECTRE*. Note de Synthese et Formation, No. 25, Direction Scientifique SEMA, Paris.

Brans, J.P. and Vincke, P.A. (1985) A preference organisation method. *Management Science*, **21** (6), 647–656.

Hokkanen, J. and Salminen, S. (1994) Choice of a Solid Waste Management System by Using the ELECTRE III Method, in *Applying MCDA for Decision to Environmental Management* (ed. M. Paruccini). Kluwer Academic Publishers, Dordrecht, The Netherlands.

Massam, B. (1988) *Spatial Search*. Pergamon Press.

Maystre, L., Pictet, J. and Simos, J. (1994) *Methodes Multicriteres ELECTRE*. Presses Polytechniques et Universitaires Romandes, Lausanne.

Rogers, M.G. and Bruen, M.P. (1995) Non-monetary decision-aid techniques in EIA – an overview. *Municipal Engineer*, **109**, 88–103.

Rogers, M.P. and Bruen, M.P. (2000) Using ELECTRE III to choose the best route option for the Dublin Port motorway. *Journal of Transportation Engineering, American Society of Civil Engineers, Urban Transport Division*, **26** (4), 313–323.

Rogers, M.G., Bruen, M.P. and Maystre, L.Y. (1999) *ELECTRE and Decision Support: Methods and Applications in Engineering and Infrastructure Investment*. Kluwer Academic Publishers, Dordrecht, The Netherlands.

Roy, B. (1968) Classement et choix en presence de points de vue multiples (la methode ELECTRE). *Revue Informatique et Recherche Operationnelle*. 2e Année, No. 8, 57–75.

Roy, B. (1978) ELECTRE III: Un algorithme de classements fonde sur une representation floue des preferences en presence de criteres multiples. *Cahiers de CERO*, **20** (1), 3–24.

Chapter 15

Concluding Comments

15.1 Introduction

The two parts within this textbook are quite different in their approach to the selection of a desired project option based on their respective expected consequences. The economic-based approaches dealt with in Part I of the book involve relatively straight-forward computations. The final output from methods of this type is expressed in units (€, $) that are readily understandable to the reader. The non-economics-based multicriteria approaches addressed in Part II can involve much more complex calculations, with the options assessed on the basis of their performance on a set of decision attributes that may, in many instances, be measured in different forms, some highly quantitatively based, others in a more qualitative manner. To express the performance of each particular option in terms of a single 'score', its performances on the various attributes must then be linked together using some direct or indirect trade-off technique. In such cases the final output obtained is much less easily understood, with the performance of each option presented either in a set of units specific to the method in question, as a ranking from first to last, or as a kernel or shortlist of preferred proposals.

15.2 Which project appraisal technique should one use?

It might be seen as logical to give added emphasis to the purely economic techniques because of their greater user accessibility. Use of a multicriteria model puts greater demands on the user in terms of gaining a basic understanding of how to use the model and of interpreting the results that output from it. Such methods can, however, be far more inclusive and the extra effort required may be more than justified in situations where the decision maker needs a method which offers a more holistic approach to the selection problem under consideration than that provided by a purely economic evaluation. In such instances the additional effort input by the decision maker will bring its reward.

Engineering Project Appraisal: The Evaluation of Alternative Development Schemes, Second Edition.
Martin Rogers and Aidan Duffy.
© 2012 John Wiley & Sons, Ltd. Published 2012 by John Wiley & Sons, Ltd.

Multicriteria methods have been proved to be of significant use in assessing complex civil engineering development proposals, allowing a set of environmental, social and/or technical concerns to be brought to light regarding a project where a purely economic evaluation was seen as inappropriate and too narrowly based. Within the United States, the main catalyst has been the National Environmental Policy Act (1970) (NEPA), which, as the pre-eminent piece of legislation responsible for requiring project planners to consider environmental and social criteria alongside the economics-based ones in order to assess a proposal's viability, has given impetus to the use of multicriteria methods within such assessments. NEPA requires US federal agencies to identify and develop methodologies which ensure that environmental factors are given appropriate consideration in decision making. The Leopold Interaction Matrix (Leopold, 1971), a multicriteria-based assessment developed in the early 1970s, was one of the earliest types of Environmental Impact Assessment methodologies. Within Europe, the 1985 Directive on Environmental Impact Assessment was the channel through which all such legislation was introduced in all the European Union member states, providing the basic format within which a multicriteria methodology could operate, with the project developer required to outline alternative proposals, identify all likely relevant impacts on the environment resulting from it and assess the magnitude of these impacts (Council of the European Communities, 1985).

The emergence and increasing use of multicriteria methodologies in recent times does not in any way imply that the importance of the economics-based methods of project appraisal, in particular cost–benefit analysis (CBA), should be underestimated or downgraded. The economic performance of a proposal remains central to its basic viability. One of the first questions asked by the project promoter before embarking on any course of action is 'what will it be worth?'. CBA remains a powerful tool for guiding decisions and further research into some of the techniques, referred to briefly in Chapter 9, which permit previously intangible costs and benefits to be assessed in money terms and thus be included in the full analysis, is continuing. Engineers and planners hold CBA in high regard, mainly for its simplicity and objectivity. However, relying on it as the lone method of appraisal leaves the decision maker open to criticisms regarding the validity of some of the assumptions at the basis of the methodology. These include its disregard of equity; the potential for criticism in whatever discount rate is chosen and the use of money as the sole measure of a proposal's worth. Despite such drawbacks, engineers may still feel drawn to placing undue emphasis on the results of the CBA. Prudent decision makers can avoid such a pitfall by supplementing the CBA with some form of multicriteria analysis as an adjunct to the purely economic analysis.

15.3   Future challenges

The two sets of techniques face different challenges in the future. For the economics-based methods, the challenge is to advance knowledge of techniques that permit the monetary evaluation of hitherto intangible costs and benefits, thereby allowing their inclusion within an economic appraisal. The ability to include environmental and social attributes within the cost–benefit analysis framework will render the process more inclusive and believable. For the multicriteria stream, experts must strive for a way of making their mathematically complex methodologies more understandable and accessible to decision-makers – this is a barrier that must be overcome. Greater effort must be made by practitioners and theorists to explain these methods to potential users.

If improvements are not strived for continually in both streams, the end effect may be that both will end up with diminished levels of public credibility and such rational decision making techniques may be sidelined and swept aside by the more political forces within the decision making system.

References

Council of the European Communities (1985) On the assessment of the effects of certain public and private projects on the environment. Official Journal L175, 28.5.85, 40–48 (85/337/EEC).

Leopold L.B. (1971) A Procedure for Evaluating Environmental Impact. Circular 645, United States Geological Survey, Washington DC.

National Environmental Policy Act (1970) Public law 91–190,91st Congress, D. 1075, United States of America, January 1.

Interest Factor Tables

Formulae for discrete cash flows with end-of-period compounding		
Name	Find / Given	Formula
Future-worth factor	F/P	$F = P(1 + i)^n$ $= P \times$ future worth factor
Present-worth factor	P/F	$P = F[1/(1 + i)^n]$ $= F \times$ present worth factor
Series present-worth factor	P/A	$P = A\{[(1 + i)^n - 1]/[i(1 + i)^n]\}$ $= A \times$ series present worth factor
Capital recovery factor	A/P	$A = P\{[i(1 + i)^n]/[(1 + i)^n - 1]\}$ $= P \times$ capital recovery factor
Series compound amount factor	F/A	$F = A\{[(1 + i)^n - 1]/i\}$ $= A \times$ series compound amount factor
Sinking fund factor	A/F	$A = F\{i/[(1 + i)^n - 1]\}$ $= F \times$ sinking fund factor
Arithmetic gradient conversion factor	A_G/G	$A = G\{[1/i] - [n /(i(1 + i)^n]\}$ $= G \times$ Arithmetic gradient conversion factor
Arithmetic gradient present worth factor	P_G/G	$P = G \times 1/i \times \{[1/I((1 + i)^n - 1)/(i(1 + i)^n)]$ $- [n /(1 + i)^n]\}$ $= G \times$ Arithmetic gradient present worth factor

Engineering Project Appraisal: The Evaluation of Alternative Development Schemes, Second Edition.
Martin Rogers and Aidan Duffy.
© 2012 John Wiley & Sons, Ltd. Published 2012 by John Wiley & Sons, Ltd.

Table 1 Discrete cash flow compound interest factors; interest rate = 1%.

1.00%								1.00%
n	F/P	P/F	A/F	A/P	F/A	P/A	A/G	P/G
1	1.010	0.990	1.000	1.010	1.000	0.990	0.000	0.000
2	1.020	0.980	0.498	0.508	2.010	1.970	0.498	0.980
3	1.030	0.971	0.330	0.340	3.030	2.941	0.993	2.921
4	1.041	0.961	0.246	0.256	4.060	3.902	1.488	5.804
5	1.051	0.951	0.196	0.206	5.101	4.853	1.980	9.610
6	1.062	0.942	0.163	0.173	6.152	5.795	2.471	14.321
7	1.072	0.933	0.139	0.149	7.214	6.728	2.960	19.917
8	1.083	0.923	0.121	0.131	8.286	7.652	3.448	26.381
9	1.094	0.914	0.107	0.117	9.369	8.566	3.934	33.696
10	1.105	0.905	0.096	0.106	10.462	9.471	4.418	41.843
11	1.116	0.896	0.086	0.096	11.567	10.368	4.901	50.807
12	1.127	0.887	0.079	0.089	12.683	11.255	5.381	60.569
13	1.138	0.879	0.072	0.082	13.809	12.134	5.861	71.113
14	1.149	0.870	0.067	0.077	14.947	13.004	6.338	82.422
15	1.161	0.861	0.062	0.072	16.097	13.865	6.814	94.481
16	1.173	0.853	0.058	0.068	17.258	14.718	7.289	107.273
17	1.184	0.844	0.054	0.064	18.430	15.562	7.761	120.783
18	1.196	0.836	0.051	0.061	19.615	16.398	8.232	134.996
19	1.208	0.828	0.048	0.058	20.811	17.226	8.702	149.895
20	1.220	0.820	0.045	0.055	22.019	18.046	9.169	165.466
21	1.232	0.811	0.043	0.053	23.239	18.857	9.635	181.695
22	1.245	0.803	0.041	0.051	24.472	19.660	10.100	198.566
23	1.257	0.795	0.039	0.049	25.716	20.456	10.563	216.066
24	1.270	0.788	0.037	0.047	26.973	21.243	11.024	234.180
25	1.282	0.780	0.035	0.045	28.243	22.023	11.483	252.894
26	1.295	0.772	0.034	0.044	29.526	22.795	11.941	272.196
27	1.308	0.764	0.032	0.042	30.821	23.560	12.397	292.070
28	1.321	0.757	0.031	0.041	32.129	24.316	12.852	312.505
29	1.335	0.749	0.030	0.040	33.450	25.066	13.304	333.486
30	1.348	0.742	0.029	0.039	34.785	25.808	13.756	355.002
31	1.361	0.735	0.028	0.038	36.133	26.542	14.205	377.039
32	1.375	0.727	0.027	0.037	37.494	27.270	14.653	399.586
33	1.389	0.720	0.026	0.036	38.869	27.990	15.099	422.629
34	1.403	0.713	0.025	0.035	40.258	28.703	15.544	446.157
35	1.417	0.706	0.024	0.034	41.660	29.409	15.987	470.158
40	1.489	0.672	0.020	0.030	48.886	32.835	18.178	596.856
45	1.565	0.639	0.018	0.028	56.481	36.095	20.327	733.704
50	1.645	0.608	0.016	0.026	64.463	39.196	22.436	879.418

Table 2 Discrete cash flow compound interest factors; interest rate = 2%.

2.00%								2.00%
n	F/P	P/F	A/F	A/P	F/A	P/A	A/G	P/G
1	1.020	0.980	1.000	1.020	1.000	0.980	0.000	0.000
2	1.040	0.961	0.495	0.515	2.020	1.942	0.495	0.961
3	1.061	0.942	0.327	0.347	3.060	2.884	0.987	2.846
4	1.082	0.924	0.243	0.263	4.122	3.808	1.475	5.617
5	1.104	0.906	0.192	0.212	5.204	4.713	1.960	9.240
6	1.126	0.888	0.159	0.179	6.308	5.601	2.442	13.680
7	1.149	0.871	0.135	0.155	7.434	6.472	2.921	18.903
8	1.172	0.853	0.117	0.137	8.583	7.325	3.396	24.878
9	1.195	0.837	0.103	0.123	9.755	8.162	3.868	31.572
10	1.219	0.820	0.091	0.111	10.950	8.983	4.337	38.955
11	1.243	0.804	0.082	0.102	12.169	9.787	4.802	46.998
12	1.268	0.788	0.075	0.095	13.412	10.575	5.264	55.671
13	1.294	0.773	0.068	0.088	14.680	11.348	5.723	64.948
14	1.319	0.758	0.063	0.083	15.974	12.106	6.179	74.800
15	1.346	0.743	0.058	0.078	17.293	12.849	6.631	85.202
16	1.373	0.728	0.054	0.074	18.639	13.578	7.080	96.129
17	1.400	0.714	0.050	0.070	20.012	14.292	7.526	107.555
18	1.428	0.700	0.047	0.067	21.412	14.992	7.968	119.458
19	1.457	0.686	0.044	0.064	22.841	15.678	8.407	131.814
20	1.486	0.673	0.041	0.061	24.297	16.351	8.843	144.600
21	1.516	0.660	0.039	0.059	25.783	17.011	9.276	157.796
22	1.546	0.647	0.037	0.057	27.299	17.658	9.705	171.379
23	1.577	0.634	0.035	0.055	28.845	18.292	10.132	185.331
24	1.608	0.622	0.033	0.053	30.422	18.914	10.555	199.630
25	1.641	0.610	0.031	0.051	32.030	19.523	10.974	214.259
26	1.673	0.598	0.030	0.050	33.671	20.121	11.391	229.199
27	1.707	0.586	0.028	0.048	35.344	20.707	11.804	244.431
28	1.741	0.574	0.027	0.047	37.051	21.281	12.214	259.939
29	1.776	0.563	0.026	0.046	38.792	21.844	12.621	275.706
30	1.811	0.552	0.025	0.045	40.568	22.396	13.025	291.716
31	1.848	0.541	0.024	0.044	42.379	22.938	13.426	307.954
32	1.885	0.531	0.023	0.043	44.227	23.468	13.823	324.403
33	1.922	0.520	0.022	0.042	46.112	23.989	14.217	341.051
34	1.961	0.510	0.021	0.041	48.034	24.499	14.608	357.882
35	2.000	0.500	0.020	0.040	49.994	24.999	14.996	374.883
40	2.208	0.453	0.017	0.037	60.402	27.355	16.889	461.993
45	2.438	0.410	0.014	0.034	71.893	29.490	18.703	551.565
50	2.692	0.372	0.012	0.032	84.579	31.424	20.442	642.361

Table 3 Discrete cash flow compound interest factors; interest rate = 3%.

3.00%								3.00%
n	F/P	P/F	A/F	A/P	F/A	P/A	A/G	P/G
1	1.030	0.971	1.000	1.030	1.000	0.971	0.000	0.000
2	1.061	0.943	0.493	0.523	2.030	1.913	0.493	0.943
3	1.093	0.915	0.324	0.354	3.091	2.829	0.980	2.773
4	1.126	0.888	0.239	0.269	4.184	3.717	1.463	5.438
5	1.159	0.863	0.188	0.218	5.309	4.580	1.941	8.889
6	1.194	0.837	0.155	0.185	6.468	5.417	2.414	13.076
7	1.230	0.813	0.131	0.161	7.662	6.230	2.882	17.955
8	1.267	0.789	0.112	0.142	8.892	7.020	3.345	23.481
9	1.305	0.766	0.098	0.128	10.159	7.786	3.803	29.612
10	1.344	0.744	0.087	0.117	11.464	8.530	4.256	36.309
11	1.384	0.722	0.078	0.108	12.808	9.253	4.705	43.533
12	1.426	0.701	0.070	0.100	14.192	9.954	5.148	51.248
13	1.469	0.681	0.064	0.094	15.618	10.635	5.587	59.420
14	1.513	0.661	0.059	0.089	17.086	11.296	6.021	68.014
15	1.558	0.642	0.054	0.084	18.599	11.938	6.450	77.000
16	1.605	0.623	0.050	0.080	20.157	12.561	6.874	86.348
17	1.653	0.605	0.046	0.076	21.762	13.166	7.294	96.028
18	1.702	0.587	0.043	0.073	23.414	13.754	7.708	106.014
19	1.754	0.570	0.040	0.070	25.117	14.324	8.118	116.279
20	1.806	0.554	0.037	0.067	26.870	14.877	8.523	126.799
21	1.860	0.538	0.035	0.065	28.676	15.415	8.923	137.550
22	1.916	0.522	0.033	0.063	30.537	15.937	9.319	148.509
23	1.974	0.507	0.031	0.061	32.453	16.444	9.709	159.657
24	2.033	0.492	0.029	0.059	34.426	16.936	10.095	170.971
25	2.094	0.478	0.027	0.057	36.459	17.413	10.477	182.434
26	2.157	0.464	0.026	0.056	38.553	17.877	10.853	194.026
27	2.221	0.450	0.025	0.055	40.710	18.327	11.226	205.731
28	2.288	0.437	0.023	0.053	42.931	18.764	11.593	217.532
29	2.357	0.424	0.022	0.052	45.219	19.188	11.956	229.414
30	2.427	0.412	0.021	0.051	47.575	19.600	12.314	241.361
31	2.500	0.400	0.020	0.050	50.003	20.000	12.668	253.361
32	2.575	0.388	0.019	0.049	52.503	20.389	13.017	265.399
33	2.652	0.377	0.018	0.048	55.078	20.766	13.362	277.464
34	2.732	0.366	0.017	0.047	57.730	21.132	13.702	289.544
35	2.814	0.355	0.017	0.047	60.462	21.487	14.037	301.627
40	3.262	0.307	0.013	0.043	75.401	23.115	15.650	361.750
45	3.782	0.264	0.011	0.041	92.720	24.519	17.156	420.632
50	4.384	0.228	0.009	0.039	112.797	25.730	18.558	477.480

Table 4 Discrete cash flow compound interest factors; interest rate = 4%.

4.00%								4.00%
n	F/P	P/F	A/F	A/P	F/A	P/A	A/G	P/G
1	1.040	0.962	1.000	1.040	1.000	0.962	0.000	0.000
2	1.082	0.925	0.490	0.530	2.040	1.886	0.490	0.925
3	1.125	0.889	0.320	0.360	3.122	2.775	0.974	2.703
4	1.170	0.855	0.235	0.275	4.246	3.630	1.451	5.267
5	1.217	0.822	0.185	0.225	5.416	4.452	1.922	8.555
6	1.265	0.790	0.151	0.191	6.633	5.242	2.386	12.506
7	1.316	0.760	0.127	0.167	7.898	6.002	2.843	17.066
8	1.369	0.731	0.109	0.149	9.214	6.733	3.294	22.181
9	1.423	0.703	0.094	0.134	10.583	7.435	3.739	27.801
10	1.480	0.676	0.083	0.123	12.006	8.111	4.177	33.881
11	1.539	0.650	0.074	0.114	13.486	8.760	4.609	40.377
12	1.601	0.625	0.067	0.107	15.026	9.385	5.034	47.248
13	1.665	0.601	0.060	0.100	16.627	9.986	5.453	54.455
14	1.732	0.577	0.055	0.095	18.292	10.563	5.866	61.962
15	1.801	0.555	0.050	0.090	20.024	11.118	6.272	69.735
16	1.873	0.534	0.046	0.086	21.825	11.652	6.672	77.744
17	1.948	0.513	0.042	0.082	23.698	12.166	7.066	85.958
18	2.026	0.494	0.039	0.079	25.645	12.659	7.453	94.350
19	2.107	0.475	0.036	0.076	27.671	13.134	7.834	102.893
20	2.191	0.456	0.034	0.074	29.778	13.590	8.209	111.565
21	2.279	0.439	0.031	0.071	31.969	14.029	8.578	120.341
22	2.370	0.422	0.029	0.069	34.248	14.451	8.941	129.202
23	2.465	0.406	0.027	0.067	36.618	14.857	9.297	138.128
24	2.563	0.390	0.026	0.066	39.083	15.247	9.648	147.101
25	2.666	0.375	0.024	0.064	41.646	15.622	9.993	156.104
26	2.772	0.361	0.023	0.063	44.312	15.983	10.331	165.121
27	2.883	0.347	0.021	0.061	47.084	16.330	10.664	174.138
28	2.999	0.333	0.020	0.060	49.968	16.663	10.991	183.142
29	3.119	0.321	0.019	0.059	52.966	16.984	11.312	192.121
30	3.243	0.308	0.018	0.058	56.085	17.292	11.627	201.062
31	3.373	0.296	0.017	0.057	59.328	17.588	11.937	209.956
32	3.508	0.285	0.016	0.056	62.701	17.874	12.241	218.792
33	3.648	0.274	0.015	0.055	66.210	18.148	12.540	227.563
34	3.794	0.264	0.014	0.054	69.858	18.411	12.832	236.261
35	3.946	0.253	0.014	0.054	73.652	18.665	13.120	244.877
40	4.801	0.208	0.011	0.051	95.026	19.793	14.477	286.530
45	5.841	0.171	0.008	0.048	121.029	20.720	15.705	325.403
50	7.107	0.141	0.007	0.047	152.667	21.482	16.812	361.164

Table 5 Discrete cash flow compound interest factors; interest rate = 5%.

5.00%								5.00%
n	F/P	P/F	A/F	A/P	F/A	P/A	A/G	P/G
1	1.050	0.952	1.000	1.050	1.000	0.952	0.000	0.000
2	1.103	0.907	0.488	0.538	2.050	1.859	0.488	0.907
3	1.158	0.864	0.317	0.367	3.153	2.723	0.967	2.635
4	1.216	0.823	0.232	0.282	4.310	3.546	1.439	5.103
5	1.276	0.784	0.181	0.231	5.526	4.329	1.903	8.237
6	1.340	0.746	0.147	0.197	6.802	5.076	2.358	11.968
7	1.407	0.711	0.123	0.173	8.142	5.786	2.805	16.232
8	1.477	0.677	0.105	0.155	9.549	6.463	3.245	20.970
9	1.551	0.645	0.091	0.141	11.027	7.108	3.676	26.127
10	1.629	0.614	0.080	0.130	12.578	7.722	4.099	31.652
11	1.710	0.585	0.070	0.120	14.207	8.306	4.514	37.499
12	1.796	0.557	0.063	0.113	15.917	8.863	4.922	43.624
13	1.886	0.530	0.056	0.106	17.713	9.394	5.322	49.988
14	1.980	0.505	0.051	0.101	19.599	9.899	5.713	56.554
15	2.079	0.481	0.046	0.096	21.579	10.380	6.097	63.288
16	2.183	0.458	0.042	0.092	23.657	10.838	6.474	70.160
17	2.292	0.436	0.039	0.089	25.840	11.274	6.842	77.140
18	2.407	0.416	0.036	0.086	28.132	11.690	7.203	84.204
19	2.527	0.396	0.033	0.083	30.539	12.085	7.557	91.328
20	2.653	0.377	0.030	0.080	33.066	12.462	7.903	98.488
21	2.786	0.359	0.028	0.078	35.719	12.821	8.242	105.667
22	2.925	0.342	0.026	0.076	38.505	13.163	8.573	112.846
23	3.072	0.326	0.024	0.074	41.430	13.489	8.897	120.009
24	3.225	0.310	0.022	0.072	44.502	13.799	9.214	127.140
25	3.386	0.295	0.021	0.071	47.727	14.094	9.524	134.228
26	3.556	0.281	0.020	0.070	51.113	14.375	9.827	141.259
27	3.733	0.268	0.018	0.068	54.669	14.643	10.122	148.223
28	3.920	0.255	0.017	0.067	58.403	14.898	10.411	155.110
29	4.116	0.243	0.016	0.066	62.323	15.141	10.694	161.913
30	4.322	0.231	0.015	0.065	66.439	15.372	10.969	168.623
31	4.538	0.220	0.014	0.064	70.761	15.593	11.238	175.233
32	4.765	0.210	0.013	0.063	75.299	15.803	11.501	181.739
33	5.003	0.200	0.012	0.062	80.064	16.003	11.757	188.135
34	5.253	0.190	0.012	0.062	85.067	16.193	12.006	194.417
35	5.516	0.181	0.011	0.061	90.320	16.374	12.250	200.581
40	7.040	0.142	0.008	0.058	120.800	17.159	13.377	229.545
45	8.985	0.111	0.006	0.056	159.700	17.774	14.364	255.315
50	11.467	0.087	0.005	0.055	209.348	18.256	15.223	277.915

Table 6 Discrete cash flow compound interest factors; interest rate = 6%.

6.00%								6.00%
n	F/P	P/F	A/F	A/P	F/A	P/A	A/G	P/G
1	1.060	0.943	1.000	1.060	1.000	0.943	0.000	0.000
2	1.124	0.890	0.485	0.545	2.060	1.833	0.485	0.890
3	1.191	0.840	0.314	0.374	3.184	2.673	0.961	2.569
4	1.262	0.792	0.229	0.289	4.375	3.465	1.427	4.946
5	1.338	0.747	0.177	0.237	5.637	4.212	1.884	7.935
6	1.419	0.705	0.143	0.203	6.975	4.917	2.330	11.459
7	1.504	0.665	0.119	0.179	8.394	5.582	2.768	15.450
8	1.594	0.627	0.101	0.161	9.897	6.210	3.195	19.842
9	1.689	0.592	0.087	0.147	11.491	6.802	3.613	24.577
10	1.791	0.558	0.076	0.136	13.181	7.360	4.022	29.602
11	1.898	0.527	0.067	0.127	14.972	7.887	4.421	34.870
12	2.012	0.497	0.059	0.119	16.870	8.384	4.811	40.337
13	2.133	0.469	0.053	0.113	18.882	8.853	5.192	45.963
14	2.261	0.442	0.048	0.108	21.015	9.295	5.564	51.713
15	2.397	0.417	0.043	0.103	23.276	9.712	5.926	57.555
16	2.540	0.394	0.039	0.099	25.673	10.106	6.279	63.459
17	2.693	0.371	0.035	0.095	28.213	10.477	6.624	69.401
18	2.854	0.350	0.032	0.092	30.906	10.828	6.960	75.357
19	3.026	0.331	0.030	0.090	33.760	11.158	7.287	81.306
20	3.207	0.312	0.027	0.087	36.786	11.470	7.605	87.230
21	3.400	0.294	0.025	0.085	39.993	11.764	7.915	93.114
22	3.604	0.278	0.023	0.083	43.392	12.042	8.217	98.941
23	3.820	0.262	0.021	0.081	46.996	12.303	8.510	104.701
24	4.049	0.247	0.020	0.080	50.816	12.550	8.795	110.381
25	4.292	0.233	0.018	0.078	54.865	12.783	9.072	115.973
26	4.549	0.220	0.017	0.077	59.156	13.003	9.341	121.468
27	4.822	0.207	0.016	0.076	63.706	13.211	9.603	126.860
28	5.112	0.196	0.015	0.075	68.528	13.406	9.857	132.142
29	5.418	0.185	0.014	0.074	73.640	13.591	10.103	137.310
30	5.743	0.174	0.013	0.073	79.058	13.765	10.342	142.359
31	6.088	0.164	0.012	0.072	84.802	13.929	10.574	147.286
32	6.453	0.155	0.011	0.071	90.890	14.084	10.799	152.090
33	6.841	0.146	0.010	0.070	97.343	14.230	11.017	156.768
34	7.251	0.138	0.010	0.070	104.184	14.368	11.228	161.319
35	7.686	0.130	0.009	0.069	111.435	14.498	11.432	165.743
40	10.286	0.097	0.006	0.066	154.762	15.046	12.359	185.957
45	13.765	0.073	0.005	0.065	212.744	15.456	13.141	203.110
50	18.420	0.054	0.003	0.063	290.336	15.762	13.796	217.457

Table 7 Discrete cash flow compound interest factors; interest rate = 7%.

7.00%								7.00%
n	F/P	P/F	A/F	A/P	F/A	P/A	A/G	P/G
1	1.070	0.935	1.000	1.070	1.000	0.935	0.000	0.000
2	1.145	0.873	0.483	0.553	2.070	1.808	0.483	0.873
3	1.225	0.816	0.311	0.381	3.215	2.624	0.955	2.506
4	1.311	0.763	0.225	0.295	4.440	3.387	1.416	4.795
5	1.403	0.713	0.174	0.244	5.751	4.100	1.865	7.647
6	1.501	0.666	0.140	0.210	7.153	4.767	2.303	10.978
7	1.606	0.623	0.116	0.186	8.654	5.389	2.730	14.715
8	1.718	0.582	0.097	0.167	10.260	5.971	3.147	18.789
9	1.838	0.544	0.083	0.153	11.978	6.515	3.552	23.140
10	1.967	0.508	0.072	0.142	13.816	7.024	3.946	27.716
11	2.105	0.475	0.063	0.133	15.784	7.499	4.330	32.466
12	2.252	0.444	0.056	0.126	17.888	7.943	4.703	37.351
13	2.410	0.415	0.050	0.120	20.141	8.358	5.065	42.330
14	2.579	0.388	0.044	0.114	22.550	8.745	5.417	47.372
15	2.759	0.362	0.040	0.110	25.129	9.108	5.758	52.446
16	2.952	0.339	0.036	0.106	27.888	9.447	6.090	57.527
17	3.159	0.317	0.032	0.102	30.840	9.763	6.411	62.592
18	3.380	0.296	0.029	0.099	33.999	10.059	6.722	67.622
19	3.617	0.277	0.027	0.097	37.379	10.336	7.024	72.599
20	3.870	0.258	0.024	0.094	40.995	10.594	7.316	77.509
21	4.141	0.242	0.022	0.092	44.865	10.836	7.599	82.339
22	4.430	0.226	0.020	0.090	49.006	11.061	7.872	87.079
23	4.741	0.211	0.019	0.089	53.436	11.272	8.137	91.720
24	5.072	0.197	0.017	0.087	58.177	11.469	8.392	96.255
25	5.427	0.184	0.016	0.086	63.249	11.654	8.639	100.676
26	5.807	0.172	0.015	0.085	68.676	11.826	8.877	104.981
27	6.214	0.161	0.013	0.083	74.484	11.987	9.107	109.166
28	6.649	0.150	0.012	0.082	80.698	12.137	9.329	113.226
29	7.114	0.141	0.011	0.081	87.347	12.278	9.543	117.162
30	7.612	0.131	0.011	0.081	94.461	12.409	9.749	120.972
31	8.145	0.123	0.010	0.080	102.073	12.532	9.947	124.655
32	8.715	0.115	0.009	0.079	110.218	12.647	10.138	128.212
33	9.325	0.107	0.008	0.078	118.933	12.754	10.322	131.643
34	9.978	0.100	0.008	0.078	128.259	12.854	10.499	134.951
35	10.677	0.094	0.007	0.077	138.237	12.948	10.669	138.135
40	14.974	0.067	0.005	0.075	199.635	13.332	11.423	152.293
45	21.002	0.048	0.003	0.073	285.749	13.606	12.036	163.756
50	29.457	0.034	0.002	0.072	406.529	13.801	12.529	172.905

Table 8 Discrete cash flow compound interest factors; interest rate = 8%.

8.00%								8.00%
n	F/P	P/F	A/F	A/P	F/A	P/A	A/G	P/G
1	1.080	0.926	1.000	1.080	1.000	0.926	0.000	0.000
2	1.166	0.857	0.481	0.561	2.080	1.783	0.481	0.857
3	1.260	0.794	0.308	0.388	3.246	2.577	0.949	2.445
4	1.360	0.735	0.222	0.302	4.506	3.312	1.404	4.650
5	1.469	0.681	0.170	0.250	5.867	3.993	1.846	7.372
6	1.587	0.630	0.136	0.216	7.336	4.623	2.276	10.523
7	1.714	0.583	0.112	0.192	8.923	5.206	2.694	14.024
8	1.851	0.540	0.094	0.174	10.637	5.747	3.099	17.806
9	1.999	0.500	0.080	0.160	12.488	6.247	3.491	21.808
10	2.159	0.463	0.069	0.149	14.487	6.710	3.871	25.977
11	2.332	0.429	0.060	0.140	16.645	7.139	4.240	30.266
12	2.518	0.397	0.053	0.133	18.977	7.536	4.596	34.634
13	2.720	0.368	0.047	0.127	21.495	7.904	4.940	39.046
14	2.937	0.340	0.041	0.121	24.215	8.244	5.273	43.472
15	3.172	0.315	0.037	0.117	27.152	8.559	5.594	47.886
16	3.426	0.292	0.033	0.113	30.324	8.851	5.905	52.264
17	3.700	0.270	0.030	0.110	33.750	9.122	6.204	56.588
18	3.996	0.250	0.027	0.107	37.450	9.372	6.492	60.843
19	4.316	0.232	0.024	0.104	41.446	9.604	6.770	65.013
20	4.661	0.215	0.022	0.102	45.762	9.818	7.037	69.090
21	5.034	0.199	0.020	0.100	50.423	10.017	7.294	73.063
22	5.437	0.184	0.018	0.098	55.457	10.201	7.541	76.926
23	5.871	0.170	0.016	0.096	60.893	10.371	7.779	80.673
24	6.341	0.158	0.015	0.095	66.765	10.529	8.007	84.300
25	6.848	0.146	0.014	0.094	73.106	10.675	8.225	87.804
26	7.396	0.135	0.013	0.093	79.954	10.810	8.435	91.184
27	7.988	0.125	0.011	0.091	87.351	10.935	8.636	94.439
28	8.627	0.116	0.010	0.090	95.339	11.051	8.829	97.569
29	9.317	0.107	0.010	0.090	103.966	11.158	9.013	100.574
30	10.063	0.099	0.009	0.089	113.283	11.258	9.190	103.456
31	10.868	0.092	0.008	0.088	123.346	11.350	9.358	106.216
32	11.737	0.085	0.007	0.087	134.214	11.435	9.520	108.857
33	12.676	0.079	0.007	0.087	145.951	11.514	9.674	111.382
34	13.690	0.073	0.006	0.086	158.627	11.587	9.821	113.792
35	14.785	0.068	0.006	0.086	172.317	11.655	9.961	116.092
40	21.725	0.046	0.004	0.084	259.057	11.925	10.570	126.042
45	31.920	0.031	0.003	0.083	386.506	12.108	11.045	133.733
50	46.902	0.021	0.002	0.082	573.770	12.233	11.411	139.593

Table 9 Discrete cash flow compound interest factors; interest rate = 9%.

9.00%								9.00%
n	F/P	P/F	A/F	A/P	F/A	P/A	A/G	P/G
1	1.090	0.917	1.000	1.090	1.000	0.917	0.000	0.000
2	1.188	0.842	0.478	0.568	2.090	1.759	0.478	0.842
3	1.295	0.772	0.305	0.395	3.278	2.531	0.943	2.386
4	1.412	0.708	0.219	0.309	4.573	3.240	1.393	4.511
5	1.539	0.650	0.167	0.257	5.985	3.890	1.828	7.111
6	1.677	0.596	0.133	0.223	7.523	4.486	2.250	10.092
7	1.828	0.547	0.109	0.199	9.200	5.033	2.657	13.375
8	1.993	0.502	0.091	0.181	11.028	5.535	3.051	16.888
9	2.172	0.460	0.077	0.167	13.021	5.995	3.431	20.571
10	2.367	0.422	0.066	0.156	15.193	6.418	3.798	24.373
11	2.580	0.388	0.057	0.147	17.560	6.805	4.151	28.248
12	2.813	0.356	0.050	0.140	20.141	7.161	4.491	32.159
13	3.066	0.326	0.044	0.134	22.953	7.487	4.818	36.073
14	3.342	0.299	0.038	0.128	26.019	7.786	5.133	39.963
15	3.642	0.275	0.034	0.124	29.361	8.061	5.435	43.807
16	3.970	0.252	0.030	0.120	33.003	8.313	5.724	47.585
17	4.328	0.231	0.027	0.117	36.974	8.544	6.002	51.282
18	4.717	0.212	0.024	0.114	41.301	8.756	6.269	54.886
19	5.142	0.194	0.022	0.112	46.018	8.950	6.524	58.387
20	5.604	0.178	0.020	0.110	51.160	9.129	6.767	61.777
21	6.109	0.164	0.018	0.108	56.765	9.292	7.001	65.051
22	6.659	0.150	0.016	0.106	62.873	9.442	7.223	68.205
23	7.258	0.138	0.014	0.104	69.532	9.580	7.436	71.236
24	7.911	0.126	0.013	0.103	76.790	9.707	7.638	74.143
25	8.623	0.116	0.012	0.102	84.701	9.823	7.832	76.926
26	9.399	0.106	0.011	0.101	93.324	9.929	8.016	79.586
27	10.245	0.098	0.010	0.100	102.723	10.027	8.191	82.124
28	11.167	0.090	0.009	0.099	112.968	10.116	8.357	84.542
29	12.172	0.082	0.008	0.098	124.135	10.198	8.515	86.842
30	13.268	0.075	0.007	0.097	136.308	10.274	8.666	89.028
31	14.462	0.069	0.007	0.097	149.575	10.343	8.808	91.102
32	15.763	0.063	0.006	0.096	164.037	10.406	8.944	93.069
33	17.182	0.058	0.006	0.096	179.800	10.464	9.072	94.931
34	18.728	0.053	0.005	0.095	196.982	10.518	9.193	96.693
35	20.414	0.049	0.005	0.095	215.711	10.567	9.308	98.359
40	31.409	0.032	0.003	0.093	337.882	10.757	9.796	105.376
45	48.327	0.021	0.002	0.092	525.859	10.881	10.160	110.556
50	74.358	0.013	0.001	0.091	815.084	10.962	10.430	114.325

Table 10 Discrete cash flow compound interest factors; interest rate = 10%.

10.00%								10.00%
n	F/P	P/F	A/F	A/P	F/A	P/A	A/G	P/G
1	1.100	0.909	1.000	1.100	1.000	0.909	0.000	0.000
2	1.210	0.826	0.476	0.576	2.100	1.736	0.476	0.826
3	1.331	0.751	0.302	0.402	3.310	2.487	0.937	2.329
4	1.464	0.683	0.215	0.315	4.641	3.170	1.381	4.378
5	1.611	0.621	0.164	0.264	6.105	3.791	1.810	6.862
6	1.772	0.564	0.130	0.230	7.716	4.355	2.224	9.684
7	1.949	0.513	0.105	0.205	9.487	4.868	2.622	12.763
8	2.144	0.467	0.087	0.187	11.436	5.335	3.004	16.029
9	2.358	0.424	0.074	0.174	13.579	5.759	3.372	19.421
10	2.594	0.386	0.063	0.163	15.937	6.145	3.725	22.891
11	2.853	0.350	0.054	0.154	18.531	6.495	4.064	26.396
12	3.138	0.319	0.047	0.147	21.384	6.814	4.388	29.901
13	3.452	0.290	0.041	0.141	24.523	7.103	4.699	33.377
14	3.797	0.263	0.036	0.136	27.975	7.367	4.996	36.800
15	4.177	0.239	0.031	0.131	31.772	7.606	5.279	40.152
16	4.595	0.218	0.028	0.128	35.950	7.824	5.549	43.416
17	5.054	0.198	0.025	0.125	40.545	8.022	5.807	46.582
18	5.560	0.180	0.022	0.122	45.599	8.201	6.053	49.640
19	6.116	0.164	0.020	0.120	51.159	8.365	6.286	52.583
20	6.727	0.149	0.017	0.117	57.275	8.514	6.508	55.407
21	7.400	0.135	0.016	0.116	64.002	8.649	6.719	58.110
22	8.140	0.123	0.014	0.114	71.403	8.772	6.919	60.689
23	8.954	0.112	0.013	0.113	79.543	8.883	7.108	63.146
24	9.850	0.102	0.011	0.111	88.497	8.985	7.288	65.481
25	10.835	0.092	0.010	0.110	98.347	9.077	7.458	67.696
26	11.918	0.084	0.009	0.109	109.182	9.161	7.619	69.794
27	13.110	0.076	0.008	0.108	121.100	9.237	7.770	71.777
28	14.421	0.069	0.007	0.107	134.210	9.307	7.914	73.650
29	15.863	0.063	0.007	0.107	148.631	9.370	8.049	75.415
30	17.449	0.057	0.006	0.106	164.494	9.427	8.176	77.077
31	19.194	0.052	0.005	0.105	181.943	9.479	8.296	78.640
32	21.114	0.047	0.005	0.105	201.138	9.526	8.409	80.108
33	23.225	0.043	0.004	0.104	222.252	9.569	8.515	81.486
34	25.548	0.039	0.004	0.104	245.477	9.609	8.615	82.777
35	28.102	0.036	0.004	0.104	271.024	9.644	8.709	83.987
40	45.259	0.022	0.002	0.102	442.593	9.779	9.096	88.953
45	72.890	0.014	0.001	0.101	718.905	9.863	9.374	92.454
50	117.391	0.009	0.001	0.101	1163.909	9.915	9.570	94.889

Table 11 Discrete cash flow compound interest factors; interest rate = 12%.

12.00%								12.00%
n	F/P	P/F	A/F	A/P	F/A	P/A	A/G	P/G
1	1.120	0.893	1.000	1.120	1.000	0.893	0.000	0.000
2	1.254	0.797	0.472	0.592	2.120	1.690	0.472	0.797
3	1.405	0.712	0.296	0.416	3.374	2.402	0.925	2.221
4	1.574	0.636	0.209	0.329	4.779	3.037	1.359	4.127
5	1.762	0.567	0.157	0.277	6.353	3.605	1.775	6.397
6	1.974	0.507	0.123	0.243	8.115	4.111	2.172	8.930
7	2.211	0.452	0.099	0.219	10.089	4.564	2.551	11.644
8	2.476	0.404	0.081	0.201	12.300	4.968	2.913	14.471
9	2.773	0.361	0.068	0.188	14.776	5.328	3.257	17.356
10	3.106	0.322	0.057	0.177	17.549	5.650	3.585	20.254
11	3.479	0.287	0.048	0.168	20.655	5.938	3.895	23.129
12	3.896	0.257	0.041	0.161	24.133	6.194	4.190	25.952
13	4.363	0.229	0.036	0.156	28.029	6.424	4.468	28.702
14	4.887	0.205	0.031	0.151	32.393	6.628	4.732	31.362
15	5.474	0.183	0.027	0.147	37.280	6.811	4.980	33.920
16	6.130	0.163	0.023	0.143	42.753	6.974	5.215	36.367
17	6.866	0.146	0.020	0.140	48.884	7.120	5.435	38.697
18	7.690	0.130	0.018	0.138	55.750	7.250	5.643	40.908
19	8.613	0.116	0.016	0.136	63.440	7.366	5.838	42.998
20	9.646	0.104	0.014	0.134	72.052	7.469	6.020	44.968
21	10.804	0.093	0.012	0.132	81.699	7.562	6.191	46.819
22	12.100	0.083	0.011	0.131	92.503	7.645	6.351	48.554
23	13.552	0.074	0.010	0.130	104.603	7.718	6.501	50.178
24	15.179	0.066	0.008	0.128	118.155	7.784	6.641	51.693
25	17.000	0.059	0.007	0.127	133.334	7.843	6.771	53.105
26	19.040	0.053	0.007	0.127	150.334	7.896	6.892	54.418
27	21.325	0.047	0.006	0.126	169.374	7.943	7.005	55.637
28	23.884	0.042	0.005	0.125	190.699	7.984	7.110	56.767
29	26.750	0.037	0.005	0.125	214.583	8.022	7.207	57.814
30	29.960	0.033	0.004	0.124	241.333	8.055	7.297	58.782
31	33.555	0.030	0.004	0.124	271.293	8.085	7.381	59.676
32	37.582	0.027	0.003	0.123	304.848	8.112	7.459	60.501
33	42.092	0.024	0.003	0.123	342.429	8.135	7.530	61.261
34	47.143	0.021	0.003	0.123	384.521	8.157	7.596	61.961
35	52.800	0.019	0.002	0.122	431.663	8.176	7.658	62.605
40	93.051	0.011	0.001	0.121	767.091	8.244	7.899	65.116
45	163.988	0.006	0.001	0.121	1358.230	8.283	8.057	66.734
50	289.002	0.003	0.000	0.120	2400.018	8.304	8.160	67.762

Table 12 Discrete cash flow compound interest factors; interest rate = 14%.

14.00%								14.00%
n	F/P	P/F	A/F	A/P	F/A	P/A	A/G	P/G
1	1.140	0.877	1.000	1.140	1.000	0.877	0.000	0.000
2	1.300	0.769	0.467	0.607	2.140	1.647	0.467	0.769
3	1.482	0.675	0.291	0.431	3.440	2.322	0.913	2.119
4	1.689	0.592	0.203	0.343	4.921	2.914	1.337	3.896
5	1.925	0.519	0.151	0.291	6.610	3.433	1.740	5.973
6	2.195	0.456	0.117	0.257	8.536	3.889	2.122	8.251
7	2.502	0.400	0.093	0.233	10.730	4.288	2.483	10.649
8	2.853	0.351	0.076	0.216	13.233	4.639	2.825	13.103
9	3.252	0.308	0.062	0.202	16.085	4.946	3.146	15.563
10	3.707	0.270	0.052	0.192	19.337	5.216	3.449	17.991
11	4.226	0.237	0.043	0.183	23.045	5.453	3.733	20.357
12	4.818	0.208	0.037	0.177	27.271	5.660	4.000	22.640
13	5.492	0.182	0.031	0.171	32.089	5.842	4.249	24.825
14	6.261	0.160	0.027	0.167	37.581	6.002	4.482	26.901
15	7.138	0.140	0.023	0.163	43.842	6.142	4.699	28.862
16	8.137	0.123	0.020	0.160	50.980	6.265	4.901	30.706
17	9.276	0.108	0.017	0.157	59.118	6.373	5.089	32.430
18	10.575	0.095	0.015	0.155	68.394	6.467	5.263	34.038
19	12.056	0.083	0.013	0.153	78.969	6.550	5.424	35.531
20	13.743	0.073	0.011	0.151	91.025	6.623	5.573	36.914
21	15.668	0.064	0.010	0.150	104.768	6.687	5.711	38.190
22	17.861	0.056	0.008	0.148	120.436	6.743	5.838	39.366
23	20.362	0.049	0.007	0.147	138.297	6.792	5.955	40.446
24	23.212	0.043	0.006	0.146	158.659	6.835	6.062	41.437
25	26.462	0.038	0.005	0.145	181.871	6.873	6.161	42.344
26	30.167	0.033	0.005	0.145	208.333	6.906	6.251	43.173
27	34.390	0.029	0.004	0.144	238.499	6.935	6.334	43.929
28	39.204	0.026	0.004	0.144	272.889	6.961	6.410	44.618
29	44.693	0.022	0.003	0.143	312.094	6.983	6.479	45.244
30	50.950	0.020	0.003	0.143	356.787	7.003	6.542	45.813
31	58.083	0.017	0.002	0.142	407.737	7.020	6.600	46.330
32	66.215	0.015	0.002	0.142	465.820	7.035	6.652	46.798
33	75.485	0.013	0.002	0.142	532.035	7.048	6.700	47.222
34	86.053	0.012	0.002	0.142	607.520	7.060	6.743	47.605
35	98.100	0.010	0.001	0.141	693.573	7.070	6.782	47.952
40	188.884	0.005	0.001	0.141	1342.025	7.105	6.930	49.238
45	363.679	0.003	0.000	0.140	2590.565	7.123	7.019	49.996
50	700.233	0.001	0.000	0.140	4994.521	7.133	7.071	50.438

Table 13 Discrete cash flow compound interest factors; interest rate = 16%.

16.00%								16.00%
n	F/P	P/F	A/F	A/P	F/A	P/A	A/G	P/G
1	1.160	0.862	1.000	1.160	1.000	0.862	0.000	0.000
2	1.346	0.743	0.463	0.623	2.160	1.605	0.463	0.743
3	1.561	0.641	0.285	0.445	3.506	2.246	0.901	2.024
4	1.811	0.552	0.197	0.357	5.066	2.798	1.316	3.681
5	2.100	0.476	0.145	0.305	6.877	3.274	1.706	5.586
6	2.436	0.410	0.111	0.271	8.977	3.685	2.073	7.638
7	2.826	0.354	0.088	0.248	11.414	4.039	2.417	9.761
8	3.278	0.305	0.070	0.230	14.240	4.344	2.739	11.896
9	3.803	0.263	0.057	0.217	17.519	4.607	3.039	14.000
10	4.411	0.227	0.047	0.207	21.321	4.833	3.319	16.040
11	5.117	0.195	0.039	0.199	25.733	5.029	3.578	17.994
12	5.936	0.168	0.032	0.192	30.850	5.197	3.819	19.847
13	6.886	0.145	0.027	0.187	36.786	5.342	4.041	21.590
14	7.988	0.125	0.023	0.183	43.672	5.468	4.246	23.217
15	9.266	0.108	0.019	0.179	51.660	5.575	4.435	24.728
16	10.748	0.093	0.016	0.176	60.925	5.668	4.609	26.124
17	12.468	0.080	0.014	0.174	71.673	5.749	4.768	27.407
18	14.463	0.069	0.012	0.172	84.141	5.818	4.913	28.583
19	16.777	0.060	0.010	0.170	98.603	5.877	5.046	29.656
20	19.461	0.051	0.009	0.169	115.380	5.929	5.167	30.632
21	22.574	0.044	0.007	0.167	134.841	5.973	5.277	31.518
22	26.186	0.038	0.006	0.166	157.415	6.011	5.377	32.320
23	30.376	0.033	0.005	0.165	183.601	6.044	5.467	33.044
24	35.236	0.028	0.005	0.165	213.978	6.073	5.549	33.697
25	40.874	0.024	0.004	0.164	249.214	6.097	5.623	34.284
26	47.414	0.021	0.003	0.163	290.088	6.118	5.690	34.811
27	55.000	0.018	0.003	0.163	337.502	6.136	5.750	35.284
28	63.800	0.016	0.003	0.163	392.503	6.152	5.804	35.707
29	74.009	0.014	0.002	0.162	456.303	6.166	5.853	36.086
30	85.850	0.012	0.002	0.162	530.312	6.177	5.896	36.423
31	99.586	0.010	0.002	0.162	616.162	6.187	5.936	36.725
32	115.520	0.009	0.001	0.161	715.747	6.196	5.971	36.993
33	134.003	0.007	0.001	0.161	831.267	6.203	6.002	37.232
34	155.443	0.006	0.001	0.161	965.270	6.210	6.030	37.444
35	180.314	0.006	0.001	0.161	1120.713	6.215	6.055	37.633
40	378.721	0.003	0.000	0.160	2360.757	6.233	6.144	38.299
45	795.444	0.001	0.000	0.160	4965.274	6.242	6.193	38.660
50	1670.704	0.001	0.000	0.160	10435.649	6.246	6.220	38.852

Table 14 Discrete cash flow compound interest factors; interest rate = 18%.

18.00%								18.00%
n	F/P	P/F	A/F	A/P	F/A	P/A	A/G	P/G
1	1.180	0.847	1.000	1.180	1.000	0.847	0.000	0.000
2	1.392	0.718	0.459	0.639	2.180	1.566	0.459	0.718
3	1.643	0.609	0.280	0.460	3.572	2.174	0.890	1.935
4	1.939	0.516	0.192	0.372	5.215	2.690	1.295	3.483
5	2.288	0.437	0.140	0.320	7.154	3.127	1.673	5.231
6	2.700	0.370	0.106	0.286	9.442	3.498	2.025	7.083
7	3.185	0.314	0.082	0.262	12.142	3.812	2.353	8.967
8	3.759	0.266	0.065	0.245	15.327	4.078	2.656	10.829
9	4.435	0.225	0.052	0.232	19.086	4.303	2.936	12.633
10	5.234	0.191	0.043	0.223	23.521	4.494	3.194	14.352
11	6.176	0.162	0.035	0.215	28.755	4.656	3.430	15.972
12	7.288	0.137	0.029	0.209	34.931	4.793	3.647	17.481
13	8.599	0.116	0.024	0.204	42.219	4.910	3.845	18.877
14	10.147	0.099	0.020	0.200	50.818	5.008	4.025	20.158
15	11.974	0.084	0.016	0.196	60.965	5.092	4.189	21.327
16	14.129	0.071	0.014	0.194	72.939	5.162	4.337	22.389
17	16.672	0.060	0.011	0.191	87.068	5.222	4.471	23.348
18	19.673	0.051	0.010	0.190	103.740	5.273	4.592	24.212
19	23.214	0.043	0.008	0.188	123.414	5.316	4.700	24.988
20	27.393	0.037	0.007	0.187	146.628	5.353	4.798	25.681
21	32.324	0.031	0.006	0.186	174.021	5.384	4.885	26.300
22	38.142	0.026	0.005	0.185	206.345	5.410	4.963	26.851
23	45.008	0.022	0.004	0.184	244.487	5.432	5.033	27.339
24	53.109	0.019	0.003	0.183	289.494	5.451	5.095	27.772
25	62.669	0.016	0.003	0.183	342.603	5.467	5.150	28.155
26	73.949	0.014	0.002	0.182	405.272	5.480	5.199	28.494
27	87.260	0.011	0.002	0.182	479.221	5.492	5.243	28.791
28	102.967	0.010	0.002	0.182	566.481	5.502	5.281	29.054
29	121.501	0.008	0.001	0.181	669.447	5.510	5.315	29.284
30	143.371	0.007	0.001	0.181	790.948	5.517	5.345	29.486
31	169.177	0.006	0.001	0.181	934.319	5.523	5.371	29.664
32	199.629	0.005	0.001	0.181	1103.496	5.528	5.394	29.819
33	235.563	0.004	0.001	0.181	1303.125	5.532	5.415	29.955
34	277.964	0.004	0.001	0.181	1538.688	5.536	5.433	30.074
35	327.997	0.003	0.001	0.181	1816.652	5.539	5.449	30.177
40	750.378	0.001	0.000	0.180	4163.213	5.548	5.502	30.527
45	1716.684	0.001	0.000	0.180	9531.577	5.552	5.529	30.701
50	3927.357	0.000	0.000	0.180	21813.094	5.554	5.543	30.786

Table 15 Discrete cash flow compound interest factors; interest rate = 20%.

20.00%								20.00%
n	F/P	P/F	A/F	A/P	F/A	P/A	A/G	P/G
1	1.200	0.833	1.000	1.200	1.000	0.833	0.000	0.000
2	1.440	0.694	0.455	0.655	2.200	1.528	0.455	0.694
3	1.728	0.579	0.275	0.475	3.640	2.106	0.879	1.852
4	2.074	0.482	0.186	0.386	5.368	2.589	1.274	3.299
5	2.488	0.402	0.134	0.334	7.442	2.991	1.641	4.906
6	2.986	0.335	0.101	0.301	9.930	3.326	1.979	6.581
7	3.583	0.279	0.077	0.277	12.916	3.605	2.290	8.255
8	4.300	0.233	0.061	0.261	16.499	3.837	2.576	9.883
9	5.160	0.194	0.048	0.248	20.799	4.031	2.836	11.434
10	6.192	0.162	0.039	0.239	25.959	4.192	3.074	12.887
11	7.430	0.135	0.031	0.231	32.150	4.327	3.289	14.233
12	8.916	0.112	0.025	0.225	39.581	4.439	3.484	15.467
13	10.699	0.093	0.021	0.221	48.497	4.533	3.660	16.588
14	12.839	0.078	0.017	0.217	59.196	4.611	3.817	17.601
15	15.407	0.065	0.014	0.214	72.035	4.675	3.959	18.509
16	18.488	0.054	0.011	0.211	87.442	4.730	4.085	19.321
17	22.186	0.045	0.009	0.209	105.931	4.775	4.198	20.042
18	26.623	0.038	0.008	0.208	128.117	4.812	4.298	20.680
19	31.948	0.031	0.006	0.206	154.740	4.843	4.386	21.244
20	38.338	0.026	0.005	0.205	186.688	4.870	4.464	21.739
21	46.005	0.022	0.004	0.204	225.026	4.891	4.533	22.174
22	55.206	0.018	0.004	0.204	271.031	4.909	4.594	22.555
23	66.247	0.015	0.003	0.203	326.237	4.925	4.647	22.887
24	79.497	0.013	0.003	0.203	392.484	4.937	4.694	23.176
25	95.396	0.010	0.002	0.202	471.981	4.948	4.735	23.428
26	114.475	0.009	0.002	0.202	567.377	4.956	4.771	23.646
27	137.371	0.007	0.001	0.201	681.853	4.964	4.802	23.835
28	164.845	0.006	0.001	0.201	819.223	4.970	4.829	23.999
29	197.814	0.005	0.001	0.201	984.068	4.975	4.853	24.141
30	237.376	0.004	0.001	0.201	1181.882	4.979	4.873	24.263
31	284.852	0.004	0.001	0.201	1419.258	4.982	4.891	24.368
32	341.822	0.003	0.001	0.201	1704.109	4.985	4.906	24.459
33	410.186	0.002	0.000	0.200	2045.931	4.988	4.919	24.537
34	492.224	0.002	0.000	0.200	2456.118	4.990	4.931	24.604
35	590.668	0.002	0.000	0.200	2948.341	4.992	4.941	24.661
40	1469.772	0.001	0.000	0.200	7343.858	4.997	4.973	24.847
45	3657.262	0.000	0.000	0.200	18281.310	4.999	4.988	24.932
50	9100.438	0.000	0.000	0.200	45497.191	4.999	4.995	24.970

Table 16 Discrete cash flow compound interest factors; interest rate = 25%.

25.00%								25.00%
n	F/P	P/F	A/F	A/P	F/A	P/A	A/G	P/G
1	1.250	0.800	1.000	1.250	1.000	0.800	0.000	0.000
2	1.563	0.640	0.444	0.694	2.250	1.440	0.444	0.640
3	1.953	0.512	0.262	0.512	3.813	1.952	0.852	1.664
4	2.441	0.410	0.173	0.423	5.766	2.362	1.225	2.893
5	3.052	0.328	0.122	0.372	8.207	2.689	1.563	4.204
6	3.815	0.262	0.089	0.339	11.259	2.951	1.868	5.514
7	4.768	0.210	0.066	0.316	15.073	3.161	2.142	6.773
8	5.960	0.168	0.050	0.300	19.842	3.329	2.387	7.947
9	7.451	0.134	0.039	0.289	25.802	3.463	2.605	9.021
10	9.313	0.107	0.030	0.280	33.253	3.571	2.797	9.987
11	11.642	0.086	0.023	0.273	42.566	3.656	2.966	10.846
12	14.552	0.069	0.018	0.268	54.208	3.725	3.115	11.602
13	18.190	0.055	0.015	0.265	68.760	3.780	3.244	12.262
14	22.737	0.044	0.012	0.262	86.949	3.824	3.356	12.833
15	28.422	0.035	0.009	0.259	109.687	3.859	3.453	13.326
16	35.527	0.028	0.007	0.257	138.109	3.887	3.537	13.748
17	44.409	0.023	0.006	0.256	173.636	3.910	3.608	14.108
18	55.511	0.018	0.005	0.255	218.045	3.928	3.670	14.415
19	69.389	0.014	0.004	0.254	273.556	3.942	3.722	14.674
20	86.736	0.012	0.003	0.253	342.945	3.954	3.767	14.893
21	108.420	0.009	0.002	0.252	429.681	3.963	3.805	15.078
22	135.525	0.007	0.002	0.252	538.101	3.970	3.836	15.233
23	169.407	0.006	0.001	0.251	673.626	3.976	3.863	15.362
24	211.758	0.005	0.001	0.251	843.033	3.981	3.886	15.471
25	264.698	0.004	0.001	0.251	1054.791	3.985	3.905	15.562
26	330.872	0.003	0.001	0.251	1319.489	3.988	3.921	15.637
27	413.590	0.002	0.001	0.251	1650.361	3.990	3.935	15.700
28	516.988	0.002	0.000	0.250	2063.952	3.992	3.946	15.752
29	646.235	0.002	0.000	0.250	2580.939	3.994	3.955	15.796
30	807.794	0.001	0.000	0.250	3227.174	3.995	3.963	15.832
31	1009.742	0.001	0.000	0.250	4034.968	3.996	3.969	15.861
32	1262.177	0.001	0.000	0.250	5044.710	3.997	3.975	15.886
33	1577.722	0.001	0.000	0.250	6306.887	3.997	3.979	15.906
34	1972.152	0.001	0.000	0.250	7884.609	3.998	3.983	15.923
35	2465.190	0.000	0.000	0.250	9856.761	3.998	3.986	15.937
40	7523.164	0.000	0.000	0.250	30088.655	3.999	3.995	15.977
45	22958.874	0.000	0.000	0.250	91831.496	4.000	3.998	15.991
50	70064.923	0.000	0.000	0.250	280255.693	4.000	3.999	15.997

Table 17 Discrete cash flow compound interest factors; interest rate = 30%.

30.00%								30.00%
n	F/P	P/F	A/F	A/P	F/A	P/A	A/G	P/G
1	1.300	0.769	1.000	1.300	1.000	0.769	0.000	0.000
2	1.690	0.592	0.435	0.735	2.300	1.361	0.435	0.592
3	2.197	0.455	0.251	0.551	3.990	1.816	0.827	1.502
4	2.856	0.350	0.162	0.462	6.187	2.166	1.178	2.552
5	3.713	0.269	0.111	0.411	9.043	2.436	1.490	3.630
6	4.827	0.207	0.078	0.378	12.756	2.643	1.765	4.666
7	6.275	0.159	0.057	0.357	17.583	2.802	2.006	5.622
8	8.157	0.123	0.042	0.342	23.858	2.925	2.216	6.480
9	10.604	0.094	0.031	0.331	32.015	3.019	2.396	7.234
10	13.786	0.073	0.023	0.323	42.619	3.092	2.551	7.887
11	17.922	0.056	0.018	0.318	56.405	3.147	2.683	8.445
12	23.298	0.043	0.013	0.313	74.327	3.190	2.795	8.917
13	30.288	0.033	0.010	0.310	97.625	3.223	2.889	9.314
14	39.374	0.025	0.008	0.308	127.913	3.249	2.969	9.644
15	51.186	0.020	0.006	0.306	167.286	3.268	3.034	9.917
16	66.542	0.015	0.005	0.305	218.472	3.283	3.089	10.143
17	86.504	0.012	0.004	0.304	285.014	3.295	3.135	10.328
18	112.455	0.009	0.003	0.303	371.518	3.304	3.172	10.479
19	146.192	0.007	0.002	0.302	483.973	3.311	3.202	10.602
20	190.050	0.005	0.002	0.302	630.165	3.316	3.228	10.702
21	247.065	0.004	0.001	0.301	820.215	3.320	3.248	10.783
22	321.184	0.003	0.001	0.301	1067.280	3.323	3.265	10.848
23	417.539	0.002	0.001	0.301	1388.464	3.325	3.278	10.901
24	542.801	0.002	0.001	0.301	1806.003	3.327	3.289	10.943
25	705.641	0.001	0.000	0.300	2348.803	3.329	3.298	10.977
26	917.333	0.001	0.000	0.300	3054.444	3.330	3.305	11.005
27	1192.533	0.001	0.000	0.300	3971.778	3.331	3.311	11.026
28	1550.293	0.001	0.000	0.300	5164.311	3.331	3.315	11.044
29	2015.381	0.000	0.000	0.300	6714.604	3.332	3.319	11.058
30	2619.996	0.000	0.000	0.300	8729.985	3.332	3.322	11.069
31	3405.994	0.000	0.000	0.300	11349.981	3.332	3.324	11.078
32	4427.793	0.000	0.000	0.300	14755.975	3.333	3.326	11.085
33	5756.130	0.000	0.000	0.300	19183.768	3.333	3.328	11.090
34	7482.970	0.000	0.000	0.300	24939.899	3.333	3.329	11.094
35	9727.860	0.000	0.000	0.300	32422.868	3.333	3.330	11.098
40	36118.865	0.000	0.000	0.300	120392.883	3.333	3.332	11.107
45	134106.817	0.000	0.000	0.300	447019.389	3.333	3.333	11.110
50	497929.223	0.000	0.000	0.300	1659760.743	3.333	3.333	11.111

Table 18 Discrete cash flow compound interest factors; interest rate = 35%.

35.00%								35.00%
n	F/P	P/F	A/F	A/P	F/A	P/A	A/G	P/G
1	1.350	0.741	1.000	1.350	1.000	0.741	0.000	0.000
2	1.823	0.549	0.426	0.776	2.350	1.289	0.426	0.549
3	2.460	0.406	0.240	0.590	4.173	1.696	0.803	1.362
4	3.322	0.301	0.151	0.501	6.633	1.997	1.134	2.265
5	4.484	0.223	0.100	0.450	9.954	2.220	1.422	3.157
6	6.053	0.165	0.069	0.419	14.438	2.385	1.670	3.983
7	8.172	0.122	0.049	0.399	20.492	2.508	1.881	4.717
8	11.032	0.091	0.035	0.385	28.664	2.598	2.060	5.352
9	14.894	0.067	0.025	0.375	39.696	2.665	2.209	5.889
10	20.107	0.050	0.018	0.368	54.590	2.715	2.334	6.336
11	27.144	0.037	0.013	0.363	74.697	2.752	2.436	6.705
12	36.644	0.027	0.010	0.360	101.841	2.779	2.520	7.005
13	49.470	0.020	0.007	0.357	138.485	2.799	2.589	7.247
14	66.784	0.015	0.005	0.355	187.954	2.814	2.644	7.442
15	90.158	0.011	0.004	0.354	254.738	2.825	2.689	7.597
16	121.714	0.008	0.003	0.353	344.897	2.834	2.725	7.721
17	164.314	0.006	0.002	0.352	466.611	2.840	2.753	7.818
18	221.824	0.005	0.002	0.352	630.925	2.844	2.776	7.895
19	299.462	0.003	0.001	0.351	852.748	2.848	2.793	7.955
20	404.274	0.002	0.001	0.351	1152.210	2.850	2.808	8.002
21	545.769	0.002	0.001	0.351	1556.484	2.852	2.819	8.038
22	736.789	0.001	0.000	0.350	2102.253	2.853	2.827	8.067
23	994.665	0.001	0.000	0.350	2839.042	2.854	2.834	8.089
24	1342.797	0.001	0.000	0.350	3833.706	2.855	2.839	8.106
25	1812.776	0.001	0.000	0.350	5176.504	2.856	2.843	8.119
26	2447.248	0.000	0.000	0.350	6989.280	2.856	2.847	8.130
27	3303.785	0.000	0.000	0.350	9436.528	2.856	2.849	8.137
28	4460.109	0.000	0.000	0.350	12740.313	2.857	2.851	8.143
29	6021.148	0.000	0.000	0.350	17200.422	2.857	2.852	8.148
30	8128.550	0.000	0.000	0.350	23221.570	2.857	2.853	8.152
31	10973.542	0.000	0.000	0.350	31350.120	2.857	2.854	8.154
32	14814.281	0.000	0.000	0.350	42323.661	2.857	2.855	8.157
33	19999.280	0.000	0.000	0.350	57137.943	2.857	2.855	8.158
34	26999.028	0.000	0.000	0.350	77137.223	2.857	2.856	8.159
35	36448.688	0.000	0.000	0.350	104136.251	2.857	2.856	8.160
40	163437.135	0.000	0.000	0.350	466960.385	2.857	2.857	8.163
45	732857.577	0.000	0.000	0.350	2093875.934	2.857	2.857	8.163
50	3286157.879	0.000	0.000	0.350	9389019.656	2.857	2.857	8.163

Table 19 Discrete cash flow compound interest factors; interest rate = 40%.

40.00%								40.00%
n	F/P	P/F	A/F	A/P	F/A	P/A	A/G	P/G
1	1.40	0.71	1.00	1.40	1.00	0.71	0.00	0.00
2	1.96	0.51	0.42	0.82	2.40	1.22	0.42	0.51
3	2.74	0.36	0.23	0.63	4.36	1.59	0.78	1.24
4	3.84	0.26	0.14	0.54	7.10	1.85	1.09	2.02
5	5.38	0.19	0.09	0.49	10.95	2.04	1.36	2.76
6	7.53	0.13	0.06	0.46	16.32	2.17	1.58	3.43
7	10.54	0.09	0.04	0.44	23.85	2.26	1.77	4.00
8	14.76	0.07	0.03	0.43	34.39	2.33	1.92	4.47
9	20.66	0.05	0.02	0.42	49.15	2.38	2.04	4.86
10	28.93	0.03	0.01	0.41	69.81	2.41	2.14	5.17
11	40.50	0.02	0.01	0.41	98.74	2.44	2.22	5.42
12	56.69	0.02	0.01	0.41	139.23	2.46	2.28	5.61
13	79.37	0.01	0.01	0.41	195.93	2.47	2.33	5.76
14	111.12	0.01	0.00	0.40	275.30	2.48	2.37	5.88
15	155.57	0.01	0.00	0.40	386.42	2.48	2.40	5.97
16	217.80	0.00	0.00	0.40	541.99	2.49	2.43	6.04
17	304.91	0.00	0.00	0.40	759.78	2.49	2.44	6.09
18	426.88	0.00	0.00	0.40	1064.70	2.49	2.46	6.13
19	597.63	0.00	0.00	0.40	1491.58	2.50	2.47	6.16
20	836.68	0.00	0.00	0.40	2089.21	2.50	2.48	6.18
21	1171.36	0.00	0.00	0.40	2925.89	2.50	2.48	6.20
22	1639.90	0.00	0.00	0.40	4097.24	2.50	2.49	6.21
23	2295.86	0.00	0.00	0.40	5737.14	2.50	2.49	6.22
24	3214.20	0.00	0.00	0.40	8033.00	2.50	2.49	6.23
25	4499.88	0.00	0.00	0.40	11247.20	2.50	2.49	6.23
26	6299.83	0.00	0.00	0.40	15747.08	2.50	2.50	6.24
27	8819.76	0.00	0.00	0.40	22046.91	2.50	2.50	6.24
28	12347.67	0.00	0.00	0.40	30866.67	2.50	2.50	6.24
29	17286.74	0.00	0.00	0.40	43214.34	2.50	2.50	6.25
30	24201.43	0.00	0.00	0.40	60501.08	2.50	2.50	6.25
31	33882.01	0.00	0.00	0.40	84702.51	2.50	2.50	6.25
32	47434.81	0.00	0.00	0.40	118584.52	2.50	2.50	6.25
33	66408.73	0.00	0.00	0.40	166019.33	2.50	2.50	6.25
34	92972.22	0.00	0.00	0.40	232428.06	2.50	2.50	6.25
35	130161.11	0.00	0.00	0.40	325400.28	2.50	2.50	6.25
40	700037.70	0.00	0.00	0.40	1750091.74	2.50	2.50	6.25
45	3764970.74	0.00	0.00	0.40	9412424.35	2.50	2.50	6.25
50	20248916.24	0.00	0.00	0.40	50622288.10	2.50	2.50	6.25

Table 20 Discrete cash flow compound interest factors; interest rate = 45%.

45.00%								45.00%
n	F/P	P/F	A/F	A/P	F/A	P/A	A/G	P/G
1	1.45	0.69	1.00	1.45	1.00	0.69	0.00	0.00
2	2.10	0.48	0.41	0.86	2.45	1.17	0.41	0.48
3	3.05	0.33	0.22	0.67	4.55	1.49	0.76	1.13
4	4.42	0.23	0.13	0.58	7.60	1.72	1.05	1.81
5	6.41	0.16	0.08	0.53	12.02	1.88	1.30	2.43
6	9.29	0.11	0.05	0.50	18.43	1.98	1.50	2.97
7	13.48	0.07	0.04	0.49	27.73	2.06	1.66	3.42
8	19.54	0.05	0.02	0.47	41.20	2.11	1.79	3.78
9	28.33	0.04	0.02	0.47	60.74	2.14	1.89	4.06
10	41.08	0.02	0.01	0.46	89.08	2.17	1.97	4.28
11	59.57	0.02	0.01	0.46	130.16	2.18	2.03	4.45
12	86.38	0.01	0.01	0.46	189.73	2.20	2.08	4.57
13	125.25	0.01	0.00	0.45	276.12	2.20	2.12	4.67
14	181.62	0.01	0.00	0.45	401.37	2.21	2.14	4.74
15	263.34	0.00	0.00	0.45	582.98	2.21	2.17	4.79
16	381.85	0.00	0.00	0.45	846.32	2.22	2.18	4.83
17	553.68	0.00	0.00	0.45	1228.17	2.22	2.19	4.86
18	802.83	0.00	0.00	0.45	1781.85	2.22	2.20	4.88
19	1164.10	0.00	0.00	0.45	2584.68	2.22	2.21	4.90
20	1687.95	0.00	0.00	0.45	3748.78	2.22	2.21	4.91
21	2447.53	0.00	0.00	0.45	5436.73	2.22	2.21	4.92
22	3548.92	0.00	0.00	0.45	7884.26	2.22	2.22	4.92
23	5145.93	0.00	0.00	0.45	11433.18	2.22	2.22	4.93
24	7461.60	0.00	0.00	0.45	16579.11	2.22	2.22	4.93
25	10819.32	0.00	0.00	0.45	24040.72	2.22	2.22	4.93
26	15688.02	0.00	0.00	0.45	34860.04	2.22	2.22	4.93
27	22747.63	0.00	0.00	0.45	50548.06	2.22	2.22	4.94
28	32984.06	0.00	0.00	0.45	73295.68	2.22	2.22	4.94
29	47826.88	0.00	0.00	0.45	106279.74	2.22	2.22	4.94
30	69348.98	0.00	0.00	0.45	154106.62	2.22	2.22	4.94
31	100556.02	0.00	0.00	0.45	223455.60	2.22	2.22	4.94
32	145806.23	0.00	0.00	0.45	324011.62	2.22	2.22	4.94
33	211419.03	0.00	0.00	0.45	469817.84	2.22	2.22	4.94
34	306557.59	0.00	0.00	0.45	681236.87	2.22	2.22	4.94
35	444508.51	0.00	0.00	0.45	987794.46	2.22	2.22	4.94
40	2849181.33	0.00	0.00	0.45	6331511.84	2.22	2.22	4.94
45	18262494.60	0.00	0.00	0.45	40583319.12	2.22	2.22	4.94
50	117057733.72	0.00	0.00	0.45	260128294.93	2.22	2.22	4.94

Table 21 Discrete cash flow compound interest factors; interest rate = 50%.

50.00%								50.00%
n	F/P	P/F	A/F	A/P	F/A	P/A	A/G	P/G
1	1.50	0.67	1.00	1.50	1.00	0.67	0.00	0.00
2	2.25	0.44	0.40	0.90	2.50	1.11	0.40	0.44
3	3.38	0.30	0.21	0.71	4.75	1.41	0.74	1.04
4	5.06	0.20	0.12	0.62	8.13	1.60	1.02	1.63
5	7.59	0.13	0.08	0.58	13.19	1.74	1.24	2.16
6	11.39	0.09	0.05	0.55	20.78	1.82	1.42	2.60
7	17.09	0.06	0.03	0.53	32.17	1.88	1.56	2.95
8	25.63	0.04	0.02	0.52	49.26	1.92	1.68	3.22
9	38.44	0.03	0.01	0.51	74.89	1.95	1.76	3.43
10	57.67	0.02	0.01	0.51	113.33	1.97	1.82	3.58
11	86.50	0.01	0.01	0.51	171.00	1.98	1.87	3.70
12	129.75	0.01	0.00	0.50	257.49	1.98	1.91	3.78
13	194.62	0.01	0.00	0.50	387.24	1.99	1.93	3.85
14	291.93	0.00	0.00	0.50	581.86	1.99	1.95	3.89
15	437.89	0.00	0.00	0.50	873.79	2.00	1.97	3.92
16	656.84	0.00	0.00	0.50	1311.68	2.00	1.98	3.95
17	985.26	0.00	0.00	0.50	1968.52	2.00	1.98	3.96
18	1477.89	0.00	0.00	0.50	2953.78	2.00	1.99	3.97
19	2216.84	0.00	0.00	0.50	4431.68	2.00	1.99	3.98
20	3325.26	0.00	0.00	0.50	6648.51	2.00	1.99	3.99
21	4987.89	0.00	0.00	0.50	9973.77	2.00	2.00	3.99
22	7481.83	0.00	0.00	0.50	14961.66	2.00	2.00	3.99
23	11222.74	0.00	0.00	0.50	22443.48	2.00	2.00	4.00
24	16834.11	0.00	0.00	0.50	33666.22	2.00	2.00	4.00
25	25251.17	0.00	0.00	0.50	50500.34	2.00	2.00	4.00
26	37876.75	0.00	0.00	0.50	75751.50	2.00	2.00	4.00
27	56815.13	0.00	0.00	0.50	113628.26	2.00	2.00	4.00
28	85222.69	0.00	0.00	0.50	170443.39	2.00	2.00	4.00
29	127834.04	0.00	0.00	0.50	255666.08	2.00	2.00	4.00
30	191751.06	0.00	0.00	0.50	383500.12	2.00	2.00	4.00
31	287626.59	0.00	0.00	0.50	575251.18	2.00	2.00	4.00
32	431439.88	0.00	0.00	0.50	862877.77	2.00	2.00	4.00
33	647159.82	0.00	0.00	0.50	1294317.65	2.00	2.00	4.00
34	970739.74	0.00	0.00	0.50	1941477.47	2.00	2.00	4.00
35	1456109.61	0.00	0.00	0.50	2912217.21	2.00	2.00	4.00
40	11057332.32	0.00	0.00	0.50	22114662.64	2.00	2.00	4.00
45	83966617.31	0.00	0.00	0.50	167933232.62	2.00	2.00	4.00
50	637621500.21	0.00	0.00	0.50	1275242998.43	2.00	2.00	4.00

Index

Engineering Project Appraisal: The Evaluation of Alternative Development Schemes, Second Edition.
Martin Rogers and Aidan Duffy.
© 2012 John Wiley & Sons, Ltd. Published 2012 by John Wiley & Sons, Ltd.